T0350170

Mesenchymal Stem Cell
DERIVED EXOSOMES

Mesenchymal Stem Cell
DERIVED EXOSOMES
The Potential for Translational Nanomedicine

Edited by

YAOLIANG TANG

Department of Medicine, Vascular Biology Center, Medical College of Georgia, Georgia Regents University, Augusta, GA, USA

BUDDHADEB DAWN

Department of Internal Medicine, Department of Biochemistry and Molecular Biology, Division of Cardiovascular Diseases, Cardiovascular Research Institute, and the Midwest Stem Cell Therapy Center, University of Kansas Medical Center, Kansas City, MO, USA

Amsterdam • Boston • Heidelberg • London • New York • Oxford
Paris • San Diego • San Francisco • Singapore • Sydney • Tokyo
Academic Press is an imprint of Elsevier

Academic Press is an imprint of Elsevier
125, London Wall, EC2Y 5AS, UK
525 B Street, Suite 1800, San Diego, CA 92101-4495, USA
225 Wyman Street, Waltham, MA 02451, USA
The Boulevard, Langford Lane, Kidlington, Oxford OX5 1GB, UK

British Library Cataloguing-in-Publication Data
A catalogue record for this book is available from the British Library

Library of Congress Cataloging-in-Publication Data
A catalog record for this book is available from the Library of Congress

ISBN: 978-0-12-800164-6

For information on all Academic Press publications
visit our website at http://store.elsevier.com/

Publisher: Mica Haley
Acquisitions Editor: Mica Haley
Editorial Project Manager: Lisa Eppich
Production Project Manager: Chris Wortley
Designer: Maria Inês Cruz
Typeset by Thomson Digital
Printed and bound in the United States of America

Working together
to grow libraries in
developing countries

www.elsevier.com • www.bookaid.org

CONTENTS

11 Diagnostic and Prognostic Applications of MicroRNA-Abundant Circulating Exosomes   223

Baron Arnone, Xiaoqi Zhao, Zhipeng Zou, Gangjian Qin, Min Cheng

LIST OF CONTRIBUTORS

Marta Adamiak
Department of Cell Biology, Faculty of Biochemistry, Biophysics and Biotechnology, Jagiellonian University, Krakow, Poland

Baron Arnone
Department of Medicine – Cardiology, Feinberg Cardiovascular Research Institute, Northwestern University Feinberg School of Medicine, Chicago, IL, USA

Muhammad Ashraf
Department of Pathology and Laboratory Medicine, University of Cincinnati Medical Center, Cincinnati, OH; Department of Pharmacology, University of Illinois College of Medicine at Chicago, Chicago, IL, USA

Sathyamoorthy Balasubramanian
Department of Internal Medicine, Cardiovascular Research Institute, University of Kansas Medical Center, Kansas City, MO, USA; Centre for Biotechnology, A.C. Technology Campus, Anna University, Chennai, India

Han Chen
Department of Cardiology, Provincial Key Cardiovascular Research Laboratory, Second Affiliated Hospital, Zhejiang University School of Medicine, Hangzhou, Zhejiang, China

Jiaquan Chen
Department of Vascular Surgery, Shanghai Jiao Tong University, School of Medicine, Renji Hospital, Shanghai, China

Tian Sheng Chen
College of Fisheries, Huazhong Agricultural University, Wuhan, Hubei Province, P.R. China

Min Cheng
Department of Cardiology, Union Hospital of Huazhong University of Science and Technology, Tongji Medical College, Wuhan, Hubei, China

William Chilian
Department of Integrative Medical Science, Northeast Ohio Medical University, Rootstown, OH, USA

Buddhadeb Dawn
Department of Internal Medicine; Division of Cardiovascular Diseases, Cardiovascular Research Institute, and the Midwest Stem Cell Therapy Center, University of Kansas Medical Center, Kansas City, MO, USA

W. Michael Dismuke
Department of Ophthalmology, Duke University School of Medicine, Durham, NC, USA

Lola DiVincenzo
Department of Integrative Medical Science, Northeast Ohio Medical University, Rootstown, OH, USA

Feng Dong
Department of Integrative Medical Science, Northeast Ohio Medical University, Rootstown, OH, USA

Kobina Essandoh
Department of Pharmacology and Cell Biophysics, University of Cincinnati College of Medicine, Cincinnati, OH, USA

Guo-Chang Fan
Department of Pharmacology and Cell Biophysics, University of Cincinnati College of Medicine, Cincinnati, OH, USA

Xiang Jiang Guo
Department of Vascular Surgery, Shanghai Jiao Tong University, School of Medicine, Renji Hospital, Shanghai, China

Mark W. Hamrick
Department of Cellular Biology and Anatomy, Medical College of Georgia, Georgia Regents University, Augusta, GA, USA

Changning Hao
Department of Vascular Surgery, Shanghai Jiao Tong University, School of Medicine, Renji Hospital, Shanghai, China

Sai Kiang Lim
Department of Surgery, Institute of Medical Biology, A*STAR, YLL School of Medicine, NUS, Singapore

Yutao Liu
Department of Cellular Biology and Anatomy, Georgia Regents University, Augusta, GA, USA

Lei Lv
Department of Vascular Surgery, Shanghai Jiao Tong University, School of Medicine, Renji Hospital, Shanghai, China

Gangjian Qin
Department of Medicine – Cardiology, Feinberg Cardiovascular Research Institute, Northwestern University Feinberg School of Medicine, Chicago, IL, USA

Johnson Rajasingh
Department of Internal Medicine, Cardiovascular Research Institute; Department of Biochemistry and Molecular Biology, University of Kansas Medical Center, Kansas City, MO, USA

Sheeja Rajasingh
Department of Internal Medicine, Cardiovascular Research Institute, University of Kansas Medical Center, Kansas City, MO, USA

Susmita Sahoo
Department of Cardiovascular Research Center, Icahn School of Medicine at Mount Sinai, New York, NY, USA

Paulomi Sanghavi
Department of Cellular Biology and Anatomy, Medical College of Georgia, Georgia
Regents University, Augusta, GA, USA

Kok Hian Tan
Department of Maternal Fetal Medicine, KK Women's and Children's Hospital, Singapore

Soon Sim Tan
Department of Surgery, Institute of Medical Biology, A*STAR, YLL School of Medicine,
NUS, Singapore

Jayakumar Thangavel
Department of Internal Medicine, Cardiovascular Research Institute, University of Kansas
Medical Center, Kansas City, MO, USA

Sunil Upadhyay
Department of Cellular Biology and Anatomy, Medical College of Georgia, Georgia
Regents University, Augusta, GA, USA

Jian'an Wang
Department of Cardiology, Provincial Key Cardiovascular Research Laboratory, Second
Affiliated Hospital, Zhejiang University School of Medicine, Hangzhou, Zhejiang, China

Shenjun Wu
Department of Vascular Surgery, Shanghai Jiao Tong University, School of Medicine, Renji
Hospital, Shanghai, China

Hui Xie
Department of Vascular Surgery, Shanghai Jiao Tong University, School of Medicine, Renji
Hospital, Shanghai, China

Meifeng Xu
Department of Pathology and Laboratory Medicine, University of Cincinnati Medical
Center, Cincinnati, OH, USA

Meng Ye
Department of Vascular Surgery, Shanghai Jiao Tong University, School of Medicine, Renji
Hospital, Shanghai, China

Liya Yin
Department of Integrative Medical Science, Northeast Ohio Medical University,
Rootstown, OH, USA

Porter Young
Department of Cellular Biology and Anatomy, Medical College of Georgia, Georgia
Regents University, Augusta, GA, USA

Bin Yu
Department of Pathology and Laboratory Medicine, University of Cincinnati Medical
Center, Cincinnati, OH, USA

Qingtan Zeng
Department of Vascular Surgery, Shanghai Jiao Tong University, School of Medicine, Renji
Hospital, Shanghai, China

Lan Zhang
Department of Vascular Surgery, Shanghai Jiao Tong University, School of Medicine, Renji Hospital, Shanghai, China

Xue Zhang
Department of Vascular Surgery, Shanghai Jiao Tong University, School of Medicine, Renji Hospital, Shanghai, China

Xiaoqi Zhao
Department of Cardiology, Union Hospital of Huazhong University of Science and Technology, Tongji Medical College, Wuhan, Hubei, China

Wei Zhu
Department of Cardiology, Provincial Key Cardiovascular Research Laboratory, Second Affiliated Hospital, Zhejiang University School of Medicine, Hangzhou, Zhejiang, China

Zhipeng Zou
Department of Neurosurgery, Wuhan Iron & Steel Corporation General Hospital, Wuhan, Hubei, China

Ewa K. Zuba-Surma
Department of Cell Biology, Faculty of Biochemistry, Biophysics and Biotechnology, Jagiellonian University, Krakow, Poland

PREFACE

Mesenchymal stem cells have emerged as one of the most promising stem cell types for angiogenesis and tissue repair. Increasing evidence indicates that therapeutic benefits of mesenchymal stem cells are largely due to paracrine effects mediated by soluble factors and exosomes. Stem cells are known to secrete exosomes, which shuttle proteins, mRNAs, long non-coding RNA and microRNAs (miRNA) between cells, and have been reported to elicit prosurvival and proangiogenic effects in damaged tissues. miRNA are evolutionarily conserved, small, noncoding RNA molecules that modulate cellular differentiation, proliferation, and apoptosis via post-transcriptional repression. Exosomes play an important role in miRNA transfer between donor stem cells and recipient tissues.

This book focuses on mesenchymal stem cell derived exosomes with contributions from internationally renowned experts in mesenchymal stem cells and exosomes. The chapters provide detailed and insightful information on comparison of methods for exosome purification, mechanisms of exosome formation and transportation, value of exosomes as biomarkers for disease diagnosis, and exosome-based therapeutics. This book is intended to benefit scientists performing basic research, as well as physicians interested in the exploitation of exosomes as biomarkers in diagnosis of disease, and formulation of novel therapeutic approaches with exosomes in nanoscale drug delivery. We would like to express our sincere gratitude to all of the contributing authors who have made this book possible. We would also like to thank Christin Minihane, Lisa Eppich, and the publisher, for their guidance, advice, and encouragement in editing this book.

Yaoliang Tang
Augusta, GA

Buddhadeb Dawn
Kansas City, MO

CHAPTER 1

Insights into the Mechanism of Exosome Formation and Secretion

Kobina Essandoh, Guo-Chang Fan
Department of Pharmacology and Cell Biophysics, University of Cincinnati College of Medicine, Cincinnati, OH, USA

Contents

1 INTRODUCTION

Exosomes are naturally generated nanoparticles from cells that range between 30 and 100 nm in size. The initial concept of exosomes can be traced to recticulocytes that secret transferrin receptor-containing tiny double membrane-bound vesicles [1]. Unlike other membrane proteins that are degraded by the endosomal pathway, transferrin receptors were released through exosomes when the reticulocytes matured into erythrocytes [2]. These exosomes were defined as products of the endosomal sorting pathway, as they originated from the intraluminal vesicles (ILVs) located in the lumen of multivesicular bodies (MVBs) [3]. Since then, numerous types of cells such as dendritic cells [4], B cells [5], T cells [6], epithelial cells [7], and mast cells [8] have been observed to release exosomes. Interestingly, exosomes collected from different cell types are rich in endosomal pathway proteins such as tumor susceptibility gene 101 (Tsg101), Alix, tetraspanins (CD9, CD63, CD81), and heat shock proteins (i.e., Hsc70 and Hsp90) [9].

Over the past few years, much of the research in the field of exosomes has centered on their role in cell-to-cell communication [3,10,11] and regulation of pathophysiological conditions such as cancer, liver disease, immune-defective disease, and neurodegenerative diseases [12–15]. In addition, research into stem cell-derived exosomes has focused on therapeutic intervention and tissue repair after injury [16–20]. While the functional role of exosomes has been widely reviewed elsewhere, fewer reviews have summarized the processes underlying the formation of exosomes. Therefore, in this chapter, we will discuss the current knowledge on the regulation of the biogenesis and release of exosomes. As products of the endocytic–endosomal pathway, biogenesis of exosomes will be discussed through endocytosis, maturation of early to late endosome, formation of MVBs, and the release of ILVs from the cell.

2 SIGNALINGS IN THE REGULATION OF ENDOCYTOSIS

The endocytic pathway was initially thought of as a mechanism to down-regulate plasma membrane receptors such as receptor tyrosine kinase (RTK) and G-protein coupled receptors (GPCRs) [21]. Later studies showed that RTKs or GPCRs might couple to extracellular signal-regulated kinase (ERK)-activating complexes to sustain signals [21,22]. Endocytic vesicles are formed from internalization of cell surface proteins. This process can be clathrin-dependent or -independent.

Clathrin-dependent endocytosis involves the formation of clathrin-coated pits along the plasma membrane, which causes inward budding of the membrane into intracellular vesicles. The clathrin–dependent pathway has been implicated in the endocytosis of RTKs, GPCRs, and transferrin receptors [23]. The formation of clathrin on the plasma membrane is preceded by the recruitment of an adaptor protein 2 (AP2) adaptor protein, which specifically binds to lipid phosphatidylinositol-4,5-bisphosphate [PtdIns(4,5)P2] at a nucleation module on the membrane or at the cytoplasmic side of the receptor [24]. At these modules, there is recruitment of complex-containing FCH domain only (F-BAR domain-containing Fer/Cip4 homology domain-only, FCHO) proteins, EGFR pathway substrate 15 (EPS15), and intersectins [24–26]. Although AP2 has binding domains for cargo, there are several other accessory adaptor proteins that provide binding for specific membrane proteins. For instance, Epsin protein possesses an epsin N-terminal homology (ENTH) domain that selects monoubiquitinated receptors for internalization [27,28]. The epidermal growth factor receptors

(EGFRs) thus undergo ligand-induced monoubiquitination that results in endocytosis and lysosomal degradation further downstream [29,30]. AP2 then recruits clathrin, which polymerizes to form interactions with curvature effector proteins (i.e., EPS15 and epsin) [31,32], which effects the bending of the cell membrane [33,34]. Subsequently, there is recruitment of a GTPase called dynamin2, which forms a collar-like structure around the neck of the inward-budded membrane wall. Through GTP hydrolysis, dynamin2 undergoes conformational changes, which initiates the dissociation of the clathrin-coated bud [35]. After release, ATPase heat shock cognate 70 (HSC70) and cyclin G-associated kinase (GAK) promote the removal of the clathrin coat, leading to the formation of endocytic vesicles in the cytosol [36].

Nonetheless, recent work has shown that endocytosis can be clathrin-independent. Such an idea originated from observations in HeLa cell lines where both clathrin- and nonclathrin-dependent endocytosis were uninhibited in the presence of a temperature-sensitive dynamin mutant [37]. The most common clathrin-independent endocytosis involves caveolae, which are subdomains of glycolipid rafts located in the plasma membrane. Caveolae, rich in cholesterol and spingolipids, possess caveolins that are necessary for the synthesis and invaginations into the cell membrane [38]. Caveolins oligomerize to form caveolin-rich rafts that coordinate with increased levels of cholesterol and spingolipid to form invaginations in the plasma membrane. Recruitment of dynamin2 results in scission to form endocytic vesicles, normally referred to as caveosomes [38,39]. For example, one study has shown that the insulin receptor, a member of RTKs, is endocytosized in a clathrin-independent (caveolae-dependent) manner in primary rat adipocytes [40]. In addition to caveolae, other factors participating in clathrin-independent endocytosis include ADP-ribosylation factor 1 (Arf1) and flotillin-1. Arf1 is believed to recruit RhoGAP domain-containing protein (ARHGAP10) to the plasma membrane, which results in the interaction of Arf1 with Cdc42 to initiate endocytosis in a dynamin-dependent manner [41]. Flotillins mediate the endocytosis of glycosylphosphatidylinositol (GPI)-anchored proteins and proteoglycans by associating with membrane lipid rafts [42,43].

Usually, electron microscopy should distinguish clathrin-dependent from clathrin-independent endocytic vesicles due to visualization of clathrin coats around endocytic vesicles [44]. It has been suggested that lower concentrations of ligand for EGFRs lead to clathrin-mediated endocytosis, whereas higher concentrations of ligand lead to clathrin-independent endocytosis and lysosomal-mediated attenuation of signal [45,46]. Most

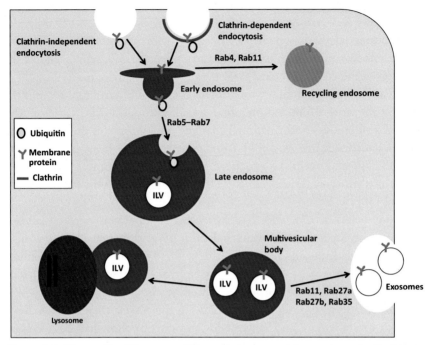

Figure 1.1 *Biogenesis of Exosomes.* Transmembrane proteins undergo both clathrin-dependent and -independent endocytosis to produce endocytic vesicles, which are incorporated into early endosomes. Early endosomes mature into late endosomes through the switch from Rab5 to Rab7 during which both early and late endosomes are subjected to sorting of cargo. After sorting of cargo, the late endosome matures into MVBs, which contains ILVs in the lumen. The MVB can then fuse with lysosome in order to degrade cargo incorporated into the ILVs. Alternatively, MVBs can be trafficked and fused with the plasma membrane with the help of Rab11, Rab27a/b, and Rab35. The MVB releases its ILVs as exosomes into extracellular space. At the early endosome stage, transmembrane proteins can be recycled back onto the plasma membrane through recycling endosomes with the help of Rab4 and Rab11.

recently, it has been reported that clathrin-coated pits account for 95% of all endocytosis [47]. Regardless of clathrin-dependent or -independent endocytosis, endocytic vesicles finally fuse and deliver their cargo to the developing early endosome (Figure 1.1).

3 SIGNALINGS IN THE REGULATION OF EARLY ENDOSOME

As described above, early endosomes receive cargo from both clathrin-dependent and -independent endocytic vesicles. Early endosomes are usually located at the outer edge of the cytoplasm along the microtubules, with proximity to the plasma membrane. Morphologically, early endosomes have a tubular domain and a vacuolar domain made up of vacuoles. The early endosome maintains a weakly acidic pH of about 6.8–5.9 [48].

Rab5 has been implicated to regulate the maturation of early endosomes to late endosomes [48]. With the help of RabGEF1 (Rabex-5) bound to internalized ubiquitinated proteins, Rab5 is recruited to early endosomes, together with its effector proteins [i.e., the early endosome antigen 1 (EEA1), vacuolar protein sorting-associated protein (Vps)15, and phosphatidylinositol 3-kinase (PI3K, Vps34)] [49]. Therefore, Rab5 can generate a complex at the cytosolic side of the early endosome. This Rab5 complex then provides the driving force to mobilize the early endosome through the microtubules [50–52].

PI3K produces phosphatidylinositol-3 phosphate [PI(3)P], which facilitates the recruitment of various effector proteins [e.g., hepatocyte growth factor-regulated tyrosine kinase substrate (Hrs), a member of the endosomal sorting complex required for transport (ESCRT)], leading to the specificity of the endosome by binding of specific cargo [53]. The association of Hrs with PI(3)P has provided some evidence that sorting of proteins may begin at the early endosomal stage [54,55]. As such, the cytosolic side of the early endosomes contains clathrin and members of ESCRT that begin to sort monoubiquitinated membrane proteins into ILVs [55].

Along the membrane of the early endosome, there are subdomains containing other Rab proteins (e.g., Rab4 and Rab11) as the early endosome matures. Receptors due to be recycled are sorted to the tubular domains and transported to the plasma membrane with the help of Rab4 [56] and Rab11 [57] (Figure 1.1). The switch from Rab5 to Rab7 regulates the maturation of early endosomes to late endosomes. At present, the most recognized model for this switch involves the conversion of Rab5-GTP to Rab5-GDP by the recruitment of Rab7 [58]. It has also been postulated that the amount of Rab5 declines overtime while Rab7 levels increase, leading to the takeover by Rab7 [59,60].

4 SIGNALINGS IN THE REGULATION OF EXOSOME CARGO LOADING TO LATE ENDOSOMES

In contrast to early endosomes, late endosomes have a round shape, signifying their maturation from the vacuole domain of the early endosomes [48]. The limiting membrane of the late endosome contains cholesterol and spingolipid rafts, clathrin, and components of ESCRT [61,62]. The limiting membrane of the late endosome undergoes invagination and budding to form ILVs, which results in the sorting of transmembrane proteins [48].

The monoubiquitination of transmembrane proteins leads to recognition and sorting by the ESCRT machinery. The ESCRT machinery consists of four complexes that lead to the formation of MVBs. The ESCRT-0 complex contains hepatocyte growth factor-regulated kinase substrate (Hrs, Vps27), signal transducing adapter molecule (STAM – STAM1 and STAM2 isoforms in human), Eps15, and clathrin. This complex recognizes monoubiquitinated proteins on the limiting membrane of late endosomes through the ubiquitin-binding domains located in Hrs and STAM [63,64]. Hrs recruits Tsg101 (Vps23), a member of ESCRT-I complex, through interaction with P(S/T)AP motif on Tsg101 [65–68]. Tsg101, which has a ubiquitin variant E2 (UEV) domain, receives the monoubiquitinated proteins from ESCRT-0 [69–71]. The ESCRT-1 complex consists of Tsg101, Vps28, Vps37, and Mvb12. Vps28 initiates binding of ESCRT-I to ESCRT-II by interacting with Vps36 of ESCRT-II [72,73]. The ESCRT-II complex consists of Vps36, Vps22, and two subunits of Vps25. Components of ESCRT 0, ESCRT-I, and ESCRT-II have ubiquitin-binding domains that transfer and sort monoubiquitinated proteins on the limiting membrane. On the other hand, ESCRT-III removes the ubiquitin tags and inward bud sorted proteins into ILVs. Vps25 of ESCRT-II binds to Vps20 of ESCRT-III, which leads to the recruitment of ESCRT-III proteins [74]. The initial members of the ESCRT-III recruited are Vps20, charged multivesicular body protein 4 (CHMP4), Vps24, and Vps2 in that order. Vps20 aids in the nucleation of CHMP4, which in turn binds to Vps24 [75]. Vps24 is then able to recruit Vps2. The addition of Vps2 brings about the recruitment of Vps4, which generates the scission of the invaginated limiting membrane into ILVs in the lumen [76]. Vps4 also has an AAA-ATPase activity, which hydrolyzes ATP and causes the breakup and recycling of ESCRT-III components [77–79].

Aside from the main components, ESCRT-III accessory protein Alix has been shown to play a major role in the ILV loading and cargo sorting. Alix interacts with CHMP4, which promotes the biogenesis of exosomes by interacting with the cytoplasmic domain of syndecan heparan sulfate proteoglycan receptors [80]. It has also been proposed that Alix binds to Tsg101, thereby inducing the interaction between ESCRT-I and ESCRT-III [81]. Alix-induced ILV formation involves only ESCRT I and III complexes, although the function of the Alix-Tsg101 interaction has not been elucidated. It is important to mention here that reduction in Alix levels in HeLa cells resulted in an increase in major histocompatibility complex (MHC) class II exosomes [82].

Recently, studies have been conducted to determine the effects of several members of the ESCRT machinery on exosome biogenesis [80–87]. Reduction in Hrs levels in HeLa-CIITA [82], head neck squamous cell carcinoma [83], HEK293 [84], and dendritic cells [85] resulted in less exosome secretion. Reduction in STAM1 levels impaired exosome secretion in HeLa-CIITA cells [82]. Reduction in Tsg101 levels in MCF-7 [80], HeLa-CIITA [82], and epithelial cells [86] resulted in less production of exosomes. However, such reduction of Tsg101 in oligodendrocytes yielded no changes in exosome secretion [87]. Downregulation of CHMP4 levels by RNAi silencing in MCF-7 cells led to reduced exosome secretion [80].

More interestingly, studies in some mammalian cell lines have demonstrated the formation of MVBs without the components of the ESCRT machinery. The inward budding of the limiting membrane of late endosomes in oligodendroglial cells was stimulated by ceramide without the help of the ESCRT machinery [87]. An inhibition of neutral sphingomyelinase, the enzyme that catalyzes the biosynthesis of ceramide, resulted in less secretion of proteolipid (PLP)-containing exosomes [87]. Inhibition of neutral sphingomyelinase with GW 4869 was also successful in blocking exosomal release from epithelial cells in the lungs of asthmatic mice [88]. Phosphatidic acid has also been shown to induce inward budding of the limiting membrane of late endosomes to promote ILV formation [89,90]. The ESCRT-dependent and ceramide (ESCRT-independent) sorting mechanisms have been summarized in Figure 1.2.

While the ESCRT machinery requires monoubiquitination of cargo, some cargos undergo sorting independent of ubiquitin tags. For example, premelanosome protein-17 (Pmel17) was sorted in an ubiquitin-independent manner to clathrin subdomains of MVBs even after the depletion of Hrs, Tsg101, and Vps4 in melanoma cells [91]. Significantly, small RNAs such as miRNAs are loaded to MVBs in mechanisms that do not require ubiquitin. Nevertheless, the association of miRNAs and other noncoding RNA molecules with ribonucleoprotein (RNP) complexes has been suggested since RNP complexes possess RNA binding motifs and are found in exosomes [92]. Recently, Villarroya-Beltri et al. revealed that miRNAs possess sequence motifs that traffic them to the protein heterogeneous nuclear ribonucleoprotein A2B1(hnRNPA2B1), which sorts miRNAs into exosomes [93]. After sorting of cargo (proteins and RNAs) into ILVs, the late endosome converts into a fully matured MVB (Figure 1.1).

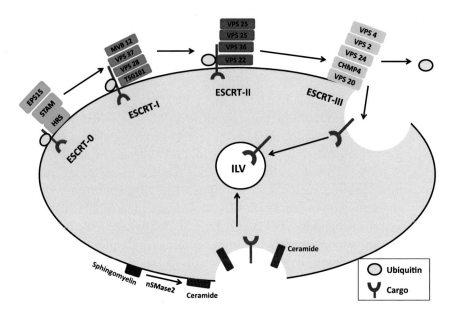

Figure 1.2 *ESCRT-Dependent and ESCRT-Independent Cargo Sorting into MVBs.* ESCRT-0 recognizes monoubiquitinated cargo through Hrs and STAM. The interaction between Hrs and Tsg101 initiates the transfer of cargo from ESCRT-0 to ESCRT-I. The cargo is transferred from ESCRT-I to ESCRT-II through the interaction between Vps28 and Vps36. Vps25 of ESCRT-II recruits Vps20, which leads to the recruitment of ESCRT-III components CHMP4, Vps24, Vps2, and Vps4 in that order. The formation of ESCRT-III causes the removal of ubiquitin and invagination and scission of sorted cargo on the limiting membrane of MVB into ILVs in the lumen. Sorting of cargo can also be initiated by ceramide, independent of ESCRT machinery. Ceramide, formed by the conversion of sphingomyelin by neutral sphingomyelinase, causes invaginations in the limiting membrane of MVBs and eventual budding into ILVs.

5 SIGNALINGS IN THE REGULATION OF THE FATE OF MULTIVESICULAR BODIES (MVBs)

Three different fates have been identified for the fully matured MVB. First, internalized receptors are sorted into MVBs and moved to lysosomes for degradation, in order to dampen ligand-induced signals (Figure 1.1). For example, most growth factor receptors are degraded by lysosomes after internalization [94]. In dendritic cells, MVBs play a role in the presentation of antigen-loaded MHC class II molecules. After the internalization of MHC class II molecules, these molecules are sorted onto MVBs to form an MHC class II compartment (MIIC). An antigen peptide is loaded onto the MHC class II molecule and is presented on the cell surface when the MIICs fuse with the plasma membrane

[95]. Alternatively, MVBs may fuse with the plasma membrane to release the ILVs as exosomes (Figure 1.1).

The current consensus in the field is that different cells have factors or conditions that promote the release of some MVBs as exosomes. Thus, there is no specific mechanism or stimuli adapted by all cells to define some MVBs for secretion from the cells. There is the possibility that cargo sorted to specific lipid-raft-enriched microdomains may be associated with ILVs destined for exosome release. Perhaps microdomains responsible for proteins and lipids encased in most exosomes [i.e., tetraspanins, cholesterol, spingolipids, and heat-shock protein 90 (HSP90)] may be responsible for the fate of such MVBs [96]. It is therefore easy to postulate that complexes associated with tetraspanins during endosomal sorting are targeted for exosomal release. For instance, tetraspanin-8 (TSPAN8) recruits exosomal proteins CD106 and CD49d in rat adenocarcinoma cells [97]. What's more, Perez-Hernandez et al. demonstrated that lymphocytes, isolated from CD81 knockout mice, released exosomes that were deficient in CD81-associated molecules such as Rac-GTPases after proteomic analysis [98]. Similarly, de Gassart et al. observed that Annexin-A2 is incorporated into cholesterol-rich lipid raft microdomains to initiate its release through exosomes and prevent degradation by lysosomes [99].

Interestingly, posttranslational modifications have been shown to direct proteins to certain microdomains on the limiting membrane of late endosomes to sort into ILVs destined for exosome release. Liang et al. recently revealed that N-glycosylation of glycoproteins is linked to sorting into exosomal MVBs [100]. Using Sk-Mel-5 cell line, the authors observed that EWI-2, a CD81 interacting protein, is recruited to exosomes and is dependent on the presence of complex N-glycans. Ske-Mel-5 cells were treated with N-glycan inhibitor 1-deoxymannojirimycin (DMJ), which resulted in less EWI-2 content in exosomes [100]. Other than glycosylation, Shen et al. have revealed that plasma membrane-derived myristoylation tag, palmitoylation tag, a PtdIns(4,5)P2-binding domain, a PtdIns(3,4,5)P3-binding domain, and a type-1 integral plasma membrane protein (CD43) target yeast protein TyA to exosomes [92,101].

6 SIGNALINGS IN THE REGULATION OF EXOSOME RELEASE

The process for the release of exosomes involves the intracellular trafficking of MVBs to the cell periphery and the fusion of MVBs with the cell membrane. Currently, several members of the Rab family of GTPases such as Rab11, Rab27a/b, and Rab35 have been shown to regulate the trafficking of MVBs to the plasma membrane.

Using time lapse confocal microscopy, Savina et al. observed that GFP-Rab11 was localized to MVBs and initiated fusion and docking of MVBs. They found that K562 cells infected with GFP-Rab11 WT generated larger MVBs and increased exosome release compared to cells expressing the mutant Q70L GFP-Rab11 [102]. More recently, the impact of Rab27a and Rab27b in regulating exosome release has been reported in HeLa B6H4 tumor cell line, where silencing of Rab27a and Rab27b reduced docking exosomes at the plasma membrane. Both Rab27a and Rab27b were localized to CD63-positve compartments at the limiting membranes of MVBs. However, silencing of Rab27a increased the size of MVBs, whereas silencing of Rab27b resulted in the localization of MVBs to the perinuclear region of the cell [103]. In oligodendrocytes, Hsu et al. showed that Rab35 and its associated GAP, TBC1 domain family, member 10B (TBC1D10B), were localized to the plasma membrane [104]. The knockdown of Rab35 by RNAi resulted in a 35% increase in accumulating late endosomes that were not released as exosomes [104]. Frühbeis et al. confirmed these observations, when inhibition of Rab35 by siRNA silencing in primary oligodendrocytes resulted in a reduction of exosomal protein markers, PLP and Alix, in the exosomes [105]. In addition, the involvement of soluble NSF attachment protein receptor (SNARE) proteins has been proposed to regulate the fusion and release of MVBs with the plasma membrane. For instance, RNAi silencing of SNARE protein, Ykt6, led to accumulation of exosomal marker GFP-CD63 in the cell [84]. Nonetheless, the exact mechanism underlying the fusion and release of MVBs with plasma membrane remains largely unknown.

7 OTHER SIGNALINGS IN THE REGULATION OF EXOSOME BIOGENESIS AND RELEASE

Aside from the traditional endocytic pathway-derived exosomes, certain molecules have been shown to induce exosome release. For example, p53 protein has been implicated to regulate the release of exosomes through a downstream factor, tumor supressor-activated pathway 6 (TSAP6) [106]. Yu et al. utilized H1299 cells, a p53-null cell line, and transfected with a construct that expressed HA-TSAP. These authors observed that exosome release was increased after p53 activation in the H1299 cells [106]. Such observations were confirmed in TSAP6-deficient mice where secretion of exosomes was reduced after p53 activation [106]. Similarly, Lespagnol et al. observed that MEF cells isolated from TSAP6−/− mice showed reduced exosomal protein content compared to wild-type controls after the cells were γ-irradiated for p53 activation [107].

In addition, changes in the concentration of intracellular calcium have been implicated to affect exosome release. For example, Savina et al. have shown that increased intracellular calcium in the K562 cell line with monensin enhanced exosome release [108]. Similarly, Krämer-Albers et al. treated oligodendrocytes with calcium ionophore ionomycin, leading to an increase in PLP-containing exosomes [109]. The role of potassium was also assessed in rat cortical neurons where exosome secretion was enhanced by K^+-dependent depolarization [110].

8 STRESS AND PATHOLOGIC CONDITIONS CAN STIMULATE EXOSOME RELEASE

In addition to factors/proteins that regulate exosome biogenesis/release, various diseases and pathologic conditions have been shown to boost exosome secretion [111,112]. Exosome release has been linked to progression of inflammation, coagulation, apoptosis, and angiogenesis [111,112]. For example, both *in vitro* and *in vivo* models exhibited increased exosome release upon stimulation with bacterial pathogens [113,114]. Elevated levels of exosomes have also been observed in tumor cells. For instance, the amount of exosomes in the urine collected from prostate cancer patients was significantly elevated compared to healthy controls [115]. It is worth noting here that exosomes can be detected in body fluids such as blood, urine, saliva, breast milk, epididymal fluid, amniotic fluid, semen, and malignant effusions [116] and may be useful for identification of new biomarkers for such disorders.

Recently, numerous studies have indicated that hypoxic and heat-stressed conditions can promote exosome release. For example, breast cancer cells cultured under severe (0.1% O_2) hypoxia conditions displayed higher expression of exosomal marker CD63 compared to normal (1% O_2) conditions. It was also revealed that hypoxia-induced exosome release was mediated by hypoxia-inducible factor α (HIF-α) [117]. Similarly, Salomon et al. observed that, under hypoxic conditions (1% and 3% O_2), placental mesenchymal stem cells (pMSCs) had a 6.7-fold increase in exosomal protein content compared to controls (8% O_2) [118]. In 3LL tumor cells, Chen et al. observed about a 2.5–3.5-fold increase in exosome release at 30 min after the induction of heat stress compared to controls [119]. Likewise, Clayton et al. observed a significant increase in the amount of exosomes released from B cells and Jurkat cells after heat stress [120].

9 CONCLUDING REMARKS

At present, much of the work in the field has been dedicated to the functional and therapeutic potential of exosomes. It appears clear that, under disease conditions, exosomes may facilitate to spread pathological information around the body. On the other hand, stem cell-derived exosomes could replace stem cells for therapeutic applications. However, the signaling pathways involved in the regulation of exosome biogenesis/release have had less attention. Indeed, better understanding of the mechanisms underlying the regulation of exosome biogenesis/release will allow us to: (1) efficiently interfere with the pathological process of disease; and (2) produce more powerful and safer exosomes for translational study and maybe clinical trials. Therefore, future study will desperately be needed to explore how to control exosome biogenesis and release in cells.

ABBREVIATIONS

AAA-ATPase	ATPases associated with diverse cellular activities
AP2	Adaptor protein 2
Arf1	ADP-ribosylation factor 1
CHMP4	Charged multivesicular body protein 4
EEA1	Early endosome antigen 1
EGFR	Epidermal growth factor receptors
ENTH	Epsin N-terminal homology
EPS15	EGFR pathway substrate 15
ESCRT	Endosomal sorting complex required for transport
FCHO	F-BAR domain-containing Fer/Cip4 homology domain-only
GAK	Cyclin G-associated kinase
GPCR	G-protein coupled receptor
GPI	Glycosylphosphatidylinositol
HIF-α	Hypoxia-inducible factor α
hnRNPA2B1	Heterogeneous nuclear ribonucleoprotein A2B1
Hrs	Hepatocyte growth factor-regulated tyrosine kinase substrate
HSP	Heat shock protein
ILVs	Intraluminal vesicles
MHC	Major histocompatibility complex
MIIC	MHC class II compartment
MVB	Multivesicular body
PI(3)P	Phosphatidylinositol-3 phosphate

PI3K	Phosphatidylinositol 3-kinase
PLP	Proteolipid
Pmel17	Premelanosome protein-17
pMSC	Placental mesenchymal stem cells
PtdIns(4,5)P2	Phosphatidylinositol-4,5-bisphosphate
RNP	Ribonucleoprotein
RTK	Receptor tyrosine kinase
SNARE	Soluble NSF attachment protein receptor
STAM	Signal transducing adapter molecule
TBC1D10B	TBC1 domain family, member 10B
TSAP6	Tumor suppressor-activated pathway 6
Tsg101	Tumor susceptibility gene 101
TSPAN8	Tetraspanin-8
UEV	Ubiquitin E2 variant
Vps	Vacuolar protein sorting-associated protein

REFERENCES

[1] Pan BT, Johnstone RM. Fate of the transferrin receptor during maturation of sheep reticulocytes *in vitro*: selective externalization of the receptor. Cell 1983;33(3):967–78.

[2] Johnstone RM, Adam M, Hammond JR, Orr L, Turbide C. Vesicle formation during reticulocyte maturation. Association of plasma membrane activities with released vesicles (exosomes). J Biol Chem 1987;262(19):9412–20.

[3] Johnstone RM, Bianchini A, Teng K. Reticulocyte maturation and exosome release: transferrin receptor containing exosomes shows multiple plasma membrane functions. Blood 1989;74(5):1844–51.

[4] Théry C, Regnault A, Garin J, Wolfers J, Zitvogel L, Ricciardi-Castagnoli P, et al. Molecular characterization of dendritic cell-derived exosomes. Selective accumulation of the heat shock protein hsc73. J Cell Biol 1999;147(3):599–610.

[5] Raposo G, Nijman HW, Stoorvogel W, Liejendekker R, Harding CV, Melief CJ, et al. B lymphocytes secrete antigen-presenting vesicles. J Exp Med 1996;183(3): 1161–72.

[6] Blanchard N, Lankar D, Faure F, Regnault A, Dumont C, Raposo G, et al. TCR activation of human T cells induces the production of exosomes bearing the TCR/CD3/zeta complex. J Immunol 2002;168(7):3235–41.

[7] van Niel G, Raposo G, Candalh C, Boussac M, Hershberg R, Cerf-Bensussan N, et al. Intestinal epithelial cells secrete exosome-like vesicles. Gastroenterology 2001;121(2):337–49.

[8] Raposo G, Tenza D, Mecheri S, Peronet R, Bonnerot C, Desaymard C. Accumulation of major histocompatibility complex class II molecules in mast cell secretory granules and their release upon degranulation. Mol Biol Cell 1997;8(12):2631–45.

[9] Mathivanan S, Ji H, Simpson RJ. Exosomes: extracellular organelles important in intercellular communication. J Proteomics 2010;73(10):1907–20.

[10] Sluijter JP, Verhage V, Deddens JC, van den Akker F, Doevendans PA. Microvesicles and exosomes for intracardiac communication. Cardiovasc Res 2014;102(2):302–11.

[11] Ribeiro MF, Zhu H, Millard RW, Fan GC. Exosomes function in pro- and anti-angiogenesis. Curr Angiogenes 2013;2(1):54–9.

[12] Buzas EI, György B, Nagy G, Falus A, Gay S. Emerging role of extracellular vesicles in inflammatory diseases. Nat Rev Rheumatol 2014;10(6):356–64.

[13] Record M, Carayon K, Poirot M, Silvente-Poirot S. Exosomes as new vesicular lipid transporters involved in cell–cell communication and various pathophysiologies. Biochim Biophys Acta 2014;1841(1):108–20.

[14] Kharaziha P, Ceder S, Li Q, Panaretakis T. Tumor cell-derived exosomes: a message in a bottle. Biochim Biophys Acta 2012;1826(1):103–11.

[15] Kalani A, Tyagi A, Tyagi N. Exosomes: mediators of neurodegeneration, neuroprotection and therapeutics. Mol Neurobiol 2014;49(1):590–600.

[16] Zhang B, Yin Y, Lai RC, Lim SK. Immunotherapeutic potential of extracellular vesicles. Front Immunol 2014;5:518.

[17] Sahoo S, Losordo DW. Exosomes and cardiac repair after myocardial infarction. Circ Res 2014;114(2):333–44.

[18] Katsuda T, Kosaka N, Takeshita F, Ochiya T. The therapeutic potential of mesenchymal stem cell-derived extracellular vesicles. Proteomics 2013;13 (10–11):1637–53.

[19] Biancone L, Bruno S, Deregibus MC, Tetta C, Camussi G. Therapeutic potential of mesenchymal stem cell-derived microvesicles. Nephrol Dial Transplant 2012;27(8):3037–42.

[20] Lai RC, Chen TS, Lim SK. Mesenchymal stem cell exosome: a novel stem cell-based therapy for cardiovascular disease. Regen Med 2011;6(4):481–92.

[21] Watson FL, Heerssen HM, Bhattacharyya A, Klesse L, Lin MZ, Segal RA. Neurotrophins use the Erk5 pathway to mediate a retrograde survival response. Nat Neurosci 2001;4(10):981–8.

[22] García-Regalado A, Guzmán-Hernández ML, Ramírez-Rangel I, Robles-Molina E, Balla T, Vazquez-Prado J, et al. G protein-coupled receptor-promoted trafficking of $G\beta_1\gamma_2$ leads to AKT activation at endosomes via a mechanism mediated by $G\beta_1\gamma_2$-Rab11a interaction. Mol Biol Cell 2008;19(10):4188–200.

[23] Lamaze C, Baba T, Redelmeier TE, Schmid SL. Recruitment of epidermal growth factor and transferrin receptors into coated pits in vitro: differing biochemical requirements. Mol Biol Cell 1993;4(7):715–27.

[24] Henne WM, Boucrot E, Meinecke M, Evergren E, Vallis Y, Mittal R, et al. FCHo proteins are nucleators of clathrin-mediated endocytosis. Science 2010;328(5983): 1281–4.

[25] Stimpson HE, Toret CP, Cheng AT, Pauly BS, Drubin DG. Early-arriving Syp1p and Ede1p function in endocytic site placement and formation in budding yeast. Mol Biol Cell 2009;20(22):4640–51.

[26] Reider A, Barker SL, Mishra SK, Im YJ, Maldonado-Báez L, Hurley JH, et al. Syp1 is a conserved endocytic adaptor that contains domains involved in cargo selection and membrane tubulation. EMBO J 2009;28(20):3103–16.

[27] Ford MG, Mills IG, Peter BJ, Vallis Y, Praefcke GJ, Evans PR, et al. Curvature of clathrin-coated pits driven by epsin. Nature 2002;419(6905):361–6.

[28] Chidambaram S, Zimmermann J, von Mollard GF. ENTH domain proteins are cargo adaptors for multiple SNARE proteins at the TGN endosome. J Cell Sci 2008;121(3):329–38.

[29] Haglund K, Sigismund S, Polo S, Szymkiewicz I, Di Fiore PP, Dikic I. Multiple monoubiquitination of RTKs is sufficient for their endocytosis and degradation. Nat Cell Biol 2003;5(5):461–6.

[30] Mosesson Y, Shtiegman K, Katz M, Zwang Y, Vereb G, Szollosi J, et al. Endocytosis of receptor tyrosine kinases is driven by monoubiquitylation, not polyubiquitylation. J Biol Chem 2003;278(24):21323–6.

[31] Motley A, Bright NA, Seaman MN, Robinson MS. Clathrin-mediated endocytosis in AP-2-depleted cells. J Cell Biol 2003;162(5):909–18.

[32] Boucrot E, Saffarian S, Zhang R, Kirchhausen T. Roles of AP-2 in clathrin-mediated endocytosis. PLoS One 2010;5(5):e10597.

[33] Tebar F, Sorkina T, Sorkin A, Ericsson M, Kirchhausen T. Eps15 is a component of clathrin-coated pits and vesicles and is located at the rim of coated pits. J Biol Chem 1996;271(46):28727–30.

[34] Saffarian S, Cocucci E, Kirchhausen T. Distinct dynamics of endocytic clathrin-coated pits and coated plaques. PLoS Biol 2009;7(9):e1000191.

[35] Kosaka T, Ikeda K. Reversible blockage of membrane retrieval and endocytosis in the garland cell of the temperature-sensitive mutant of *Drosophila melanogaster*, shibirets1. J Cell Biol 1983;97(2):499–507.

[36] Schlossman DM, Schmid SL, Braell WA, Rothman JE. An enzyme that removes clathrin coats: purification of an uncoating ATPase. J Cell Biol 1984;99(2):723–33.

[37] Damke H, Baba T, van der Bliek AM, Schmid SL. Clathrin-independent pinocytosis is induced in cells overexpressing a temperature-sensitive mutant of dynamin. J Cell Biol 1995;131(1):69–80.

[38] Hansen CG, Nichols BJ. Exploring the caves: cavins, caveolins and caveolae. Trends Cell Biol 2010;20(4):177–86.

[39] Oh P, McIntosh DP, Schnitzer JE. Dynamin at the neck of caveolae mediates their budding to form transport vesicles by GTP-driven fission from the plasma membrane of endothelium. J Cell Biol 1998;141(1):101–14.

[40] Fagerholm S, Ortegren U, Karlsson M, Ruishalme I, Strålfors P. Rapid insulin-dependent endocytosis of the insulin receptor by caveolae in primary adipocytes. PLoS One 2009;4(6):e5985.

[41] Kumari S, Mayor S. ARF1 is directly involved in dynamin-independent endocytosis. Nat Cell Biol 2008;10(1):30–41.

[42] Aït-Slimane T, Galmes R, Trugnan G, Maurice M. Basolateral internalization of GPI-anchored proteins occurs via a clathrin-independent flotillin-dependent pathway in polarized hepatic cells. Mol Biol Cell 2009;20(17):3792–800.

[43] Payne CK, Jones SA, Chen C, Zhuang X. Internalization and trafficking of cell surface proteoglycans and proteoglycan-binding ligands. Traffic 2007;8(4):389–401.

[44] McMahon HT, Boucrot E. Molecular mechanism and physiological functions of clathrin-mediated endocytosis. Nat Rev Mol Cell Biol 2011;12(8):517–33.

[45] Sigismund S, Argenzio E, Tosoni D, Cavallaro E, Polo S, Di Fiore PP. Clathrin-mediated internalization is essential for sustained EGFR signaling but dispensable for degradation. Dev Cell 2008;15(2):209–19.

[46] Sigismund S, Woelk T, Puri C, Maspero E, Tacchetti C, Transidico P, et al. Clathrin-independent endocytosis of ubiquitinated cargos. Proc Natl Acad Sci USA 2005;102(8):2760–5.

[47] Bitsikas V, Corrêa IR Jr, Nichols BJ. Clathrin-independent pathways do not contribute significantly to endocytic flux. Elife 2014;3:e03970.

[48] Huotari J, Helenius A. Endosome maturation. EMBO J 2011;30(17):3481–500.

[49] Mattera R, Bonifacino JS. Ubiquitin binding and conjugation regulate the recruitment of Rabex-5 to early endosomes. EMBO J 2008;27(19):2484–94.

[50] Li G, D'Souza-Schorey C, Barbieri MA, Roberts RL, Klippel A, Williams LT, et al. Evidence for phosphatidylinositol 3-kinase as a regulator of endocytosis via activation of Rab5. Proc Natl Acad Sci USA 1995;92(22):10207–11.

[51] Christoforidis S, McBride HM, Burgoyne RD, Zerial M. The Rab5 effector EEA1 is a core component of endosome docking. Nature 1999;397(6720):621–5.

[52] Nielsen E, Severin F, Backer JM, Hyman AA, Zerial M. Rab5 regulates motility of early endosomes on microtubules. Nat Cell Biol 1999;1(6):376–82.

[53] Petiot A, Faure J, Stenmark H, Gruenberg J. PI3P signaling regulates receptor sorting but not transport in the endosomal pathway. J Cell Biol 2003;162(6):971–9.

[54] Raiborg C, Bache KG, Gillooly DJ, Madshus IH, Stang E, Stenmark H. Hrs sorts ubiquitinated proteins into clathrin-coated microdomains of early endosomes. Nat Cell Biol 2002;4(5):394–8.

[55] Sachse M, Urbé S, Oorschot V, Strous GJ, Klumperman J. Bilayered clathrin coats on endosomal vacuoles are involved in protein sorting toward lysosomes. Mol Biol Cell 2002;13(4):1313–28.

[56] de Wit H, Lichtenstein Y, Kelly RB, Geuze HJ, Klumperman J, van der Sluijs P. Rab4 regulates formation of synaptic-like microvesicles from early endosomes in PC12 cells. Mol Biol Cell 2001;12(11):3703–15.

[57] Peden AA, Schonteich E, Chun J, Junutula JR, Scheller RH, Prekeris R. The RCP-Rab11 complex regulates endocytic protein sorting. Mol Biol Cell 2004;15(8): 3530–41.

[58] Rink J, Ghigo E, Kalaidzidis Y, Zerial M. Rab conversion as a mechanism of progression from early to late endosomes. Cell 2005;122(5):735–49.

[59] Lakadamyali M, Rust MJ, Zhuang X. Ligands for clathrin-mediated endocytosis are differentially sorted into distinct populations of early endosomes. Cell 2006;124(5):997–1009.

[60] Del Conte-Zerial P, Brusch L, Rink JC, Collinet C, Kalaidzidis Y, Zerial M, et al. Membrane identity and GTPase cascades regulated by toggle and cut-out switches. Mol Sys Biol 2008;4:206.

[61] Mukherjee S, Soe TT, Maxfield FR. Endocytic sorting of lipid analogues differing solely in the chemistry of their hydrophobic tails. J Cell Biol 1999;144(6):1271–84.

[62] Mukherjee S, Maxfield FR. Membrane domains. Annu Rev Cell Dev Biol 2004;20:839–66.

[63] Stringer DK, Piper RC. A single ubiquitin is sufficient for cargo protein entry into MVBs in the absence of ESCRT ubiquitination. J Cell Biol 2011;192(2):229–42.

[64] Ren J, Kee Y, Huibregtse JM, Piper RC. Hse1, a component of the yeast Hrs-STAM ubiquitin-sorting complex, associates with ubiquitin peptidases and a ligase to control sorting efficiency into multivesicular bodies. Mol Biol Cell 2007;18(1):324–35.

[65] Pornillos O, Higginson DS, Stray KM, Fisher RD, Garrus JE, Payne M, et al. HIV Gag mimics the Tsg101-recruiting activity of the human Hrs protein. J Cell Biol 2003;162(3):425–34.

[66] Lu Q, Hope LW, Brasch M, Reinhard C, Cohen SN. TSG101 interaction with HRS mediates endosomal trafficking and receptor down-regulation. Proc Natl Acad Sci USA 2003;100(13):7626–31.

[67] Katzmann DJ, Stefan CJ, Babst M, Emr SD. Vps27 recruits ESCRT machinery to endosomes during MVB sorting. J Cell Biol 2003;162(3):413–23.

[68] Bilodeau PS, Winistorfer SC, Kearney WR, Robertson AD, Piper RC. Vps27-Hse1 and ESCRT-I complexes cooperate to increase efficiency of sorting ubiquitinated proteins at the endosome. J Cell Biol 2003;163(2):237–43.

[69] Teo H, Veprintsev DB, Williams RL. Structural insights into endosomal sorting complex required for transport (ESCRT-I) recognition of ubiquitinated proteins. J Biol Chem 2004;279(27):28689–96.

[70] Sundquist WI, Schubert HL, Kelly BN, Hill GC, Holton JM, Hill CP. Ubiquitin recognition by the human TSG101 protein. Mol Cell 2004;13(6):783–9.

[71] Katzmann DJ, Babst M, Emr SD. Ubiquitin-dependent sorting into the multivesicular body pathway requires the function of a conserved endosomal protein sorting complex ESCRT-I. Cell 2001;106(2):145–55.

[72] Kostelansky MS, Schluter C, Tam YY, Lee S, Ghirlando R, Beach B, et al. Molecular architecture and functional model of the complete yeast ESCRT-I heterotetramer. Cell 2007;129(3):485–98.

[73] Gill DJ, Teo H, Sun J, Perisic O, Veprintsev DB, Emr SD, et al. Structural insight into the ESCRT-I/-II link and its role in MVB trafficking. EMBO J 2007;26(2):600–12.

[74] Im YJ, Wollert T, Boura E, Hurley JH. Structure and function of the ESCRT-II-III interface in multivesicular body biogenesis. Dev Cell 2009;17(2):234–43.

[75] Teis D, Saksena S, Emr SD. Ordered assembly of the ESCRT-III complex on endosomes is required to sequester cargo during MVB formation. Dev Cell 2008;15(4): 578–89.

[76] Wollert T, Wunder C, Lippincott-Schwartz J, Hurley JH. Membrane scission by the ESCRT-III complex. Nature 2009;458(7235):172–7.

[77] Gonciarz MD, Whitby FG, Eckert DM, Kieffer C, Heroux A, Sundquist WI, et al. Biochemical and structural studies of yeast Vps4 oligomerization. J Mol Biol 2008;384(4):878–95.

[78] Xiao J, Xia H, Yoshino-Koh K, Zhou J, Xu Z. Structural characterization of the ATPase reaction cycle of endosomal AAA protein Vps4. J Mol Biol 2007;374(3):655–70.

[79] Scott A, Chung HY, Gonciarz-Swiatek M, Hill GC, Whitby FG, Gaspar J, et al. Structural and mechanistic studies of VPS4 proteins. EMBO J 2005;24(20):3658–69.

[80] Baietti MF, Zhang Z, Mortier E, Melchior A, Degeest G, Geeraerts A, et al. Syndecan-syntenin-ALIX regulates the biogenesis of exosomes. Nat Cell Biol 2012;14(7): 677–85.

[81] Odorizzi G. The multiple personalities of Alix. J Cell Sci 2006;119(15):3025–32.

[82] Colombo M, Moita C, van Niel G, Kowal J, Vigneron J, Benaroch P, et al. Analysis of ESCRT functions in exosome biogenesis, composition and secretion highlights the heterogeneity of extracellular vesicles. J Cell Sci 2013;126(24):5553–65.

[83] Hoshino D, Kirkbride KC, Costello K, Clark ES, Sinha S, Grega-Larson N, et al. Exosome secretion is enhanced by invadopodia and drives invasive behavior. Cell Rep 2013;5(5):1159–68.

[84] Gross JC, Chaudhary V, Bartscherer K, Boutros M. Active Wnt proteins are secreted on exosomes. Nat Cell Biol 2012;14(10):1036–45.

[85] Tamai K, Tanaka N, Nakano T, Kakazu E, Kondo Y, Inoue J, et al. Exosome secretion of dendritic cells is regulated by Hrs, an ESCRT-0 protein. Biochem Biophys Res Commun 2010;399(3):384–90.

[86] Abrami L, Brandi L, Moayeri M, Brown MJ, Krantz BA, Leppla SH, et al. Hijacking multivesicular bodies enables long-term and exosome-mediated long-distance action of anthrax toxin. Cell Rep 2013;5(4):986–96.

[87] Trajkovic K, Hsu C, Chiantia S, Rajendran L, Wenzel D, Wieland F, et al. Ceramide triggers budding of exosome vesicles into multivesicular endosomes. Science 2008;319(5867):1244–7.

[88] Kulshreshtha A, Ahmad T, Agrawal A, Ghosh B. Proinflammatory role of epithelial cell-derived exosomes in allergic airway inflammation. J Allergy Clin Immunol 2013;131(4):1194–203. 1203.e1-14.

[89] Ghossoub R, Lembo F, Rubio A, Gaillard CB, Bouchet J, Vitale N, et al. Syntenin-ALIX exosome biogenesis and budding into multivesicular bodies are controlled by ARF6 and PLD2. Nat Commun 2014;5:3477.

[90] Géminard C, De Gassart A, Blanc L, Vidal M. Degradation of AP2 during reticulocyte maturation enhances binding of hsc70 and Alix to a common site on TFR for sorting into exosomes. Traffic 2004;5(3):181–93.

[91] Theos AC, Truschel ST, Tenza D, Hurbain I, Harper DC, Berson JF, et al. A lumenal domain-dependent pathway for sorting to intralumenal vesicles of multivesicular endosomes involved in organelle morphogenesis. Dev Cell 2006;10(3):343–54.

[92] Yang JM, Gould SJ. The cis-acting signals that target proteins to exosomes and microvesicles. Biochem Soc Trans 2013;41(1):277–82.

[93] Villarroya-Beltri C, Gutiérrez-Vázquez C, Sánchez-Cabo F, Pérez-Hernández D, Vázquez J, Martin-Cofreces N, et al. Sumoylated hnRNPA2B1 controls the sorting of miRNAs into exosomes through binding to specific motifs. Nat Commun 2013;4:2980.

[94] Ceresa BP, Schmid SL. Regulation of signal transduction by endocytosis. Curr Opin Cell Biol 2000;12(2):204–10.

[95] Kleijmeer MJ, Escola JM, UytdeHaag FG, Jakobson E, Griffith JM, Osterhaus AD, et al. Antigen loading of MHC class I molecules in the endocytic tract. Traffic 2001;2(2):124–37.

[96] Wubbolts R, Leckie RS, Veenhuizen PT, Schwarzmann G, Möbius W, Hoernschemeyer J, et al. Proteomic and biochemical analyses of human B cell–derived exosomes. Potential implications for their function and. multivesicular body formation. J Biol Chem 2003;278(13):10963–72.

[97] Nazarenko I, Rana S, Baumann A, McAlear J, Hellwig A, Trendelenburg M, et al. Cell surface tetraspanin Tspan8 contributes to molecular pathways of exosome-induced endothelial cell activation. Cancer Res 2010;70(4):1668–78.

[98] Perez-Hernandez D, Gutiérrez-Vázquez C, Jorge I, López-Martín S, Ursa A, Sánchez-Madrid F, et al. The intracellular interactome of tetraspanin-enriched microdomains reveals their function as sorting machineries toward exosomes. J Biol Chem 2013;288(17):11649–61.

[99] Valapala M, Vishwanatha JK. Lipid raft endocytosis and exosomal transport facilitate extracellular trafficking of annexin A2. J Biol Chem 2011;286(35):30911–25.

[100] Liang Y, Eng WS, Colquhoun DR, Dinglasan RR, Graham DR, Mahal LK. Complex N-linked glycans serve as a determinant for exosome/microvesicle cargo recruitment. J Biol Chem 2014;289(47):32526–37.

[101] Shen B, Wu N, Yang JM, Gould SJ. Protein targeting to exosomes/microvesicles by plasma membrane anchors. J Biol Chem 2011;286(16):14383–95.

[102] Savina A, Fader CM, Damiani MT, Colombo MI. Rab11 promotes docking and fusion of multivesicular bodies in a calcium-dependent manner. Traffic 2005;6(2):131–43.

[103] Ostrowski M, Carmo NB, Krumeich S, Fanget I, Raposo G, Savina A, et al. Rab27a and Rab27b control different steps of the exosome secretion pathway. Nat Cell Biol 2010;12(1):19–30.

[104] Hsu C, Morohashi Y, Yoshimura S, Manrique-Hoyos N, Jung S, Lauterbach MA, et al. Regulation of exosome secretion by Rab35 and its GTPase-activating proteins TBC1D10A-C. J Cell Biol 2010;189(2):223–32.

[105] Frühbeis C, Fröhlich D, Kuo WP, Amphornrat J, Thilemann S, Saab AS, et al. Neurotransmitter-triggered transfer of exosomes mediates oligodendrocyte-neuron communication. PLoS Biol 2013;11(7):e1001604.

[106] Yu X, Harris SL, Levine AJ. The regulation of exosome secretion: a novel function of the p53 protein. Cancer Res 2006;66(9):4795–801.

[107] Lespagnol A, Duflaut D, Beekman C, Blanc L, Fiucci G, Marine JC, et al. Exosome secretion, including the DNA damage-induced p53-dependent secretory pathway, is severely compromised in TSAP6/Steap3-null mice. Cell Death Differ 2008;15(11):1723–33.

[108] Savina A, Furlán M, Vidal M, Colombo MI. Exosome release is regulated by a calcium-dependent mechanism in K562 cells. J Biol Chem 2003;278(22):20083–90.

[109] Krämer-Albers EM, Bretz N, Tenzer S, Winterstein C, Möbius W, Berger H, et al. Oligodendrocytes secrete exosomes containing major myelin and stress-protective proteins: trophic support for axons? Proteomics Clin Appl 2007;1(11):1446–61.

[110] Fauré J, Lachenal G, Court M, Hirrlinger J, Chatellard-Causse C, Blot B, et al. Exosomes are released by cultured cortical neurones. Mol Cell Neurosci 2006;31(4):642–8.

[111] Ailawadi S, Wang X, Gu H, Fan GC. Pathologic function and therapeutic potential of exosomes in cardiovascular disease. Biochim Biophys Acta 2015;1852(1):1–11.

[112] Janowska-Wieczorek A, Wysoczynski M, Kijowski J, Marquez-Curtis L, Machalinski B, Ratajczak J, et al. Microvesicles derived from activated platelets induce metastasis and angiogenesis in lung cancer. Int J Cancer 2005;113(5):752–60.

[113] Bhatnagar S, Shinagawa K, Castellino FJ, Schorey JS. Exosomes released from macrophages infected with intracellular pathogens stimulate a proinflammatory response *in vitro* and *in vivo*. Blood 2007;110(9):3234–44.

[114] Bhatnagar S, Schorey JS. Exosomes released from infected macrophages contain *Mycobacterium avium* glycopeptidolipids and are proinflammatory. J Biol Chem 2007;282(35):25779–89.

[115] Mitchell PJ, Welton J, Staffurth J, Court J, Mason MD, Tabi Z, et al. Can urinary exosomes act as treatment response markers in prostate cancer? J Transl Med 2009;7:4.

[116] Yamada T, Inoshima Y, Matsuda T, Ishiguro N. Comparison of methods for isolating exosomes from bovine milk. J Vet Med Sci 2012;74(11):1523–5.

[117] King HW, Michael MZ, Gleadle JM. Hypoxic enhancement of exosome release by breast cancer cells. BMC Cancer 2012;12:421.

[118] Salomon C, Ryan J, Sobrevia L, Kobayashi M, Ashman K, Mitchell M, et al. Exosomal signaling during hypoxia mediates microvascular endothelial cell migration and vasculogenesis. PLoS One 2013;8(7):e68451.

[119] Chen T, Guo J, Yang M, Zhu X, Cao X. Chemokine-containing exosomes are released from heat-stressed tumor cells via lipid raft-dependent pathway and act as efficient tumor vaccine. J Immunol 2011;186(4):2219–28.

[120] Clayton A, Turkes A, Navabi H, Mason MD, Tabi Z. Induction of heat shock proteins in B-cell exosomes. J Cell Sci 2005;118(16):3631–8.

CHAPTER 2

An Overview of the Proteomic and miRNA Cargo in MSC-Derived Exosomes

Soon Sim Tan*, Tian Sheng Chen**, Kok Hian Tan†, Sai Kiang Lim*,‡
*Department of Surgery, Institute of Medical Biology, A*STAR, YLL School of Medicine, NUS, Singapore
**College of Fisheries, Huazhong Agricultural University, Wuhan, Hubei Province, P.R. China
†Department of Maternal Fetal Medicine, KK Women's and Children's Hospital, Singapore
‡Department of Surgery, YLL School of Medicine, NUS, Singapore

Contents

1 BACKGROUND

1.1 Mesenchymal Stem Cells

Mesenchymal stem cells (MSCs) were first described in 1968 as multipotent fibroblast-like cells from bone marrow [1] with a potential to differentiate into osteocytes, chondrocytes, adipocytes, and myoblasts [2,3]. MSCs have now been isolated from diverse sources of tissues such adipose tissue [4,5], liver [6], muscle [7], amniotic fluid [8], placenta [9,10], umbilical cord blood [4], dental pulp [11,12], and other sources [6,13]. Despite the difference in tissue sources, MSCs share several common characteristics: plastic adherence, expression of

CD105, CD73, and CD90 but not CD45, CD34, CD14 or CD11b, CD79α or CD19 and HLA-DR surface molecules, and an *in vitro* differentiation potential of osteogenesis, adipogenesis, and chondrogenesis [14].

MSCs are currently the most evaluated stem cells with more than 118 registered clinical trials (http://www.clinicaltrials.gov/, January 2014). The attractiveness of MSCs could be attributed to their regenerative potential, ease of isolation from adult tissues such as bone marrow and adipose tissue, and a large *ex vivo* expansion capacity (reviewed in [15–17]).

1.2 Mechanism of Action of MSCs: Paracrine Secretion

The use of MSCs as therapeutics was predicated on the hypothetical potential of transplanted MSCs to home in engraft and differentiate to repair injured tissues. However, it has been estimated that <1% of transplanted cells actually reached the target tissue with most of the cells being trapped in the liver, spleen, and lung [18] and few of the appropriately homed and engrafted cells demonstrated unambiguous evidence of differentiation [19–21]. Instead, the therapeutic efficacy of MSC therapy was observed to be independent of the engraftment or differentiation of transplanted MSCs at the site of injury [22–26]. Hence, it was proposed that MSCs exert their therapeutic effects through the secretion of growth factors and cytokines [27]. This proposal provides for a more all-encompassing mechanism of action for MSC efficacy in a wide range of disease indications and injuries [28–35].

1.3 Secreted MSC Mediators

The search for soluble mediators of MSC therapy has focused initially on chemokines, cytokines, or secreted proteins. For example, MSC that was approved for treatment of pediatric graft-versus-host disease in Canada and New Zealand was postulated to modulate regulatory T cells or Tregs, a subpopulation of T cells through the secretion of soluble mediators known to enhance Treg expansion. Many candidates such as transforming growth factor beta, prostaglandin E2, human leukocyte antigen G, interleukin-10, and indoleamine-pyrrole 2,3-dioxygenase were proposed (reviewed in [36]). However, none of the soluble mediators identified to date was sufficient in mediating the MSC immune modulatory effect [37].

To conduct an unbiased search for a secreted mediator, we took a first principle approach by profiling the proteome of MSC secretion, which is essentially culture medium conditioned by MSC. At that time, proteomic analysis of MSC-conditioned culture medium was not amenable to high-throughput unbiased mass spectrometry as MSC culture was still dependent on serum, which contains high abundance serum proteins such as albumin and immunoglobulins that obscure the detection of low abundance secreted proteins. To circumvent this, MSCs were cultured for 3 days in serum-free Dulbecco's modified eagle's medium supplemented with five peptides, namely insulin, transferrin, selenoprotein, fibroblast growth factor 2 with bovine serum albumin carrier, and platelet-derived growth factor AB [38]. Under this culture condition, MSCs remain viable for at least a week and when returned to serum-containing medium, they regain their proliferative activity without significant loss in their differentiation potential (unpublished observation). Using this defined medium, the conditioned medium (CM) was highly amenable to sensitive unbiased high-throughput mass spectrometry analysis to elucidate MSC secretome. As mass spectrometry is relatively insensitive in detecting small peptides such as cytokines, chemokines, and growth factors, mass spectrometry analysis was complemented with commercially available antibody arrays to detect small proteins that had commercially available antibodies. Using these two different analytical approaches, we detected 201 proteins secreted by MSCs. Of the 33 proteins that were previously reported to be secreted by MSCs, 29 proteins were present in the 201 proteins leaving 172 of the 201 proteins not known to be secreted by MSCs [38]. Among the 172 proteins, a significant proportion was cytoplasmic proteins that were not known to be secreted. Based on this proteome, we hypothesized that MSC-conditioned culture medium could be cardioprotective [38]. This was confirmed in a pig and mouse model of acute myocardial ischemia/reperfusion injury from [39] and a pig model of chronic myocardial ischemia [40].

1.4 Active Agent in MSC Secretion: Exosome

To identify the active therapeutic agent/s in the complex MSC secretion, we first estimated the size range of the agent by ultrafiltration using membranes with different pore sizes, and then testing the filtrate or retentate for cardioprotective activity. Using this method, we established the size range of the active agent as 50–200 ηm, which was then confirmed visually by electron microscopy [39,41]. Subsequent biochemical and biophysical analysis identified these agents as lipid membrane vesicles with a detergent-sensitive flotation density range of 1.10–1.18 g/mL in sucrose, enriched in exosome-associated protein markers such as CD9, CD81, and Alix, and membrane lipids such as cholesterol, sphingomyelin, and phosphatidylcholine [41].

When these membrane vesicles were isolated by size exclusion high performance liquid chromatography (HPLC), they constituted a population of homogeneously sized particles with a hydrodynamic radius of 55–65 ηm and were as efficacious as CM in reducing infarct size in a mouse model. Therefore, the cardioprotective agent in MSC secretion is an exosome. Following our report, exosomes were implicated as the agent mediating the biological activity of MSCs in attenuating type 1 diabetes and multiple myeloma progression [42,43].

2 CARGO OF MSC EXOSOMES

The implication of exosome as the active agent underpinning therapeutic efficacy of MSC converts MSC-based therapy from a cell to a biologic therapy. However, an exosome is far more complex than a biologic, and is more akin to a miniature cell. Exosomes are bilipid membrane vesicles that carry not only proteins but also RNAs. This latter adds an important dimension to the role of exosomes as a conveyer of intercellular communication by expanding the repertoire of communication to include genetic materials. However, the physiological role of exosomes and the underlying mechanism have remained relatively unknown, as exosomes themselves are relatively uncharacterized. Intensive efforts to isolate and define their biophysical and biochemical properties and their cargos are under way.

Studies on the protein and RNA cargo of MSC exosomes have been reported. The RNAs of MSC exosomes are found to be encapsulated within cholesterol-rich phospholipid vesicles such that they are susceptible to enzymatic RNase digestion only in the presence of an sodium dodecyl sulfate (SDS)-based lysis buffer, cyclodextrin, a cholesterol chelator, and phospholipase A2. The RNAs also have a buoyant density of a lipid vesicle, i.e., 1.11–1.15 g/mL that was sensitive to SDS, phospholipase A2, and cyclodextrin [44]. Like RNAs, proteins are also somewhat protected from enzymatic degradation. However, unlike RNAs, proteins localized in the lumen of the exosomes such as superoxide dismutase 1 are resistant to enzymatic degradation while others such as membrane-bound CD9, are partially resistant to trypsin digestion. The trypsin resistance of exosomal protein is abolished by pretreatment with a detergent-based cell lysis buffer [41].

2.1 RNA Cargo of MSC Exosomes

MSC exosomal RNAs when stained with ethidium bromide and visualized on agarose or polyacrylamide gels were mainly short RNAs of less than 300 nt [44]. 18S, or 28S RNAs were not visible [45,46].

Microarray hybridization of MSC exosomal RNA against probes for 151 miRNAs revealed the presence of 60 miRNAs and at least one ribosomal RNA degradative product [44]. When the composition of MSC exosomal miRNAs was compared against their cellular miRNAs, 106 of the miRNAs that were present in MSCs were not secreted, suggesting that MSCs secrete a select population of miRNAs in a regulated process. In addition, many passenger miRNA sequences were also found to be present in MSC exosome. As passenger miRNA sequences are only found in the primary and pre-miRNAs, and not mature miRNAs, the presence of passenger miRNA suggests the presence of primary and pre-miRNAs. Consistent with this, it was observed that RNase III digestion of the exosome RNA preferentially degraded 50–100 nt RNAs with a corresponding increase in <30 nt RNAs. As this shift in RNA sizes could be visually detected when the RNAs were resolved in ethidium bromide-stained gels, this suggested that primary and pre-miRNAs constituted a significant fraction of exosomal miRNAs. Semi-quantitative real time RT-PCR assay for specific miRNAs such as hsa-let-7b and hsa-let-7g, and their corresponding primary and pre-miRNAs [47], revealed that only mature miRNAs and their pre-miRNAs but not the primary miRNAs were present in MSC exosomal RNAs. Furthermore, the relative ratio of the mature miRNAs to their pre-miRNAs was significantly lower than those in the MSC source, and RNase III treatment increased this ratio through a reduction in precursor miRNA level and a concomitant increase in mature miRNA level. This preferential sequestering of pre-miRNAs in exosomes further demonstrated the high degree of regulation involved in the loading of miRNAs into exosomes. A rationale for this selective sequestration of pre-miRNAs could be that unlike others [48], Dicer and protein argonaute-2, i.e., RNA-induced silencing complex (RISC) components were not detected in MSC exosomes. Since miRNAs are biologically functional only when associated with RISC, and only pre-miRNA could be loaded onto RISCs, this absence of RISC in MSC exosomes implied that any miRNA-induced biological effects of MSC exosomes [49] would have to be mediated by pre-miRNA and not the mature miRNA.

Although miRNA and pre-miRNAs were readily detected in MSC exosomal RNA, it is obvious from the molecular weight distribution of RNAs in ethidium bromide-stained gels that they probably represent a fraction of total MSC exosomal RNA. While MSC exosome-mediated activities were reportedly underpinned by miRNAs, it is highly probable that other RNA species also mediate MSC exosomal activities. Hence, it is important to map MSC exosomal RNA extensively beyond miRNA profiling. At present, deep sequencing represents the state of art for non-biased DNA and RNA analysis, and deep sequencing has been reportedly used to map exosomal RNA from plasma or neuronal cell secretion [50,51]. These deep sequencing studies revealed the complexity of RNAs that are present in exosomal RNA and provide a harbinger for the complexity of MSC exosomal RNA. In plasma exosomes, miRNA sequences represent 76.2% of mappable reads. The remaining sequences included ribosomal RNA, long noncoding RNA, piwi-interacting RNA, transfer RNA, small nuclear RNA, and small nucleolar RNA. In contrast, 50% of the sequences in neuronal cell exosomes are the RNA repeat regions with some rRNA. The next largest class of RNA or 15% of the sequences map to small RNA sequences such as miRNA, tRNA, small nucleolar RNA (snoRNA), small nuclear RNA (snRNA), silencing RNA (siRNA), and small cytoplasmic RNA. Interestingly, it was observed that while prion infection did not have a qualitative effect on the distribution of RNAs across different classes, it changed the miRNA signature of the neuronal cell exosomes and many of the affected miRNAs have been implicated in neuronal disease or dysfunction. Based on these observations, we expect that MSC exosomes would have an equally complex and diverse RNA cargo that could be altered by its differentiation potency or state and microenvironment.

2.2 Proteome of MSC Exosomes

The proteome of MSC exosomes has been characterized using standard proteomic analysis tools such as mass spectrometry. Our group has previously described the proteome of three independently prepared batches of HPLC purified HuES9.E1 exosomes [52,53]. The proteome was elucidated using two complementary approaches, mass spectrometry and antibody arrays [38]. Although mass spectrometry is currently the state of the art for unbiased high-throughput identification of proteins, it is limited to relatively large proteins, and small proteins such as the cytokines are generally not amenable to detection by mass spectrometry. One approach that we have taken to circumvent this limitation is to use antibody arrays. The use of these complementary approaches enables a more comprehensive coverage of the proteome.

A critical issue in preparing exosomes from conditioned culture medium for mass spectrometric analysis is the need to eliminate serum from the culture medium as serum itself contains exosomes and highly abundant proteins such as albumin and immunoglobulins. In our proteomic and RNA analysis of MSC exosomes, the exosomes were prepared from serum-free chemically defined culture medium that was conditioned by MSCs. To ensure a high fidelity in the mass spectrometric analysis, analysis was performed on multiple batches of exosomes purified by HPLC from independent preparations of CM. Exosomes were tryptic digested and desalted, and then analyzed by multidimensional protein identification technology with an LC–MS/MS system. Proteins were identified by searching the mass spectra against the International Protein Index (IPI) human protein database via an in-house Mascot server using the search parameters: a maximum of three missed cleavages using trypsin; fixed modification, carbamidomethylation; and variable modification, oxidation of methionine with mass tolerances of 2.0 and 0.8 Da for peptide precursor and fragmentations, respectively. A protein was accepted as a true positive if two or more different peptides from the same protein had ion scores greater than their Mascot identity score of 41 with significance threshold $p < 0.05$. The IPI identifier of each protein was converted to the gene symbol using the protein cross reference table.

The other approach to identify exosomal proteins was by screening against commercially available arrays of antibodies against known proteins, particularly small proteins such as the cytokines and growth factors. Altogether, the combined analysis using mass spectrometry and antibody arrays identified a total of 866 gene products in MSC exosomes (www.exocarta. org). Of these, 101 proteins were identified by antibody array and 10 of these 101 proteins were also detected by mass spectrometry analysis.

Of the 866 proteins identified in MSC exosomes, 320 were found in the 739 proteins previously identified in the unfractionated CM, and 419 proteins found in the CM were not present in the exosome [13,41]. Together, these observations indicate that MSCs secrete proteins via several routes. While some, as represented by the 320 proteins, are secreted as both soluble and membrane-encapsulated proteins, others, such as the 419 proteins or the remaining 539 proteins detected only in the exosome, were secreted either exclusively as soluble proteins or membrane-encapsulated proteins, respectively. Most importantly, among the proteins that were detected in the exosome preparations were 17 proteins, namely glyceraldehyde 3-phosphate dehydrogenase (GAPDH), pyruvate kinase (PK), eukaryotic translation elongation factor 1A1, milk fat globule EGF factor 8 protein, tetraspanins, 14-3-3 proteins, Gα proteins, clathrin, Alix (PDCD6IP), MHC class 1, annexins (ANX), Rab proteins, ezrin (VIL2), radixin (RDX) and moesin (MSN) (ERM), actin, and tubulin; heat shock protein 70 and heat shock protein 90 were found to be present in at least 50% of the exosomes that have been characterized to date [54]. In addition, the detection of endosome-associated proteins such as Alix (PDCD6IP), Tsg101, and Rab is indicative of an endosomal origin.

2.2.1 Computational Analysis of the MSC Exosome Proteome

A useful first-level analysis of the MSC exosome proteome is a functional clustering of proteins based on the gene ontology classification system using publicly available algorithms such as PANTHER (Protein ANalysis THrough Evolutionary Relationships) analytical software [55,56] or commercially available software. When the 866 proteins were clustered into biological processes using PANTHER, there were 32 biological processes in which the frequency of genes from the exosome proteome was significantly greater than the reference frequency of genes in the NCBI database ($p < 0.001$), and three processes where the frequency was significantly lower than the reference, i.e., under-represented in the MSC exosome proteome.

These 32 biological processes could be broadly classified into several distinct activities, namely communication, structure and mechanics, inflammation, exosome biogenesis, tissue repair and regeneration, and metabolism [52,53]. The three processes that were under-represented in the MSC exosome proteome were mRNA transcription, nucleoside, nucleotide and nucleic acid metabolism, and mRNA transcription regulation. As the latter three processes are essentially processes for gene expression, their absence implied that the mechanism of action for MSC exosome-mediated activities is not through regulation of gene expression.

The concentration of MSC exosome-mediated biological processes in communication, structure and mechanics, inflammation, exosome biogenesis, tissue repair and regeneration, and metabolism resonates with the known biology of exosomes in general and also relates specifically to the therapeutic efficacy of MSC secretion in tissue repair and regeneration.

Exosomes are generally viewed as vehicles for intercellular communication and morphogen signaling [57,58]. The clustering of exosomal proteins in important cellular communication processes such as signal transduction, ligand-mediated signaling and intracellular signaling cascade, etc., corroborates the current perception of exosomes as a vehicle for intercellular communication. Similarly, processes such as intracellular protein trafficking, general vesicle transport, endocytosis, receptor-mediated endocytosis, other protein targeting and localization, and exocytosis are highly pertinent to the current view on the biogenesis of exosome as an endosomal process involving ESCRT-mediated sorting and the formation of multivesicular bodies, and fusion with the plasma membrane. On the other hand, the concentration of MSC exosomal proteome-mediated processes in the area of structure and mechanics, inflammation, exosome biogenesis, tissue repair and regeneration, and metabolism appears to be more specific to the therapeutic efficacy of MSC exosomes and could potentially provide a molecular underpinning for this efficacy.

2.2.2 Biochemical Activity of the MSC Proteome

The elucidation of the RNA and protein cargo of MSC exosome based on partial RNA or protein sequences represents an important initial step to understand the biological importance of the MSC exosome. However, these partial sequences alone are limited in evaluating or predicting the integrity or activity of a biological molecule. In particular, the biochemical activity of a protein is especially sensitive to a multitude of factors such as pH and structural conformation notwithstanding the integrity of its sequence. Therefore, the elucidation of the MSC exosomal proteome would be inconsequential without a corresponding assessment of its biochemical potency.

As the proteome of MSC exosomes includes enzymes or enzyme complexes, inhibitors, and immune modulators, MSC exosomes were assessed for activity in each of these three categories, namely enzymatic, inhibitory, and immune modulatory activities.

2.2.3 Enzymes

To date, we have evaluated the enzymatic activity of several enzymes such as NT5E (or ecto-5′-nucleotidase or CD73), three glycolytic enzymes, GAPDH, phosphoglycerate kinase (PGK), and pyruvate kinase muscle isoenzyme 2 (PKm2), and the 20S proteasome.

CD73 is an ecto-5′-nucleotidase [59] and the only known extracellular ecto-5′-nucleotidase. It is primarily responsible for the dephosphorylation of extracellular adenosine monophosphate (AMP) released by distressed cells into adenosine, an important activator of Ras/Raf/mitogen-activated protein kinase /extracellular signal-regulated kinases mitogen-activated protein kinases and phosphatase and tensin homolog /phosphoinositide 3-kinase (PI3K)/protein kinase B (Akt)/mammalian target of rapamycin [60]. The integrity of CD73 on MSC exosomes was evidenced by the capacity of the MSC exosome to dephosphorylate AMP to inorganic phosphate and adenosine. In the presence of AMP, MSC exosomes activate adenosine receptors and phosphorylate the PI3K/Akt pathway in H9C2 cardiomyocytes [52]. More importantly, MSC exosome treatment elicited an immediate and significant increase in cardiac Akt and GSK3 phosphorylation within an hour after reperfusion in a mouse model of myocardial ischemia/reperfusion [61].

Proteomic analysis of MSC exosome revealed the presence of all five glycolytic enzymes, GAPDH, PGK, PGM, enolase, and PKm2 that respectively catalyze the five consecutive reactions in the adenosine triphosphate (ATP)-generating stage of glycolysis. Of these, the three enzymes namely, GAPDH, PGK, and PKm2, that generate either ATP or NADH were confirmed to be present in MSC exosomes by immunoblotting and determined to enzymatically active *in vitro* [52]. The presence of the five glycolytic enzymes is consistent with close subcellular localization of glycolytic enzymes in red blood cells [62], possibly to produce ATP efficiently. It is possible that as in the red blood cells, these five enzymes also colocalized in close proximity in the exosomes to increase the efficiency of glycolysis. The presence of these five enzymes indicates a potential to generate glycolytic ATP. MSC exosomes increased the intracellular ATP level of cells pretreated with oligomycin to inhibit mitochondrial ATPase and oxidative phosphorylation [52], and ATP level in the cardiac tissues of a mouse model of myocardial ischemia/reperfusion [61].

A significant observation during proteomic analysis of MSC exosome was the presence of the full complement of all seven α and seven β chains that made up the 20S proteasome as well as the additional three beta subunits of "immunoproteasome" [53]. These observations are indicative of fully assembled and functional 20S proteasome and immunoproteasome. Consistent with this indication, the MSC exosome exhibits proteolytic activity that is inhibited by lactacystin, a proteasome-specific inhibitor, and reduces misfolded proteins in a mouse model of myocardial ischemia/reperfusion injury [53]. The presence of 20S proteasome in exosomes could be the configuration of extracellular, circulating proteasomes that have been implicated in important physiological and pathological functions [63–67].

2.2.4 Inhibitors

An example of a protein inhibitor in MSC exosomes is the glycosylphosphatidyl inositol (GPI)-anchored membrane CD59, commonly known as protectin. Both membrane-bound and soluble recombinant CD59 are known to inhibit MAC-mediated cell lysis in a cis or trans manner [68–70]. MAC stands for membrane attack complex, which is a transmembrane channel that causes cell lysis and its formation is initiated by the activation of the complement system (reviewed in [71]). MSC exosomes have been reported to inhibit complement-mediated lysis of sheep red blood cells in a CD59-dependent manner [53].

2.2.5 Immune Modulators

The proteomes of MSC exosomes carry several candidate endogenous Toll-like receptor 4 (TLR4) ligands, namely fibronectin 1 (FN1), the heat shock proteins, and fibrinogens [72,73]. Although TLRs are present in MSCs [74], they were not detectable in the exosomes. FN1 is a family of high molecular weight alternatively spliced glycosylated products of a single gene and can be classified largely into cellular and plasma FN1. Plasma FN1s, which are produced by liver cells, do not contain a specific alternatively spliced domain known as extradomain A (EDA) found in cellular FN1s, which are produced by many cell types in response to injury. EDA-containing FN1 is the first and best characterized endogenous TLR4 ligand [75]. The FN1 in MSC exosomes has EDA. Also, IST-9, an EDA-neutralizing monoclonal antibody [76], abolished 60% of exosome-induced secreted embryonic alkaline phosphatase in HEK-Blue-hTLR4 indicating that 60% of TLR4 activation by exosomes was mediated by EDA-containing FN1 and the remaining 40% by other endogenous ligands such as the heat shock proteins or fibrinogens.

3 CONCLUSIONS

MSC exosome encapsulates a complex protein and RNA cargo that represents a highly selective fraction of the cellular contents in MSC. The enrichment of pre-miRNAs and the *in vitro* biochemical activity of the proteins provide a molecular rationale for the therapeutic efficacy of MSC exosomes in animal models of human diseases. In support of this, MSC exosomes elicit a physiological and biochemical response in animals that are consistent with the biochemical potential of its proteome and RNA cargo.

ABBREVIATIONS

Akt	Protein kinase B
AMP	adenosine monophosphate
CM	Conditioned medium
EDA	Extradomain A
GAPDH	Glyceraldehyde 3-phosphate dehydrogenase
HPLC	High-performance liquid chromatography
MSC	Mesenchymal stem cells
PANTHER	Protein analysis through evolutionary relationships
PGK	Phosphoglycerate kinase
PI3K	Phosphoinositide 3-kinase
PKm2	Pyruvate kinase muscle isoenzyme 2
RISC	RNA-induced silencing complex
SDS	Sodium dodecyl sulfate
TLR4	Toll-like receptor 4

REFERENCES

[1] Friedenstein AJ, Deriglasova UF, Kulagina NN, Panasuk AF, Rudakowa SF, Luria EA, et al. Precursors for fibroblasts in different populations of hematopoietic cells as detected by the *in vitro* colony assay method. Exp Hematol 1974;2(2):83–92.

[2] Prockop DJ. Marrow stromal cells as stem cells for nonhematopoietic tissues. Science 1997;276(5309):71–4.

[3] Pittenger MF, Mackay AM, Beck SC, Jaiswal RK, Douglas R, Mosca JD, et al. Multilineage potential of adult human mesenchymal stem cells. Science 1999;284(5411):143–7.

[4] Kern S, Eichler H, Stoeve J, Kluter H, Bieback K. Comparative analysis of mesenchymal stem cells from bone marrow, umbilical cord blood, or adipose tissue. Stem Cells 2006;24(5):1294–301.

[5] Banas A, Teratani T, Yamamoto Y, Tokuhara M, Takeshita F, Quinn G, et al. Adipose tissue-derived mesenchymal stem cells as a source of human hepatocytes. Hepatology 2007;46(1):219–28.

[6] In 't Anker PS, Noort WA, Scherjon SA, Kleijburg-Van der Keur C, Kruisselbrink AB, Van Bezooijen RL, et al. Mesenchymal stem cells in human second-trimester bone marrow, liver, lung, and spleen exhibit a similar immunophenotype but a heterogeneous multilineage differentiation potential. Haematologica 2003;88(8):845–52.

[7] Young HE, Steele TA, Bray RA, Hudson J, Floyd JA, Hawkins K, et al. Human reserve pluripotent mesenchymal stem cells are present in the connective tissues of skeletal muscle and dermis derived from fetal, adult, and geriatric donors. Anat Rec 2001;264(1):51–62.

[8] Roubelakis MG, Pappa KI, Bitsika V, Zagoura D, Vlahou A, Papadaki HA, et al. Molecular and proteomic characterization of human mesenchymal stem cells derived from amniotic fluid: comparison to bone marrow mesenchymal stem cells. Stem Cells Dev 2007;16(6):931–52.

[9] Fukuchi Y, Nakajima H, Sugiyama D, Hirose I, Kitamura T, Tsuji K. Human placenta-derived cells have mesenchymal stem/progenitor cell potential. Stem Cells 2004;22(5):649–58.

[10] Miao Z, Jin J, Chen L, Zhu J, Huang W, Zhao J, et al. Isolation of mesenchymal stem cells from human placenta: comparison with human bone marrow mesenchymal stem cells. Cell Biol Int 2006;30(9):681–7.

[11] Jo YY, Lee HJ, Kook SY, Choung HW, Park JY, Chung JH, et al. Isolation and characterization of postnatal stem cells from human dental tissues. Tissue Eng 2007;13(4): 767–73.

[12] Huang GTJ, Gronthos S, Shi S. Critical reviews in oral biology & medicine: mesenchymal stem cells derived from dental tissues vs. those from other sources: their biology and role in regenerative medicine. J Dent Res 2009;88(9):792–806.

[13] Lai RC, Arslan F, Tan SS, Tan B, Choo A, Lee MM, et al. Derivation and characterization of human fetal MSCs: an alternative cell source for large-scale production of cardioprotective microparticles. J Mol Cell Cardiol 2010;48(6):1215–24.

[14] Dominici M, Le Blanc K, Mueller I, Slaper-Cortenbach I, Marini F, Krause D, et al. Minimal criteria for defining multipotent mesenchymal stromal cells. The International Society for Cellular Therapy position statement. Cytotherapy 2006;8(4):315–17.

[15] Bernardo ME, Pagliara D, Locatelli F. Mesenchymal stromal cell therapy: a revolution in regenerative medicine? Bone Marrow Transpl 2011;47:164–71.

[16] Brooke G, Cook M, Blair C, Han R, Heazlewood C, Jones B, et al. Therapeutic applications of mesenchymal stromal cells. Semin Cell Dev Biol 2007;18(6):846–58.

[17] Salem HK, Thiemermann C. Mesenchymal stromal cells: current understanding and clinical status. Stem Cells 2010;28(3):585–96.

[18] Phinney DG, Prockop DJ. Concise review: mesenchymal stem/multipotent stromal cells: the state of transdifferentiation and modes of tissue repair – current views. Stem Cells 2007;25(11):2896–902.

[19] Ferrand J, Noel D, Lehours P, Prochazkova-Carlotti M, Chambonnier L, Menard A, et al. Human bone marrow-derived stem cells acquire epithelial characteristics through fusion with gastrointestinal epithelial cells. PLoS One 2011;6(5):e19569.

[20] Spees JL, Olson SD, Ylostalo J, Lynch PJ, Smith J, Perry A, et al. Differentiation, cell fusion, and nuclear fusion during *ex vivo* repair of epithelium by human adult stem cells from bone marrow stroma. Proc Natl Acad Sci USA 2003;100(5):2397–402.

[21] Vassilopoulos G, Wang PR, Russell DW. Transplanted bone marrow regenerates liver by cell fusion. Nature 2003;422(6934):901–4.

[22] Prockop DJ. "Stemness" does not explain the repair of many tissues by mesenchymal stem/multipotent stromal cells (MSCs). Clin Pharmacol Ther 2007;82(3):241–3.

[23] da Silva Meirelles L, Caplan AI, Nardi NB. In search of the *in vivo* identity of mesenchymal stem cells. Stem Cells 2008;26(9):2287–99.

[24] Dai W, Hale SL, Martin BJ, Kuang JQ, Dow JS, Wold LE, et al. Allogeneic mesenchymal stem cell transplantation in postinfarcted rat myocardium: short- and long-term effects. Circulation 2005;112(2):214–23.

[25] Noiseux N, Gnecchi M, Lopez-Ilasaca M, Zhang L, Solomon SD, Deb A, et al. Mesenchymal stem cells overexpressing Akt dramatically repair infarcted myocardium and improve cardiac function despite infrequent cellular fusion or differentiation. Mol Ther 2006;14(6):840–50.

[26] Iso Y, Spees JL, Serrano C, Bakondi B, Pochampally R, Song YH, et al. Multipotent human stromal cells improve cardiac function after myocardial infarction in mice without long-term engraftment. Biochem Biophys Res Commun 2007;354(3):700–6.

[27] Caplan AI, Dennis JE. Mesenchymal stem cells as trophic mediators. J Cell Biochem 2006;98(5):1076–84.

[28] Meirelles Lda S, Fontes AM, Covas DT, Caplan AI. Mechanisms involved in the therapeutic properties of mesenchymal stem cells. Cytokine Growth Factor Rev 2009;20(5–6):419–27.

[29] Chen L, Tredget EE, Wu PY, Wu Y. Paracrine factors of mesenchymal stem cells recruit macrophages and endothelial lineage cells and enhance wound healing. PLoS One 2008;3(4):e1886.

[30] Hung SC, Pochampally RR, Chen SC, Hsu SC, Prockop DJ. Angiogenic effects of human multipotent stromal cell conditioned medium activate the PI3K-Akt pathway

in hypoxic endothelial cells to inhibit apoptosis, increase survival, and stimulate angiogenesis. Stem Cells 2007;25(9):2363–70.

[31] Kinnaird T, Stabile E, Burnett MS, Shou M, Lee CW, Barr S, et al. Local delivery of marrow-derived stromal cells augments collateral perfusion through paracrine mechanisms. Circulation 2004;109(12):1543–9.

[32] Li L, Zhang S, Zhang Y, Yu B, Xu Y, Guan Z. Paracrine action mediate the anti-fibrotic effect of transplanted mesenchymal stem cells in a rat model of global heart failure. Mol Biol Rep 2009;36(4):725–31.

[33] Lin YC, Ko TL, Shih YH, Lin MY, Fu TW, Hsiao HS, et al. Human umbilical mesenchymal stem cells promote recovery after ischemic stroke. Stroke 2011;42(7):2045–53.

[34] Togel F, Weiss K, Yang Y, Hu Z, Zhang P, Westenfelder C. Vasculotropic, paracrine actions of infused mesenchymal stem cells are important to the recovery from acute kidney injury. Am J Physiol Renal Physiol 2007;292(5):F1626–35.

[35] van Poll D, Parekkadan B, Cho CH, Berthiaume F, Nahmias Y, Tilles AW, et al. Mesenchymal stem cell-derived molecules directly modulate hepatocellular death and regeneration in vitro and in vivo. Hepatology 2008;47(5).1634–43.

[36] Burr SP, Dazzi F, Garden OA. Mesenchymal stromal cells and regulatory T cells: the Yin and Yang of peripheral tolerance? Immunol Cell Biol 2013;91(1):12–18.

[37] Ghannam S, Bouffi C, Djouad F, Jorgensen C, Noel D. Immunosuppression by mesenchymal stem cells: mechanisms and clinical applications. Stem Cell Res Ther 2010;1(1):2.

[38] Sze SK, de Kleijn DP, Lai RC, Khia Way Tan E, Zhao H, Yeo KS, et al. Elucidating the secretion proteome of human embryonic stem cell-derived mesenchymal stem cells. Mol Cell Proteomics 2007;6(10):1680–9.

[39] Timmers L, Lim S-K, Arslan F, Armstrong JS, Hoefler IE, Doevendans PA, et al. Reduction of myocardial infarct size by human mesenchymal stem cell conditioned medium. Stem Cell Res 2008;1:129–37.

[40] Timmers L, Lim SK, Hoefer IE, Arslan F, Lai RC, van Oorschot AAM, et al. Human mesenchymal stem cell-conditioned medium improves cardiac function following myocardial infarction. Stem Cell Res 2011;6(3):206–14.

[41] Lai RC, Arslan F, Lee MM, Sze NS, Choo A, Chen TS, et al. Exosome secreted by MSC reduces myocardial ischemia/reperfusion injury. Stem Cell Res 2010;4:214–22.

[42] Rahman MJ, Regn D, Bashratyan R, Dai YD. Exosomes released by islet-derived mesenchymal stem cells trigger autoimmune responses in NOD mice. Diabetes 2013;63:1008–20.

[43] Roccaro AM, Sacco A, Maiso P, Azab AK, Tai Y-T, Reagan M, et al. BM mesenchymal stromal cell-derived exosomes facilitate multiple myeloma progression. J Clin Invest 2013;123(4):1542–55.

[44] Chen TS, Lai RC, Lee MM, Choo AB, Lee CN, Lim SK. Mesenchymal stem cell secretes microparticles enriched in pre-microRNAs. Nucleic Acids Res 2010;38(1):215–24.

[45] Smalheiser NR. Exosomal transfer of proteins and RNAs at synapses in the nervous system. Biol Direct 2007;2:35.

[46] Valadi H, Ekstrom K, Bossios A, Sjostrand M, Lee JJ, Lotvall JO. Exosome-mediated transfer of mRNAs and microRNAs is a novel mechanism of genetic exchange between cells. Nat Cell Biol 2007;9(6):654–9.

[47] Chen TS, Lim SK. Measurement of precursor miRNA in exosomes from human ESC-derived mesenchymal stem cells. Method Mol Biol 2013;1024:69–86.

[48] Gibbings DJ, Ciaudo C, Erhardt M, Voinnet O. Multivesicular bodies associate with components of miRNA effector complexes and modulate miRNA activity. Nat Cell Biol 2009;11(9):1143–9.

[49] Xin H, Li Y, Buller B, Katakowski M, Zhang Y, Wang X, et al. Exosome-mediated transfer of miR-133b from multipotent mesenchymal stromal cells to neural cells contributes to neurite outgrowth. Stem Cells 2012;30(7):1556–64.

[50] Bellingham SA, Coleman BM, Hill AF. Small RNA deep sequencing reveals a distinct miRNA signature released in exosomes from prion-infected neuronal cells. Nucleic Acids Res 2012;40:10937–49.

[51] Huang X, Yuan T, Tschannen M, Sun Z, Jacob H, Du M, et al. Characterization of human plasma-derived exosomal RNAs by deep sequencing. BMC Genomics 2013;14(1):319.

[52] Lai RC, Yeo RWY, Tan SS, Zhang B, Yin Y, Sze NSK, et al. Mesenchymal stem cell exosomes: the future MSC-based therapy? Mesenchymal Stem Cell Ther 2013;39.

[53] Lai RC, Tan SS, Teh BJ, Sze SK, Arslan F, de Kleijn DP, et al. Proteolytic potential of the MSC exosome proteome: implications for an exosome-mediated delivery of therapeutic proteasome. Int J Proteomics 2012;2012:971907.

[54] Thery C, Ostrowski M, Segura E. Membrane vesicles as conveyors of immune responses. Nat Rev Immunol 2009;9(8):581–93.

[55] Thomas PD, Campbell MJ, Kejariwal A, Mi H, Karlak B, Daverman R, et al. PANTHER: a library of protein families and subfamilies indexed by function. Genome Res 2003;13(9):2129–41.

[56] Thomas PD, Kejariwal A, Guo N, Mi H, Campbell MJ, Muruganujan A, et al. Applications for protein sequence-function evolution data: mRNA/protein expression analysis and coding SNP scoring tools. Nucleic Acids Res 2006;34(Web server issue): W645–50.

[57] Simons M, Raposo G. Exosomes – vesicular carriers for intercellular communication. Curr Opin Cell Biol 2009;21(4):575–81.

[58] Keller S, Sanderson MP, Stoeck A, Altevogt P. Exosomes: from biogenesis and secretion to biological function. Immunol Lett 2006;107(2):102–8.

[59] Colgan SP, Eltzschig HK, Eckle T, Thompson LF. Physiological roles for ecto-5'-nucleotidase (CD73). Purinergic Signal 2006;2(2):351–60.

[60] Steelman LS, Chappell WH, Abrams SL, Kempf RC, Long J, Laidler P, et al. Roles of the Raf/MEK/ERK and PI3K/PTEN/Akt/mTOR pathways in controlling growth and sensitivity to therapy-implications for cancer and aging. Aging (Albany, NY) 2011;3(3):192–222.

[61] Arslan F, Lai RC, Smeets MB, Akeroyd L, Choo A, Aguor EN, et al. Mesenchymal stem cell-derived exosomes increase ATP levels, decrease oxidative stress and activate PI3K/Akt pathway to enhance myocardial viability and prevent adverse remodeling after myocardial ischemia/reperfusion injury. Stem Cell Res 2013;10(3):301–12.

[62] Campanella ME, Chu H, Low PS. Assembly and regulation of a glycolytic enzyme complex on the human erythrocyte membrane. Proc Natl Acad Sci USA 2005;102(7):2402–7.

[63] Zoeger A, Blau M, Egerer K, Feist E, Dahlmann B. Circulating proteasomes are functional and have a subtype pattern distinct from 20S proteasomes in major blood cells. Clin Chem 2006;52(11):2079–86.

[64] Jakob C, Egerer K, Liebisch P, Turkmen S, Zavrski I, Kuckelkorn U, et al. Circulating proteasome levels are an independent prognostic factor for survival in multiple myeloma. Blood 2007;109(5):2100–5.

[65] Sixt SU, Peters J. Extracellular alveolar proteasome: possible role in lung injury and repair. Proc Am Thorac Soc 2010;7(1):91–6.

[66] Sixt SU, Adamzik M, Spyrka D, Saul B, Hakenbeck J, Wohlschlaeger J, et al. Alveolar extracellular 20S proteasome in patients with acute respiratory distress syndrome. Am J Respir Crit Care Med 2009;179(12):1098–106.

[67] Ma W, Kantarjian H, Bekele B, Donahue AC, Zhang X, Zhang ZJ, et al. Proteasome enzymatic activities in plasma as risk stratification of patients with acute myeloid leukemia and advanced-stage myelodysplastic syndrome. Clin Cancer Res 2009;15(11): 3820–6.

[68] Davies A, Simmons DL, Hale G, Harrison RA, Tighe H, Lachmann PJ, et al. CD59, an LY-6-like protein expressed in human lymphoid cells, regulates the action of the complement membrane attack complex on homologous cells. J Exp Med 1989;170(3):637–54.

[69] Li LM, Li JB, Zhu Y, Fan GY. Soluble complement receptor type 1 inhibits complement system activation and improves motor function in acute spinal cord injury. Spinal Cord 2010;48(2):105–11.

[70] Gandhi J, Cashman SM, Kumar-Singh R. Soluble CD59 expressed from an adenovirus *in vivo* is a potent inhibitor of complement deposition on murine liver vascular endothelium. PLoS One 2011;6(6):e21621.

[71] Zipfel PF, Skerka C. Complement regulators and inhibitory proteins. Nat Rev Immunol 2009;9(10):729–40.

[72] Lai RC, Wang XD, Zhang X, Lin WQ, Rong TH. Heart fatty acid-binding protein may not be an early biomarker for anthracycline-induced cardiotoxicity in rabbits. Med Oncol 2012;29(3):2303–8.

[73] Lai RC, Yeo RW, Tan SS, Zhang B, Yin Y, Sze NS, et al. Mesenchymal stem cell exosomes: the future MSC-based therapy? In: Chase LG, Vemuri MC, editors. Mesenchymal stem cell therapy. New York: Humana Press; 2012.

[74] Tomchuck SL, Zwezdaryk KJ, Coffelt SB, Waterman RS, Danka ES, Scandurro AB. Toll-like receptors on human mesenchymal stem cells drive their migration and immunomodulating responses. Stem Cells 2008;26(1):99–107.

[75] Okamura Y, Watari M, Jerud ES, Young DW, Ishizaka ST, Rose J, et al. The extra domain A of fibronectin activates Toll-like receptor 4. J Biol Chem 2001;276(13): 10229–333.

[76] Serini G, Bochaton-Piallat ML, Ropraz P, Geinoz A, Borsi L, Zardi L, et al. The fibronectin domain ED-A is crucial for myofibroblastic phenotype induction by transforming growth factor-beta1. J Cell Biol 1998;142(3):873–81.

CHAPTER 3

Exosome Function in miRNA-Mediated Paracrine Effects

Sathyamoorthy Balasubramanian*,†, Sheeja Rajasingh*, Jayakumar Thangavel*, Buddhadeb Dawn*, Johnson Rajasingh*,**
*Department of Internal Medicine, Cardiovascular Research Institute, University of Kansas Medical Center, Kansas City, MO, USA
**Department of Biochemistry and Molecular Biology, University of Kansas Medical Center, Kansas City, MO, USA
†Centre for Biotechnology, A.C. Technology Campus, Anna University, Chennai, India

Contents

1 INTRODUCTION

Most cells, including stem cells, are capable of secreting a variety of substances into the extracellular space in the form of vesicles with different size, morphology, and contents [1]. These vesicles are ultra-small membrane vesicles, which range from 30 nm to 100 nm in size [2–4]. The size can be measured by electron microscopy, dynamic light scattering analysis [5], and nanoparticle tracking analysis [6]. These vesicles are generally known as microvesicular bodies (MVBs), ectosomes, shedding vesicles, and extracellular vesicles (EVs)

[7–12]. The term exosome means "exfoliated membrane vesicle" [13] and was first discovered as prostasome in 1978 [14], and in 1987, the term exosome was coined [15]. Exosomes are secreted by many cell types and cell lines, including mast cells, B lymphocyte cell lines, dendritic cells, stem cells, endothelial cells, smooth muscle cells, neuronal cells, and cancer cells [5,16–20]. Exosomes are released from viable cells or diseased cell types either constitutively or upon activation into interstitial spaces as well as into body fluids. Interestingly, these exosomes are also released from cells *in vitro* [21]. Studies have also shown that exosomes can be used as a cell-free vaccine for various diseases [3,22].

Recently, the role of exosomes has been well documented both in normal physiological conditions such as lactation, immune response, and neuronal function and in pathophysiological conditions such as development and progression of liver diseases, neurodegenerative diseases, and cancer [23]. Based on the cell origin, exosomes attribute morphogen transporters in the creation of polarity during development and differentiation [24]. Exosomes contain proteins, lipids, RNAs, nontranscribed RNAs, microRNAs (miRNAs), and small RNAs [23] as well as DNA [25]. Interestingly, DNA and RNA containing exosomes are known to bud from the endosomal compartment and the RNA present in the exosomes has an active biological role in cells. Recent studies have shown that exosomes function as a transporter, which can deliver their cargos to the recipient cells. Because of its cellular function between the cells, exosomes are also called intercellular communicators. This cell-to-cell communication, which showed the possibility of transferring genetic material via exosomes, has been reported [26]. A study has reported that embryonic stem cell-derived small vesicles containing messenger RNAs (mRNAs) are functional [27]. Recently, a novel way of interneuronal communication was reported among neurons depending on the synaptic activities mediated through exosomes [28]. Various cell types, including stem cell-derived exosomes and their biological roles have been well documented. Numerous studies have shown that exosomes are the most important and well-accepted small vesicles for cell-to-cell interactions and communications. There is still a lot to learn about the exosome formation mechanism, cargo-binding formation, and signaling pathways and their functions. This chapter will focus on the exosome's function in miRNA-mediated paracrine effects.

2 PHYSICAL AND MOLECULAR COMPOSITIONS

Studies have shown that various types of cell-based exosome are present and its physical and molecular nature, and biological applications have recently appreciated. Most recently, exosomes generated special attention regarding various diseases. It is difficult to observe exosomes under a phase contrast microscope because of their sizes; instead they can be seen under the electron microscope or photon microscope [12]. A study has shown that exosomes derived from neoplastic cells possessed $5'$-nucleotidase activity [13]. The secretions of exosomes are regulated by two Rab GTPase proteins: Rab27a and Rab27b. It is possible to manipulate the exosome secretions by altering the Rab GTPase [29]. Exosomes are found in the peripheral circulation at higher concentrations during pregnancy and cancer [30].

2.1 Isolation of Exosomes

The isolation of exosomes remains a big challenge and a hot topic in the field of exosome research and is described in Table 3.1. The biological companies have designed a variety of reagents to isolate exosomes, such as Exo-Quick (System Biosciences), Exo Test Kits (HansaBioMed), and ExoMir Kit (Bio Scientifics). In most cases, the exosomes are purified by four methods, e.g., immune affinity capture, 0.22 μm cut-off filtration [31], size exclusion, and high-speed ultracentrifugation. Filtrating through a 0.22 μm pore size, although yielding a higher number of exosomes, may compromise its purity. The relatively purified exosomes have been obtained by using immune affinity-specific antibodies such as CD9, CD63, and CD81. This antibody-specific isolation of exosomes has advantages over the other methods, but the yields are often low [31]. A better method is to follow a well-defined series of centrifugation to remove dead cells, followed by ultracentrifugation and sucrose density gradient isolation [4,32]. The sucrose density gradient technique is used to achieve a purer form of exosomes because it floats at a density of 1.13–1.19 g/mL [12]. Most microvesicles range from 100 nm to 1000 nm and can be pelleted at 10,000 to 20,000 g whereas

Table 3.1 Different exosomes isolation protocols

S. no.	Methods	References
1	Ultracentrifugation	[4]
2	Ultracentrifugation + 30% sucrose cushion	[12]
3	Ultracentrifugation + 0.22 μm filtration	[31]
5	Standard exosome precipitation-ExoQuick-TC; System Biosciences	[2]
6	Exo Test Kits (HansaBioMed)	Aiavilja, Estonia
7	ExoMir Kit (Bio Scientifics)	Austin, Texas
8	Immunoprecipitation using CD9 antibody or prefiltration using 0.2 μm cut-off filters	[31]

exosomes require 100,000 g. However, small plasma membrane vesicles are also recovered along with exosomes during ultracentrifugation but have a lighter density than exosomes [33]. Thus, the advantage of this method of isolation is that the exosomes have a lesser density than apoptotic vesicles (1.24–1.28 g) and a higher density than plasma membrane vesicles (1.04–1.07 g). The flow cytometry study is used as an alternative method to purify the exosomes according to their size and cell-surface marker expressions [34]. Recently, a detailed review identified that several other methods, such as microfluidic devices, antibody-coated magnetic beads, and precipitation-based isolation, have been developed [35]. Even though different methods are available, the optimal method of exosome isolation and analysis has not been achieved.

2.2 Biogenesis of Exosomes

The foundation for understanding and defining the biogenesis of exosomes has led to further insight into their functions over the past decades. Various types of extracellular vesicle are released by exocytosis and the details are shown in Table 3.2 in different cell types [36,37]. Among these extracellular vesicles, exosomes have received significant attention in recent years [24,38]. Exosomes are lipid bilayer vesicles present in varying diameter and after negative staining with 2% uranyl acetate, show cup or round-shaped structure under transmission and cryoelectron microscopy, respectively [39]. This arises within endosomal compartments and buds internally to form

Table 3.2 Types of microvesicles or microparticles

S. no.	Microparticle	Size	Markers	Contents
1	Apoptotic bodies	≥1000 platelet size	Exposure of PS	Fragmented DNA
2	Exosomes	30–120 nm	CD63, CD81, CD9, Tsg101, Alix, Hsp60, Hsp70, Hsp90 (low exposure of PS)	Proteins, lipids, mRNAs, miRNA
3	Shedding vesicles	100–1000 nm	Lipid raft-associated molecules – TF, flotillin	Proteins, lipids, mRNAs, miRNAs

intraluminal vesicles. These vesicles on fusion with the plasma membrane release their contents into the extracellular fluid but they are not released by plasma membrane shedding. These free forms of vesicles in the extracellular fluid are referred to as exosomes [40,41] (Figure 3.1). They are very stable in extracellular fluids because of a rigid membrane, which helps to prevent degradation by hydrolytic enzymes. A recent study using nanotracking analysis indicated that an average of 500 single exosomes is released from a cell. In addition, some of the exosomes can naturally circulate as aggregates [42]. Two types of secretion have been identified and are constitutive or inducible, depending on the cell type [33]. The constitutive secretions are the proteins destined to the cell surface or to the extracellular medium, which

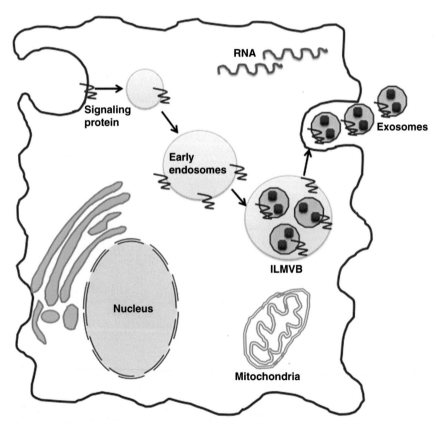

Figure 3.1 *Biosynthesis and Release of Exosomes.* Exosome synthesis begins with the inward budding of plasma membrane along with the cellular contents. Because of their contents, exosomes might be considered as cell status updaters. Exosomes are released from the cells through the process of exocytosis. ILMVB: Intraluminal MVB.

can be routed from the trans–Golgi network but do not require any stimulus. During this constitutive secretion, exosomes are transported within the vesicles but not transmitted through the MVB. Various cell activation processes are available to control the inducible or regulated secretion pathway [42,43]. The regulators are calcium [44], potassium, cross-linking receptor, small GTPase Rab11, the citron kinase, RhoA effector [45,46], and Rab27 family proteins [29]. Apart from the Rab27 family, other Rab family members, Rab35 and Rab11, have also been shown to regulate the secretion of exosomes [23] (Figure 3.2).

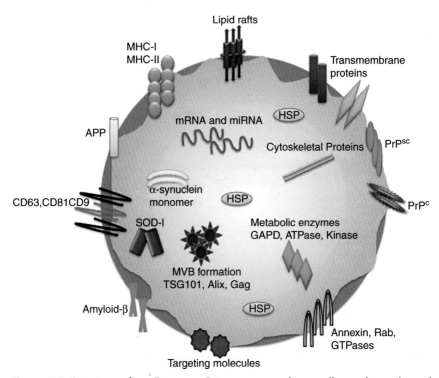

Figure 3.2 *Structure of an Exosome.* Exosomes are ultra-small membrane-bound vesicles sharing a similar structure to the plasma membrane. The inner contents are mRNAs, miRNAs, cytoskeletal proteins, metabolic enzymes, proteins involved in MVB formation, α-synuclein, and HSP. The surface proteins are tetraspanins (CD9, CD63, and CD81), annexins and Rab proteins, MHC 1 and II, lipid rafts (flotillin, cholesterol, sphingomyelin), transmembrane proteins, adhesion molecules (ICAM), β-amyloid, cellular form of prion protein (PrP^c) and sole or infectious prion protein (PrP^{sc}), and amyloid precursor protein (APP).

2.3 Characterization of Exosomes

The exosomal vesicles always carry and share their own cell-specific membranes and cell components. Therefore a number of features should be taken into account for the characterization of exosomes. Apart from morphology, the unique protein and lipid composition help us to identify the exosomes. These are annexins, Rab GTPases, flotillin, CD63, CD81, CD82, and CD9, and heat shock proteins (HSP) like Hsp60, Hsp70, and Hsp90 [38,40]. The common molecular markers are tetraspanins, CD9, CD63, CD81, CD82, CD151, intercellular adhesion molecule-1 (ICAM1), CD51, CD61, Alix, externalized phosphatidylserine, milk fat globule-E8/lactoferrin, CD80, CD86, CD96, Rab5b, and major histo-compatibility complex (MHC) class, I and II [47,48]. Recently, the presence of exosomes in plasma samples of healthy donors has been identified [49]. Besides blood, exosomes also have been identified in urine, cerebrospinal fluid (CSF), saliva, milk, semen, amniotic fluid, and were already summarized in a review by Simpson et al [50]. Stability of individual exosomes containing miRNAs depends on the storage period and temperature. Different storage conditions, such as 4, −20, and −80°C, may alter the exosomes' profile and miRNAs [32].

2.4 Functions of Exosomes

Exosome is one of the family members of bioactive microvesicles. These vesicles are described during the maturation of reticulocytes into erythrocytes. Surprisingly, these vesicles have receptors and are called transferrin, which downregulate mature erythrocytes [51]. These exosomes have both beneficial physiological and detrimental pathological roles during their biological processes. Studies have shown that the exosomes function as a scavenger in elimination of unnecessary proteins or unwanted molecules from the cells. This also involves the exchange of materials between cells, intercellular communication, propagation of pathogens, functions of the immune system (both stimulatory and inhibitory), antigen presentation, and protein secretion [12]. Interestingly, the role of exosomes has been reported in atherosclerosis, calcification, and kidney diseases [52]. The exosomes released from the immune cells play an important role in maintaining the immune system and also have immunomodulatory effects [53]. Mesenchymal stem cell (MSC)-derived exosomes have the capacity to promote tissue repair and mediate regeneration. They also can repair acute and chronic injuries in kidney because of paracrine mediation through miRNAs and exosomes or microvesicles [40]. The most important function of exosomes has been documented on natural killer (NK) cell activation in immune-competent mice and NK cell-dependent antitumor effects. Although it has many important functions, the pathophysiological stimuli-mediated production of exosomes, mechanism of release, and its uptake has yet to be elucidated clearly [52].

3 MicroRNAs (miRNAs) IN EXOSOMES

miRNAs or miRs are a new class of highly conserved, 21–25 nucleotide, genetically encoded, endogenous, noncoding regulatory single-stranded RNAs that upregulate or downregulate gene expression at the posttranscriptional level by binding to the 3′ untranslated region (3′ UTR). These miRNAs mediate and balance cell proliferations and differentiation during tumorigenesis and organ development by translational suppression or degradation of their target mRNAs [54,55]. During the past decade, numerous research and articles have shown a wide knowledge on the basic mechanism of miRNAs, biogenesis, and their molecular function in our body [56–58]. These noncoding RNAs are found within intergenic or intronic regions of host genes, and make up approximately 10% of the human genome.

The presence of miRNAs within exosomes was described for the first time in 2007 [25]. Recently, studies have proven that miRNAs are found in microvesicles and they are released by different types of cells, including human renal and cancer stem cells [59] and tumor-associated macrophages [60]. miRNAs containing exosomes are described as a great biomarker for diseases. The vital role of miRNAs has been reported in maturation and differentiation of regulatory T cells (Tregs) in the thymus [61,62]. Importantly, the importance of miR-155 has been underlined by the development of the Tregs' lineage [63]. Recent studies have shown that the miRNAs are considered as a key modulator of lung and cardiovascular diseases [64]. miRNA-specific knockdown studies have shown its involvement in various diseases. In many conditions, miRNAs are aberrantly expressed either upregulated or downregulated depending on the physiological state of the cells. Understanding the roles of these molecular modulators and their signaling pathways with their target genes is a great challenge in future research.

Cellular communication is one of the most important physiological events for maintaining homeostasis. In particular, miRNAs play a vital role in ovarian follicle growth and maturation that could develop an embryo through cell-to-cell communication. A study has shown that the presence of both mRNA and miRNA in exosomes from mast cells has a paracrine effect on neighboring cells [65]. In recent years, the measurement of serum miRNAs in an extracellular environment has been exploited to assess cell function at distance. Activated $CD4^+$ T cells release a large amount of exosomes, containing miRNA; however, it is unsure whether this phenomenon is reflected in modulation of serum miRNAs. Interestingly, miRNA signatures of exosome-released $CD4^+$ T cells are substantially different from intracellular miRNA signatures of the same cells [66].

3.1 Paracrine Effects of miRNAs

An epidemiological study showed that raw cow's milk consumption in the first year of life protects against atopic diseases and increases regulatory T-cell numbers ($CD4^+$, $CD25^+$, and $FoxP3^+$). This benefit reduces the unusual sensitization and asthma in children younger than 4.5 years old. Milk transfers exosomal miRNAs including miRNA-155, which is important for the development of the immune system and thereby regulating FoxP3 expression, IL-4 signaling, and class switching of immunoglobulins during immune response [67]. The active forms of miRNAs are degraded when they are boiled and, surprisingly, miRNA-155 and miRNA-146 are absent in human breast milk [68,69]. miRNA-155 plays a vital role in the development of Tregs and further miRNA-155 deficient mice showed a reduction of Tregs numbers both in the thymus and periphery [63]. The membrane-bound exosomal miRNAs are important for horizontal miRNA transfer [67]. In the human immune system, cell-to-cell communication and interactions are being done by exosomes, which carry genetic information. More recently, the roles of exosomal miRNAs have been reported and considered as diagnostic biomarkers for lung cancer, ovarian cancer, and cardiovascular diseases [70,71]. The exosomes containing miRNA-21 and miRNA-146a are upregulated in cervical cancers, and are considered potential novel biomarkers for cervical cancer diagnosis [72].

4 EXOSOMES AND DISEASES

MSCs are multipotent stem cells, derived from many sources and are capable of differentiating into many cell types. Exosomes can be isolated from MSCs of different origins and are well characterized by their specific markers. Both normal and engineered MSC (En. MSC)-derived exosomes are used for therapy. Intravenous administration and transplantation of exosomes are involved in anticancer, immune suppressive response, and regenerative medicine for brain, heart, kidney, liver, and pulmonary (Figure 3.3).

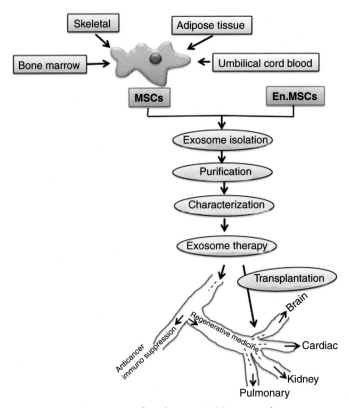

Figure 3.3 *MSC-Derived Exosomes for Therapy*. MSCs are multipotent stem cells derived from many sources and are capable of differentiating into many cell types. Exosomes can be isolated from MSCs of different origins and are well characterized by their specific markers. Both normal and engineered MSCs (En.MSCs)-derived exosomes are used for therapy. Intravenous administration and transplantation of exosomes are involved in anticancer, immune suppressive response, and regenerative medicine for brain, heart, kidney, liver, and pulmonary.

4.1 Cardiovascular Diseases

Ischemic heart disease resulting in loss or dysfunction of cardiomyocytes (CMCs) is the leading cause of death worldwide [73]. The survivors of acute myocardial infarction (MI) eventually develop chronic heart failure, with over 5 million cases in the United States alone [74]. A fundamental goal of regenerative cardiology is to successfully repair injured hearts with functional cell therapy. Cell therapy has emerged as a promising therapeutic option for the treatment of MI or heart failure. However, an ideal cell source remains elusive, as all current cell sources have shortcomings such as limited differentiation potential, ethical issues, immunologic incompatibility, or teratoma formation [75–78]. Recent studies have reported the potential of specific cardiac transcription factors or a combination of miRNAs for reprogramming fibroblasts into cardiomyopathy (CM)-like cells, providing a new option for cell therapy [79–81]. Despite significant improvement in the prognoses of patients with acute MI after treatment with available therapies, including thrombolysis and urgent revascularization, mortality remains significant and a substantial number of survivors develop heart failure. The identification in adult animals of stem cells that can contribute to tissue repair and regeneration has raised the possibility that cell therapy could be employed for repair of the damaged myocardium. In particular, bone marrow (BM)-derived cells have been mobilized or transplanted into animal models of MI and shown to incorporate into the infarct and peril-infarct zone, where they induce myogenesis and angiogenesis, thereby augmenting cardiac function and survival [82–84]. A recent systematic review and meta-analysis study showed that the transplantation of adult bone marrow progenitor cells (BMCs) improves left ventricle function, infarct size, and remodeling in patients with ischemic heart disease compared with standard therapy, and these benefits persist during long-term follow-up. BMC transplantation also reduces the incidence of death, recurrent MI, and stent thrombosis in patients with ischemic heart disease [85].

Several factors may limit the utility of autologous BM-derived or peripheral-blood stem/progenitor cells for treatment of myocardial conditions. Peripheral-blood endothelial progenitor cells are often less prevalent and/or less functionally active in patients with risk factors for coronary artery disease [86–89] or diabetes [90], and EPC-mediated re-endothelialization is impaired in diabetic mice [91]. Concerns have also been expressed about the heterogeneity of BM-derived stem-cell populations, the incomplete understanding of how they function, and their limited potential for transdifferentiation into various cell lineages. In addition, the ability of transdifferentiated cells to maintain their newly acquired phenotype and the heritability of such changes is unknown. Thus, the development of autologous stem cells with enhanced lineage plasticity for subsequent testing in the setting of myocardial ischemia is an integral part of stem-cell research.

The clinical application of exosomes in regenerative medicine is just beginning, and the results are encouraging. Studies have shown that exosomes play an important role in promoting angiogenesis. The *in vitro* and *in vivo* studies by using exosome derived from human $CD34^+$ cells have displayed an angiogenic activity [5]. This study further suggests that the benefit of $CD34^+$ cell therapy on functional recovery after ischemic injury could be induced primarily through the exosome-mediated transfer of angiogenic factors to surrounding cells [5]. Several studies suggested that MSC-derived extracellular vesicles (MSC-EVs) promote angiogenesis activity, but therapeutic mechanism of MSC-EVs on an ischemic heart is unclear. A study has recently reported that the role of human BM MSC-derived extracellular vesicles promoted angiogenesis in a rat MI model [92].

It has been shown that MSC-derived exosomes secretes paracrine factor, which ameliorates increased oxidative stress and cell death and preserves left ventricular function in ischemic/reperfused myocardium [93]. It has also been shown that an increase in the number of peripheral white blood cells (WBCs) is an indicator for large infarct size. Treatment with exosomes in the ischemia/reperfusion injury animal model reduces the WBC count further than in the untreated saline control animals [94]. miRNA-208a is identified and determined as a potential marker for the early stage of MI. Dysfunctional endothelial cells release microvesicles and it is kind of biomarker for endothelial damage [95]. The microvesicles have been found in the blood of various cardiovascular diseases such as acute coronary syndrome [96] and severe hypertension [97].

4.2 Neurodegenerative Diseases

The central nervous system is one of the most important systems, which coordinates and maintains the homeostasis of the human body. In the nervous system, mainly glia cells play a vital role in many aspects such as myelin synthesis, metabolic supports, and immune responses. It has been reported that exosomes can be found in both adult and embryonic animal CSF [98,99]. Recently, exosomes have emerged as a prominent mediator of neurodegenerative diseases where they have been shown to carry disease particles such as β-amyloid and prions from their cells of origin to other cells. In a neuronal disease, the involvement of exosomes in association with pathophysiological functions needs to be studied. Exosomes play an important role in synchronizing the intercellular communication between the glia and neurons [100]. Exosome-mediated interactions between neurons and the glia induce neurite outgrowth, which supports the survival of neurons [101,102].

MSC-derived exosomes containing miRNA-133b support the induction of neurite outgrowth and survival of neurons [103]. The exosome-mediated transfer of miRNAs in neuron-to-astrocyte signaling has been shown recently. Exosomes isolated from the neuronal-conditioned medium contain abundant miRNAs and small RNAs. These exosomes can be directly internalized into astrocytes and increase astrocyte miRNA-124a and GLT1 protein levels, indicating that within the central nervous system cell-to-cell interaction and/or communication is also due to exosomes [104]. Recent studies have shown various biological roles of exosomes in brain malignancies. In glioblastoma, the cells are capable of secreting exosomes loaded with mRNAs and miRNAs, which are utilized by the recipient tumor cells to contribute tumor proliferation and angiogenesis [105]. An electron microscopic study showed that in the mammalian nervous system, exosomes are released from the differentiated neurons of somatodentric compartments. These exosome secretions are regulated by glutamatergic activity-induced increase in postsynaptic calcium levels [106,107]. Accumulating evidence suggests that exosomes play an important role in intercellular signaling. Intercellular transfer and paracrine signaling of exosomes is associated with pathology in several neurodegenerative diseases, pathogenic proteins, β-amyloid peptide, and superoxide dismutase; synuclein and tau are released from cells in association with EVs [100]. In amyotrophic lateral sclerosis, Parkinson's disease (PD), and Alzheimer's disease (AD), the superoxide dismutase α-synuclein in association with exosomes plays an important role in disease progression [108]. Mainly, exosomes are involved in intercellular communication and cleaning for cellular waste products. Exosomes released by A549 cells during DDP exposure decreased the sensitivity of other A549 cells to cisplatin, which may be mediated by miRNA and mRNA exchange by exosomes via cell-to-cell communication [109].

Neuronal cell death is a hallmark of neurodegenerative disorders. AD, Huntington's disease, PD, Niemann-Pick disease, frontotemporal dementia, and amyotrophic lateral sclerosis are the important neurodegenerative diseases. All these diseases have a common cellular and molecular mechanism, which involves protein aggregation and formation of inclusion bodies in the selected region of the nervous system. Interestingly, the degradation of these unwanted proteins inside the cells is important for neuronal health and is mediated either by endosomal pathways of exosomes or lysosomal degradation [110].

The role of MVBs in AD patients was reported in 1970 and HD in 1997 [108,111]. Studies from human AD brain samples and transgenic AD mouse models identified miRNAs, which are significantly reduced during the disease process [112]. MVBs accumulated with amyloid-β 42 a peptide fragment are the main constituents of plague and characteristics of AD. In AD, the exosomes play an important role in both degradation and accumulation of amyloid-β 42 toxic peptides from the neuronal cells [108,113].

PD is the second most common neurodegenerative disorder after AD, affecting the population over 50 years old. PD is characterized by damage or degeneration of dopaminergic neurons, but the exact mechanism for disease onset and progression is not clear. In PD, exosomes also play a vital role by transferring the toxic form of α-synuclein to other cells. The transferred α-synuclein tends to form aggregates in the recipient cells [114]. Further understanding of cell-to-cell transferring of α-synuclein and inflammatory signaling may provide a novel mechanism of onset and progression of PD.

4.3 Renal Diseases

There has been increasing interest and attention towards newly recognized exosomes over the past several years by the nephrology community. The exosomes from urine provide an attractive means of noninvasively detecting and monitoring the function of kidney diseases. It was thought that the major role exosomes present in the urinary tract is to eliminate the senescent proteins by exocytosis from the cells. Another important function of exosomes in the urinary tract is to regulate the cofunctioning of different parts of the nephron through secretion of mRNAs and miRNAs that may affect the functions of their own or the recipient cells [52]. Alteration of miRNA profile and its expression have been observed in exosomes associated with polycystic kidney diseases [115], renal cell carcinoma [116], and allograft rejection [117]. It has been shown that in the rat model of acute kidney injury induced by cisplatin, the rats treated with human umbilical cord MSC-derived exosomes showed decreased acute kidney injury as well as blood urea nitrogen and creatinine levels, apoptosis, necrosis, and oxidative stress. This study also reported that exosomes from MSCs are capable of reducing cell death signaling by Bax and increasing survival signaling by Bcl-2 thereby modulating apoptosis and proliferation, respectively, in the kidney during injury. This suggests that MSC-derived exosomes have therapeutic potential [118]. It has been shown that stem cell-derived exosomes having the role of repairing injured renal and cardiac tissues compared to direct usage of stem cells [119]. Recently, immunogold staining and quantitative reverse transcription-polymerase chain reaction data showed reduce levels of miRNA-29 and miRNA-200 in chronic kidney disease patients compared to normal controls. This suggests that the exosomes derived from the urinary tract contain members of the miR-29 and miRNA-200 family and could be considered as a potential noninvasive biomarker for renal fibrosis [120,121].

4.4 Exosomes in Cancer

The expressions of individual miRNAs and their specific signatures have now been associated with the diagnosis and prognosis of many human cancers. Studies have shown that the miRNAs present in the serum exosome are considered a promising potential biomarker for cancer [121–123]. In breast cancer, MSC-derived exosomes suppress angiogenesis by down-regulating vascular endothelial cell growth factor expression both *in vivo* and *in vitro* [3]. In contrast, exosomes derived from tumor have angiogenic properties and were demonstrated by an *in vitro* endothelial spheroid formation model in matrigel [124]. It was also demonstrated that glioblastoma exosomes have angiogenic properties and stimulating tube formation [105]. Exosomal miRNA-21 has been upregulated in the serum of esophageal squamous-cell cancer patient samples [125]. For cancer, exosome-based immunotherapeutic approaches are increasing the hope of cancer therapy. A study on ovarian cancer tissues demonstrated that the specific miRNA-141, miRNA-200a, miRNA-200b and miRNA-200c are significantly over-expressed in ovarian cancer tissues in comparison with the normal tissues [126]. Tumor-derived exosomes, which are found in the plasma, could be a good biomarker for cancers [30]. Marrow stromal cell-derived exosomes have a role for facilitating multiple myeloma (MM) progressions. A confocal microscopic study characterized the exosome by the expression of surface markers CD63 and CD81 in MM cells [127].

4.5 Pulmonary Diseases

The understanding of miRNA biogenesis and its function in determining the change in the structural component of the pulmonary vasculature during pulmonary hypertension (PH) becomes obvious when it is shown that the miRNAs engage in the cellular process of change in PH progress. Further, the expressions of miRNAs were altered in the experimental model of PH. The expression of miRNA-21 was found to be increased in lung tissue of the chronic hypoxia model [128] and monocrotaline model [129]. In the miRNA profiling study of the monocrotaline experimental model of PH, data suggest that the expression of miRNA-17, miRNA-19b, and miRNA-92a is significantly increased [128,130]. This experimental study also demonstrates the intravenous administration of anti-miRNA-17 to sequester the increased level of miRNA-17 in the lung tissue and reduce the severity of PH. It has been shown that exosomal miRNAs isolated from bronchoalveolar lavage fluid of asthmatic patients are different from healthy individuals [131]. It has been shown that MSC-derived exosomes mediate the cryoprotective beneficial effect on hypoxia-induced PH. In this study, it was identified that the miRNA-17 family of miRNAs is the key effector for MSC paracrine function [132]. Another study showed that the marrow's stromal cell-derived exosomes are expressing miR-146b to inhibit glioma growth [133]. These exosomes act as biological delivery vehicles [134] because there is no immune rejection and risk-free tumor formation [135].

4.6 Stem Cells and Exosomes

MSCs or mesenchymal stromal cells are multipotent stem cells that can differentiate into various cell types [136] such as bone, cartilage, muscle, ligament, tendon, adipose, and stromal cells. A study has reported that MSC derivative exosomes possesssalient features of the lipid raft, a microdomain with an effective endocytic activity in the plasma membrane [137]. BM-derived MSCs are one of the most promising cells for treating cardiovascular diseases [138]. In our recent electron microscopic study, the MSCs treated with bacterial protein lipopolysaccharide (LPS) showed an increased number of exosomes compared to the unstimulated resting MSCs (Figure 3.4). Further studies are in progress to identify and characterize whether these are beneficial or nonbeneficial exosomes. Therapeutic strategies of MSCs are differentiated into either reparative or replacement cell types [139]. These reparations are mediated by paracrine factors, which are secreted by MSCs [140]. MSCs also can regulate the immune response in pathological conditions, including musculoskeletal and cardiovascular diseases [136]. Studies have shown that the MSCs in tissue repair mechanisms are based on homing capacity and differentiation within the damaged or diseased cells into specialized cells, but the integration and survival rate in host tissues are a very small proportion [136] (Figure 3.5).

miRNAs influence multiple cellular functions, including cell survival, metabolism, proliferation, development, and differentiation. A report on combined expression of Akt and angiopoietin-1 on MSCs has shown that miR-NA-143 is the critical regulator for in vitro proliferation and has an enhanced angiomyogenic capacity in the infarcted rat's heart [141]. Moreover, it has been shown that knockdown of miRNAs from stem cells was unable to differentiate [142,143] demonstrating that miRNAs may contribute significantly to the differentiation of MSC toward CMCs. Studies have shown that coculturing of MSC with CMCs increases their transdifferentiation potential into CM or cardiac-like cells [144–146]. Although a coculture-based preprogramming strategy has been extensively studied, the underlying mechanism remains unclear. More recent studies report that exchange of cellular materials, including

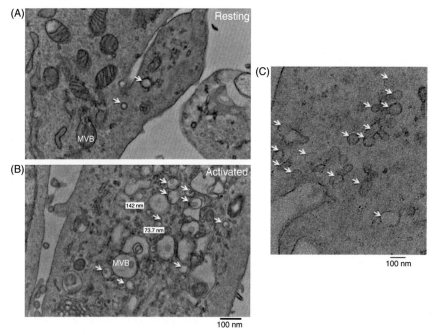

Figure 3.4 *Electron Microscopic Image of Exosomes and Multivesicular Bodies (MVBs) in MSCs.* (A) Resting MSC (10,000×), (B) LPS-induced MSC (10,000×), and (C) activated MSC (25,000×). The exosomes are indicated by white arrows. The LPS-activated MSC contains an increased number of exosomes compared to the resting MSC.

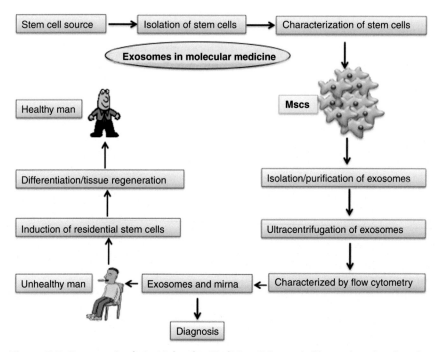

Figure 3.5 *Exosomes' role in Molecular Medicine.* Schematic illustration showing the perspective of the stem cell-derived exosomes for tissue regeneration.

miRNAs, occurs between MSC and CM in coculture conditions [147,148]. MSC possesses the potential of converting into different cell types depending upon interventions, and may be the more relevant cell type for cell therapy.

5 FUTURE PERSPECTIVE

Growing evidence suggests that exosomes play an important role in intercellular communication and carry signaling molecules to activate the distance molecules during disease progression and prevention. However, the field of regenerative medicine, which is currently in its infancy, needs to address several unanswered questions and challenges. Some of the issues that remain are: (i) how are beneficial exosomes separated from the cellular fluid for transplantation? (ii) what are the signaling pathways targeted by exosomes during injury? (iii) are the cells present in the injured region release similar or different type of exosomes both qualitatively and quantitatively? (iv) how can the therapeutic potential of exosomes be improved? (v) what are the sources of beneficial exosomes? (vi) do the exosomes have a direct effect on the residential stem cells?

6 CONCLUSIONS

In the past few years there have been many studies that aim to understand the biology and signaling pathways of exosomes in regenerative medicine. However, we are at an early stage of exploring the basic functions that have been unraveled in exosome biology.

ABBREVIATIONS

AD	Alzheimer's disease
ALS	Amyotrophic lateral sclerosis
APP	Amyloid precursor protein
BM	Bone marrow
BMC	Bone marrow progenitor cells
CMC	Cardiomyocytes
CSF	Cerebrospinal fluid
HSP	Heat shock proteins
LPS	Lipopolysaccharide
MHC	Major histocompatibility complex
MI	Myocardial infarction
MM	Multiple myeloma
miRNA	Microribonucleic acid
mRNA	Messenger ribonucleic acid

MSCs	Mesenchymal stem cells
MVBs	Microvesicular bodies
NK	Natural killer
PD	Parkinson's disease
PH	Pulmonary hypertension
PrP^c	Cellular form of prion protein
Tregs	Regulatory T-cells
WBCs	White blood cells

REFERENCES

[1] Ge Q, Zhou Y, Lu J, Bai Y, Xie X, Lu Z. miRNA in plasma exosome is stable under different storage conditions. Molecules (Basel, Switzerland) 2014;19(2):1568–75.

[2] Schageman J, Zeringer E, Li M, Barta T, Lea K, Gu J, et al. The complete exosome workflow solution: from isolation to characterization of RNA cargo. Biomed Res Int 2013;2013:253957.

[3] Lee JK, Park SR, Jung BK, Jeon YK, Lee YS, Kim MK, et al. Exosomes derived from mesenchymal stem cells suppress angiogenesis by down-regulating VEGF expression in breast cancer cells. PloS One 2013;8(12):e84256.

[4] Raposo G, Nijman HW, Stoorvogel W, Liejendekker R, Harding CV, Melief CJ, et al. B lymphocytes secrete antigen-presenting vesicles. J Exp Med 1996;183(3):1161–72.

[5] Sahoo S, Klychko E, Thorne T, Misener S, Schultz KM, Millay M, et al. Exosomes from human CD34(+) stem cells mediate their proangiogenic paracrine activity. Circ Res 2011;109(7):724–8.

[6] van der Pol E, Coumans F, Varga Z, Krumrey M, Nieuwland R. Innovation in detection of microparticles and exosomes. J Thromb Haemost 2013;11(Suppl 1):36–45.

[7] Stoorvogel W, Kleijmeer MJ, Geuze HJ, Raposo G. The biogenesis and functions of exosomes. Traffic (Copenhagen, Denmark) 2002;3(5):321–30.

[8] Holme PA, Solum NO, Brosstad F, Roger M, Abdelnoor M. Demonstration of platelet-derived microvesicles in blood from patients with activated coagulation and fibrinolysis using a filtration technique and western blotting. Thromb Haemost 1994;72(5):666–71.

[9] Hess C, Sadallah S, Hefti A, Landmann R, Schifferli JA. Ectosomes released by human neutrophils are specialized functional units. J Immunol (Baltimore, Md.: 1950) 1999;163(8):4564–73.

[10] Cocucci E, Racchetti G, Meldolesi J. Shedding microvesicles: artefacts no more. Trends Cell Biol 2009;19(2):43–51.

[11] Raposo G, Stoorvogel W. Extracellular vesicles: exosomes, microvesicles, and friends. J Cell Biol 2013;200(4):373–83.

[12] Thery C, Zitvogel L, Amigorena S. Exosomes: composition, biogenesis and function. Nat Rev Immunol 2002;2(8):569–79.

[13] Trams EG, Lauter CJ, Salem N Jr, Heine U. Exfoliation of membrane ecto-enzymes in the form of micro-vesicles. Biochim Biophys Acta 1981;645(1):63–70.

[14] Ronquist G, Brody I. The prostasome: its secretion and function in man. Biochim Biophys Acta 1985;822(2):203–18.

[15] Johnstone RM, Adam M, Hammond JR, Orr L, Turbide C. Vesicle formation during reticulocyte maturation. Association of plasma membrane activities with released vesicles (exosomes). J Biol Chem 1987;262(19):9412–20.

[16] Laulagnier K, Motta C, Hamdi S, Roy S, Fauvelle F, Pageaux JF, et al. Mast cell- and dendritic cell-derived exosomes display a specific lipid composition and an unusual membrane organization. Biochem J 2004;380(Pt 1):161–71.

[17] Thery C. Exosomes: secreted vesicles and intercellular communications. F1000 Biol Rep 2011;3:15.

[18] Hergenreider E, Heydt S, Tréguer K, Boettger T, Horrevoets AJ, Zeiher AM, et al. Atheroprotective communication between endothelial cells and smooth muscle cells through miRNAs. Nat Cell Biol 2012;14(3):249–56.

[19] Bang C, Thum T. Exosomes: new players in cell-cell communication. Int J Biochem Cell Biol 2012;44(11):2060–4.

[20] Lai CP, Breakefield XO. Role of exosomes/microvesicles in the nervous system and use in emerging therapies. Front Physiol 2012;3:228.

[21] Zhang HG, Grizzle WE. Exosomes: a novel pathway of local and distant intercellular communication that facilitates the growth and metastasis of neoplastic lesions. Am J Pathol 2014;184(1):28–41.

[22] Viaud S, Ullrich E, Zitvogel L, Chaput N. Exosomes for the treatment of human malignancies. Horm Metab Res 2008;40(2):82–8.

[23] Beach A, Zhang HG, Ratajczak MZ, Kakar SS. Exosomes: an overview of biogenesis, composition and role in ovarian cancer. J Ovarian Res 2014;7(1):14.

[24] Lakkaraju A, Rodriguez-Boulan E. Itinerant exosomes: emerging roles in cell and tissue polarity. Trends Cell Biol 2008;18(5):199–209.

[25] Valadi H, Ekström K, Bossios A, Sjöstrand M, Lee JJ, Lötvall JO, et al. Exosome-mediated transfer of mRNAs and microRNAs is a novel mechanism of genetic exchange between cells. Nat Cell Biol 2007;9(6):654–9.

[26] Camussi G, Cantaluppi V, Deregibus MC, Gatti E, Tetta C. Role of microvesicles in acute kidney injury. Contrib Nephrol 2011;174:191–9.

[27] Ratajczak J, Miekus K, Kucia M, Zhang J, Reca R, Dvorak P, et al. Embryonic stem cell-derived microvesicles reprogram hematopoietic progenitors: evidence for horizontal transfer of mRNA and protein delivery. Leukemia: official journal of the Leukemia Society of America, Leukemia Research Fund, UK 2006;20(5):847–56.

[28] Chivet M, Hemming F, Pernet-Gallay K, Fraboulet S, Sadoul R. Emerging role of neuronal exosomes in the central nervous system. Front Physiol 2012;3:145.

[29] Ostrowski M, Carmo NB, Krumeich S, Fanget I, Raposo G, Savina A, et al. Rab27a and Rab27b control different steps of the exosome secretion pathway. Nat Cell Biol 2010;12(1):19–30. sup pp 11-13.

[30] Taylor DD, Gercel-Taylor C. MicroRNA signatures of tumor-derived exosomes as diagnostic biomarkers of ovarian cancer. Gynecol Oncol 2008;110(1):13–21.

[31] Jansen FH, Krijgsveld J, van Rijswijk A, van den Bemd GJ, van den Berg MS, van Weerden WM, et al. Exosomal secretion of cytoplasmic prostate cancer xenograft-derived proteins. MolCell Prot 2009;8(6):1192–205.

[32] Thery C, Amigorena S, Raposo G, Clayton A. Isolation and characterization of exosomes from cell culture supernatants and biological fluids. Curr Prot Cell Biol 2006;Chapter 3:Unit 3 22.

[33] Thery C, Ostrowski M, Segura E. Membrane vesicles as conveyors of immune responses. Nat Rev Immunol 2009;9(8):581–93.

[34] Nielsen MH, Beck-Nielsen H, Andersen MN, Handberg A. A flow cytometric method for characterization of circulating cell-derived microparticles in plasma. J Extracell Ves 2014;3.

[35] Momen-Heravi F, Balaj L, Alian S, Mantel PY, Halleck AE, Trachtenberg AJ, et al. Current methods for the isolation of extracellular vesicles. Biol Chem 2013;394(10):1253–62.

[36] Ronquist G. Prostasomes are mediators of intercellular communication: from basic research to clinical implications. J Intern Med 2012;271(4):400–13.

[37] Camussi G, Deregibus MC, Bruno S, Cantaluppi V, Biancone L. Exosomes/microvesicles as a mechanism of cell-to-cell communication. Kidney Int 2010;78(9):838–48.

[38] Simons M, Raposo G. Exosomes – vesicular carriers for intercellular communication. Curr Opin Cell Biol 2009;21(4):575–81.

[39] Conde-Vancells J, Rodriguez-Suarez E, Embade N, Gil D, Matthiesen R, Valle M, et al. Characterization and comprehensive proteome profiling of exosomes secreted by hepatocytes. J Proteome Res 2008;7(12):5157–66.

[40] Biancone L, Bruno S, Deregibus MC, Tetta C, Camussi G. Therapeutic potential of mesenchymal stem cell-derived microvesicles. Nephrol Dial Transplant: official publication of the European Dialysis and Transplant Association – European Renal Association 2012;27(8):3037–42.

[41] Yellon DM, Davidson SM. Exosomes: nanoparticles involved in cardioprotection? Circ Res 2014;114(2):325–32.

[42] Record M, Carayon K, Poirot M, Silvente-Poirot S. Exosomes as new vesicular lipid transporters involved in cell-cell communication and various pathophysiologies. Biochim Biophys Acta 2014;1841(1):108–20.

[43] Record M. Exosome-like nanoparticles from food: protective nanoshuttles for bioactive cargo. Mol Ther: the journal of the American Society of Gene Therapy 2013;21(7):1294–6.

[44] Savina A, Furlan M, Vidal M, Colombo MI. Exosome release is regulated by a calcium-dependent mechanism in K562 cells. J Biol Chem 2003;278(22):20083–90.

[45] Loomis RJ, Holmes DA, Elms A, Solski PA, Der CJ, Su L. Citron kinase, a RhoA effector, enhances HIV-1 virion production by modulating exocytosis. Traffic (Copenhagen, Denmark) 2006;7(12):1643–53.

[46] Savina A, Fader CM, Damiani MT, Colombo MI. Rab11 promotes docking and fusion of multivesicular bodies in a calcium-dependent manner. Traffic (Copenhagen, Denmark) 2005;6(2):131–43.

[47] Thery C, Boussac M, Véron P, Ricciardi-Castagnoli P, Raposo G, Garin J, et al. Proteomic analysis of dendritic cell-derived exosomes: a secreted subcellular compartment distinct from apoptotic vesicles. J Immunol (Baltimore, Md.: 1950) 2001;166(12): 7309–18.

[48] Escola JM, Kleijmeer MJ, Stoorvogel W, Griffith JM, Yoshie O, Geuze HJ. Selective enrichment of tetraspan proteins on the internal vesicles of multivesicular endosomes and on exosomes secreted by human B-lymphocytes. J Biol Chem 1998;273(32):20121–7.

[49] Caby MP, Lankar D, Vincendeau-Scherrer C, Raposo G, Bonnerot C. Exosomal-like vesicles are present in human blood plasma. Int Immunol 2005;17(7):879–87.

[50] Simpson RJ, Lim JW, Moritz RL, Mathivanan S. Exosomes: proteomic insights and diagnostic potential. Expert Rev Prot 2009;6(3):267–83.

[51] Pan BT, Johnstone RM. Fate of the transferrin receptor during maturation of sheep reticulocytes in vitro: selective externalization of the receptor. Cell 1983;33(3):967–78.

[52] Fang DY, King HW, Li JY, Gleadle JM. Exosomes and the kidney: blaming the messenger. Nephrology (Carlton, Vic.) 2013;18(1):1–10.

[53] Chaput N, Thery C. Exosomes: immune properties and potential clinical implementations. Semin Immunopathol 2011;33(5):419–40.

[54] Kloosterman WP, Plasterk RH. The diverse functions of microRNAs in animal development and disease. Dev Cell 2006;11(4):441–50.

[55] Calin GA, Croce CM. MicroRNA signatures in human cancers. Nat Rev Cancer 2006;6(11):857–66.

[56] Bartel DP. MicroRNAs: target recognition and regulatory functions. Cell 2009;136(2):215–33.

[57] Carthew RW, Sontheimer EJ. Origins and mechanisms of miRNAs and siRNAs. Cell 2009;136(4):642–55.

[58] Chekulaeva M, Filipowicz W. Mechanisms of miRNA-mediated post-transcriptional regulation in animal cells. Curr Opin Cell Biol 2009;21(3):452–60.

[59] Camussi G, Deregibus MC, Bruno S, Grange C, Fonsato V, Tetta C. Exosome/microvesicle-mediated epigenetic reprogramming of cells. Am J Cancer Res 2011;1(1):98–110.

[60] Yang M, Chen J, Su F, Yu B, Su F, Lin L, et al. Microvesicles secreted by macrophages shuttle invasion-potentiating microRNAs into breast cancer cells. Mol Cancer 2011;10:117.

[61] Xiao C, Rajewsky K. MicroRNA control in the immune system: basic principles. Cell 2009;136(1):26–36.

[62] Josefowicz SZ, Lu LF, Rudensky AY. Regulatory T cells: mechanisms of differentiation and function. Annu Rev Immunol 2012;30:531–64.

[63] Kohlhaas S, Garden OA, Scudamore C, Turner M, Okkenhaug K, Vigorito E. Cutting edge: the Foxp3 target miR-155 contributes to the development of regulatory T cells. J Immunol (Baltimore Md.: 1950) 2009;182(5):2578–82.

[64] van Rooij E, Purcell AL, Levin AA. Developing microRNA therapeutics. Circ Res 2012;110(3):496–507.

[65] Lotvall J, Valadi H. Cell to cell signalling via exosomes through esRNA. Cell Adhes Migr 2007;1(3):156–8.

[66] de Candia P, Torri A, Pagani M, Abrignani S. Serum microRNAs as biomarkers of human lymphocyte activation in health and disease. Front Immunol 2014;5:43.

[67] Melnik BC, John SM, Schmitz G. Milk: an exosomal microRNA transmitter promoting thymic regulatory T cell maturation preventing the development of atopy? J Transl Med 2014;12(1):43.

[68] Zhou Q, Li M, Wang X, Li Q, Wang T, Zhu Q, et al. Immune-related microRNAs are abundant in breast milk exosomes. Int J Biol Sci 2012;8(1):118–23.

[69] Munch EM, Harris RA, Mohammad M, Benham AL, Pejerrey SM, Showalter L, et al. Transcriptome profiling of microRNA by Next-Gen deep sequencing reveals known and novel miRNA species in the lipid fraction of human breast milk. PloS One 2013;8(2):e50564.

[70] Gallo A, Tandon M, Alevizos I, Illei GG. The majority of microRNAs detectable in serum and saliva is concentrated in exosomes. PloS One 2012;7(3):e30679.

[71] Kuwabara Y, Ono K, Horie T, Nishi H, Nagao K, Kinoshita M, et al. Increased microRNA-1 and microRNA-133a levels in serum of patients with cardiovascular disease indicate myocardial damage. Circ Cardiovasc Genet 2011;4(4):446–54.

[72] Liu J, Sun H, Wang X, Yu Q, Li S, Yu X, et al. Increased exosomal microRNA-21 and microRNA-146a levels in the cervicovaginal lavage specimens of patients with cervical cancer. Int J Mol Sci 2014;15(1):758–73.

[73] Lopez AD, Mathers CD, Ezzati M, Jamison DT, Murray CJ. Global and regional burden of disease and risk factors, 2001: systematic analysis of population health data. Lancet 2006;367(9524):1747–57.

[74] Roger VL, Go AS, Lloyd-Jones DM, Benjamin EJ, Berry JD, Borden WB, et al. Heart disease and stroke statistics – 2012 update: a report from the American Heart Association. Circulation 2012;125(1):e2–e220.

[75] Blum B, Benvenisty N. The tumorigenicity of human embryonic stem cells. Adv Cancer Res 2008;100:133–58.

[76] Ben-David U, Benvenisty N. The tumorigenicity of human embryonic and induced pluripotent stem cells. Nat Rev Cancer 2011;11(4):268–77.

[77] Nussbaum J, Minami E, Laflamme MA, Virag JA, Ware CB, Masino A, et al. Transplantation of undifferentiated murine embryonic stem cells in the heart: teratoma formation and immune response. FASEB J: official publication of the Federation of American Societies for Experimental Biology 2007;21(7):1345–57.

[78] Lee AS, Tang C, Cao F, Xie X, van der Bogt K, Hwang A, et al. Effects of cell number on teratoma formation by human embryonic stem cells. Cell Cycle (Georgetown, Tex.) 2009;8(16):2608–12.

[79] Ieda M, Fu JD, Delgado-Olguin P, Vedantham V, Hayashi Y, Bruneau BG, et al. Direct reprogramming of fibroblasts into functional cardiomyocytes by defined factors. Cell 2010;142(3):375–86.

[80] Qian L, Huang Y, Spencer CI, Foley A, Vedantham V, Liu L, et al. In vivo reprogramming of murine cardiac fibroblasts into induced cardiomyocytes. Nature 2012;485(7400):593–8.

[81] Song K, Nam YJ, Luo X, Qi X, Tan W, Huang GN, et al. Heart repair by reprogramming non-myocytes with cardiac transcription factors. Nature 2012;485(7400):599–604.

[82] Losordo DW, Dimmeler S. Therapeutic angiogenesis and vasculogenesis for ischemic disease. Part I: Angiogenic cytokines. Circulation 2004;109(21):2487–91.

[83] Yoon YS, Uchida S, Masuo O, Cejna M, Park JS, Gwon HC, et al. Progressive attenuation of myocardial vascular endothelial growth factor expression is a seminal event in diabetic cardiomyopathy: restoration of microvascular homeostasis and recovery of cardiac function in diabetic cardiomyopathy after replenishment of local vascular endothelial growth factor. Circulation 2005;111(16):2073–85.

[84] Rajasingh J, Thangavel J, Siddiqui MR, Gomes I, Gao XP, Kishore R, et al. Improvement of cardiac function in mouse myocardial infarction after transplantation of epigenetically-modified bone marrow progenitor cells. PloS One 2011;6(7):e22550.

[85] Jeevanantham V, Butler M, Saad A, Abdel-Latif A, Zuba-Surma EK, Dawn B. Adult bone marrow cell therapy improves survival and induces long-term improvement in cardiac parameters: a systematic review and meta-analysis. Circulation 2012;126(5):551–68.

[86] Vasa M, Fichtlscherer S, Aicher A, Adler K, Urbich C, Martin H, et al. Number and migratory activity of circulating endothelial progenitor cells inversely correlate with risk factors for coronary artery disease. Circ Res 2001;89(1):E1–7.

[87] Walter DH, Haendeler J, Reinhold J, Rochwalsky U, Seeger F, Honold J, et al. Impaired CXCR4 signaling contributes to the reduced neovascularization capacity of endothelial progenitor cells from patients with coronary artery disease. Circ Res 2005;97(11):1142–51.

[88] Schmidt-Lucke C, Rössig L, Fichtlscherer S, Vasa M, Britten M, Kämper U, et al. Reduced number of circulating endothelial progenitor cells predicts future cardiovascular events: proof of concept for the clinical importance of endogenous vascular repair. Circulation 2005;111(22):2981–7.

[89] Urbich C, Dimmeler S. Risk factors for coronary artery disease, circulating endothelial progenitor cells, and the role of HMG-CoA reductase inhibitors. Kidney Int 2005;67(5):1672–6.

[90] Tepper OM, Galiano RD, Capla JM, Kalka C, Gagne PJ, Jacobowitz GR, et al. Human endothelial progenitor cells from type II diabetics exhibit impaired proliferation, adhesion, and incorporation into vascular structures. Circulation 2002;106(22):2781–6.

[91] Ii M, Takenaka H, Asai J, Ibusuki K, Mizukami Y, Maruyama K, et al. Endothelial progenitor thrombospondin-1 mediates diabetes-induced delay in reendothelialization following arterial injury. Circ Res 2006;98(5):697–704.

[92] Bian S, Zhang L, Duan L, Wang X, Min Y, Yu H. Extracellular vesicles derived from human bone marrow mesenchymal stem cells promote angiogenesis in a rat myocardial infarction model. J Mol Med (Berlin, Germany) 2013;92:387–97.

[93] Arslan F, Lai RC, Smeets MB, Akeroyd L, Choo A, Aguor EN, et al. Mesenchymal stem cell-derived exosomes increase ATP levels, decrease oxidative stress and activate PI3K/Akt pathway to enhance myocardial viability and prevent adverse remodeling after myocardial ischemia/reperfusion injury. Stem Cell Res 2013;10(3):301–12.

[94] Chia S, Nagurney JT, Brown DF, Raffel OC, Bamberg F, Senatore F, et al. Association of leukocyte and neutrophil counts with infarct size, left ventricular function and outcomes after percutaneous coronary intervention for ST-elevation myocardial infarction. Am J Cardiol 2009;103(3):333–7.

[95] Horstman LL, Jy W, Jimenez JJ, Ahn YS. Endothelial microparticles as markers of endothelial dysfunction. Front Biosci: a journal and virtual library 2004;9:1118–35.

[96] Bernal-Mizrachi L, Jy W, Jimenez JJ, Pastor J, Mauro LM, Horstman LL, et al. High levels of circulating endothelial microparticles in patients with acute coronary syndromes. Am Heart J 2003;145(6):962–70.

[97] Preston RA, Jy W, Jimenez JJ, Mauro LM, Horstman LL, Valle M, et al. Effects of severe hypertension on endothelial and platelet microparticles. Hypertension 2003;41(2):211–17.

[98] Vella LJ, Greenwood DL, Cappai R, Scheerlinck JP, Hill AF. Enrichment of prion protein in exosomes derived from ovine cerebral spinal fluid. Vet Immunol Immunop 2008;124(3-4):385–93.

[99] Bachy I, Kozyraki R, Wassef M. The particles of the embryonic cerebrospinal fluid: how could they influence brain development? Brain Res Bull 2008;75(2–4):289–94.

[100] Fruhbeis C, Frohlich D, Kuo WP, Kramer-Albers EM. Extracellular vesicles as mediators of neuron-glia communication. Front Cell Neurosci 2013;7:182.

[101] Braccioli L, van Velthoven C, Heijnen CJ. Exosomes: a new weapon to treat the central nervous system. Mol Neurobiol 2014;49(1):113–9.

[102] Wang S, Cesca F, Loers G, Schweizer M, Buck F, Benfenati F, et al. Synapsin I is an oligomannose-carrying glycoprotein, acts as an oligomannose-binding lectin, and promotes neurite outgrowth and neuronal survival when released via glia-derived exosomes. J Neurosci 2011;31(20):7275–90.

[103] Xin H, Li Y, Buller B, Katakowski M, Zhang Y, Wang X, et al. Exosome-mediated transfer of miR-133b from multipotent mesenchymal stromal cells to neural cells contributes to neurite outgrowth. Stem Cells (Dayton, Ohio) 2012;30(7):1556–64.

[104] Morel L, Regan M, Higashimori H, Ng SK, Esau C, Vidensky S, et al. Neuronal exosomal miRNA-dependent translational regulation of astroglial glutamate transporter GLT1. J Biol Chem 2013;288(10):7105–16.

[105] Skog J, Würdinger T, van Rijn S, Meijer DH, Gainche L, Sena-Esteves M, et al. Glioblastoma microvesicles transport RNA and proteins that promote tumour growth and provide diagnostic biomarkers. Nat Cell Biol 2008;10(12):1470–6.

[106] Lachenal G, Pernet-Gallay K, Chivet M, Hemming FJ, Belly A, Bodon G, et al. Release of exosomes from differentiated neurons and its regulation by synaptic glutamatergic activity. Mol Cell Neurosci 2011;46(2):409–18.

[107] Chivet M, Javalet C, Hemming F, Pernet-Gallay K, Laulagnier K, Fraboulet S, et al. Exosomes as a novel way of interneuronal communication. Biochem Soc Trans 2013;41(1):241–4.

[108] Yuyama K, Sun H, Mitsutake S, Igarashi Y. Sphingolipid-modulated exosome secretion promotes clearance of amyloid-beta by microglia. J Biol Chem 2012;287(14): 10977–89.

[109] Xiao X, Yu S, Li S, Wu J, Ma R, Cao H, et al. Exosomes: decreased sensitivity of lung cancer a549 cells to cisplatin. PloS One 2014;9(2):e89534.

[110] Kalani A, Tyagi A, Tyagi N. Exosomes: mediators of neurodegeneration, neuroprotection and therapeutics. Mol Neurobiol 2014;49(1):590–600.

[111] Aronin N, Kim M, Laforet G, DiFiglia M. Are there multiple pathways in the pathogenesis of Huntington's disease? Philos Trans R Soc Lond B Biol Sci 1999;354(1386): 995–1003.

[112] Bellingham SA, Guo BB, Coleman BM, Hill AF. Exosomes: vehicles for the transfer of toxic proteins associated with neurodegenerative diseases? Front Physiol 2012;3:124.

[113] Rajendran L, Honsho M, Zahn TR, Keller P, Geiger KD, Verkade P, et al. Alzheimer's disease beta-amyloid peptides are released in association with exosomes. Proc Natl Acad Sci USA 2006;103(30):11172–7.

[114] Desplats P, Lee HJ, Bae EJ, Patrick C, Rockenstein E, Crews L, et al. Inclusion formation and neuronal cell death through neuron-to-neuron transmission of alpha-synuclein. Proc Natl Acad Sci USA 2009;106(31):13010–15.

[115] Pandey P, Brors B, Srivastava PK, Bott A, Boehn SN, Groene HJ, et al. Microarray-based approach identifies microRNAs and their target functional patterns in polycystic kidney disease. BMC Genomics 2008;9:624.

[116] Zhang Y, Luo CL, He BC, Zhang JM, Cheng G, Wu XH. Exosomes derived from IL-12-anchored renal cancer cells increase induction of specific antitumor response *in vitro*: a novel vaccine for renal cell carcinoma. Int J Oncol 2010;36(1):133–40.

[117] Anglicheau D, Sharma VK, Ding R, Hummel A, Snopkowski C, Dadhania D, et al. MicroRNA expression profiles predictive of human renal allograft status. Proc Natl Acad Sci USA 2009;106(13):5330–5.

[118] Zhou Y, Xu H, Xu W, Wang B, Wu H, Tao Y, et al. Exosomes released by human umbilical cord mesenchymal stem cells protect against cisplatin-induced renal oxidative stress and apoptosis *in vivo* and *in vitro*. Stem Cell Res Ther 2013;4(2):34.

[119] Dorronsoro A, Robbins PD. Regenerating the injured kidney with human umbilical cord mesenchymal stem cell-derived exosomes. Stem Cell Res Ther 2013;4(2):39.

[120] Lorenzen JM, Haller H, Thum T. MicroRNAs as mediators and therapeutic targets in chronic kidney disease. Nat Rev. Nephrol 2011;7(5):286–94.

[121] Lv LL, Cao YH, Ni HF, Xu M, Liu D, Liu H, et al. MicroRNA-29c in urinary exosome/microvesicle as a biomarker of renal fibrosis. Am J Physiol Renal Physiol 2013;305(8):F1220–1227.

[122] Kosaka N, Iguchi H, Ochiya T. Circulating microRNA in body fluid: a new potential biomarker for cancer diagnosis and prognosis. Cancer Sci 2010;101(10):2087–92.

[123] Yang C, Chalasani G, Ng YH, Robbins PD. Exosomes released from *Mycoplasma* infected tumor cells activate inhibitory B cells. PloS One 2012;7(4):e36138.

[124] Hood JL, Pan H, Lanza GM, Wickline SA. Paracrine induction of endothelium by tumor exosomes. Lab Invest 2009;89(11):1317–28.

[125] Tanaka Y, Kamohara H, Kinoshita K, Kurashige J, Ishimoto T, Iwatsuki M, et al. Clinical impact of serum exosomal microRNA-21 as a clinical biomarker in human esophageal squamous cell carcinoma. Cancer 2013;119(6):1159–67.

[126] Iorio MV, Visone R, Di Leva G, Donati V, Petrocca F, Casalini P, et al. MicroRNA signatures in human ovarian cancer. Cancer Res 2007;67(18):8699–707.

[127] Roccaro AM, Sacco A, Maiso P, Azab AK, Tai YT, Reagan M, et al. BM mesenchymal stromal cell-derived exosomes facilitate multiple myeloma progression. J Clin Invest 2013;123(4):1542–55.

[128] Caruso P, MacLean MR, Khanin R, McClure J, Soon E, Southgate M, et al. Dynamic changes in lung microRNA profiles during the development of pulmonary hypertension due to chronic hypoxia and monocrotaline. Arterioscler Thromb Vasc Biol 2010;30(4):716–23.

[129] Parikh VN, Jin RC, Rabello S, Gulbahce N, White K, Hale A, et al. MicroRNA-21 integrates pathogenic signaling to control pulmonary hypertension: results of a network bioinformatics approach. Circulation 2012;125(12):1520–32.

[130] Pullamsetti SS, Doebele C, Fischer A, Savai R, Kojonazarov B, Dahal BK, et al. Inhibition of microRNA-17 improves lung and heart function in experimental pulmonary hypertension. Am J Respir Crit Care Med 2012;185(4):409–19.

[131] Levanen B, Bhakta NR, Torregrosa Paredes P, Barbeau R, Hiltbrunner S, Pollack JL, et al. Altered microRNA profiles in bronchoalveolar lavage fluid exosomes in asthmatic patients. J Allergy Clin Immunol 2013;131(3):894–903.

[132] Zhu YG, Hao Q, Monsel A, Feng XM, Lee JW. Adult stem cells for acute lung injury: remaining questions and concerns. Respirology (Carlton, Vic.) 2013;18(5):744–56.

[133] Katakowski M, Buller B, Zheng X, Lu Y, Rogers T, Osobamiro O, et al. Exosomes from marrow stromal cells expressing miR-146b inhibit glioma growth. Cancer Lett 2013;335(1):201–4.

[134] Hu G, Drescher KM, Chen XM. Exosomal miRNAs: biological properties and therapeutic potential. Front Genet 2012;3:56.

[135] Chen TS, Arslan F, Yin Y, Tan SS, Lai RC, Choo AB, et al. Enabling a robust scalable manufacturing process for therapeutic exosomes through oncogenic immortalization of human ESC-derived MSCs. J Transl Med 2011;9:47.

[136] Baglio SR, Pegtel DM, Baldini N. Mesenchymal stem cell secreted vesicles provide novel opportunities in (stem) cell-free therapy. Front Physiol 2012;3:359.

[137] Xin H, Li Y, Cui Y, Yang JJ, Zhang ZG, Chopp M. Systemic administration of exosomes released from mesenchymal stromal cells promote functional recovery and neurovascular plasticity after stroke in rats. J Cereb Blood Flow Metab 2013;33(11):1711–15.

[138] Lai RC, Arslan F, Lee MM, Sze NS, Choo A, Chen TS, et al. Exosome secreted by MSC reduces myocardial ischemia/reperfusion injury. Stem Cell Res 2010;4(3):214–22.

[139] Minguell JJ, Erices A. Mesenchymal stem cells and the treatment of cardiac disease. Exp Biol Med (Maywood, N.J.) 2006;231(1):39–49.

[140] Schafer R, Northoff H. Cardioprotection and cardiac regeneration by mesenchymal stem cells. Panminerva Medica 2008;50(1):31–9.

[141] Lai VK, Ashraf M, Jiang S, Haider K. MicroRNA-143 is a critical regulator of cell cycle activity in stem cells with co-overexpression of Akt and angiopoietin-1 via transcriptional regulation of Erk5/cyclin D1 signaling. Cell Cycle (Georgetown, Tex.) 2012;11(4):767–77.

[142] Hosoda T, Zheng H, Cabral-da-Silva M, Sanada F, Ide-Iwata N, Ogórek B, et al. Human cardiac stem cell differentiation is regulated by a mircrine mechanism. Circulation 2011;123(12):1287–96.

[143] Winter J, Jung S, Keller S, Gregory RI, Diederichs S. Many roads to maturity: microRNA biogenesis pathways and their regulation. Nat Cell Biol 2009;11(3):228–34.

[144] Pijnappels DA, Schalij MJ, Ramkisoensing AA, van Tuyn J, de Vries AA, van der Laarse A, et al. Forced alignment of mesenchymal stem cells undergoing cardiomyogenic differentiation affects functional integration with cardiomyocyte cultures. Circ Res 2008;103(2):167–76.

[145] He XQ, Chen MS, Li SH, Liu SM, Zhong Y, McDonald Kinkaid HY, et al. Co-culture with cardiomyocytes enhanced the myogenic conversion of mesenchymal stromal cells in a dose-dependent manner. Mol Cell Biochem 2010;339(1-2):89–98.

[146] Rota M, Kajstura J, Hosoda T, Bearzi C, Vitale S, Esposito G, et al. Bone marrow cells adopt the cardiomyogenic fate in vivo. Proc Natl Acad Sci USA 2007;104(45):17783–8.

[147] Kim HW, Jiang S, Ashraf M, Haider KH. Stem cell-based delivery of Hypoxamir-210 to the infarcted heart: implications on stem cell survival and preservation of infarcted heart function. J Mol Med (Berlin, Germany) 2012;90(9):997–1010.

[148] Acquistapace A, Bru T, Lesault PF, Figeac F, Coudert AE, le Coz O, et al. Human mesenchymal stem cells reprogram adult cardiomyocytes toward a progenitor-like state through partial cell fusion and mitochondria transfer. Stem Cells (Dayton, Ohio) 2011;29(5):812–24.

CHAPTER 4

Current Methods to Purify and Characterize Exosomes

W. Michael Dismuke*, Yutao Liu**

*Department of Ophthalmology, Duke University School of Medicine, Durham, NC, USA
**Department of Cellular Biology and Anatomy, Georgia Regents University, Augusta, GA, USA

Contents

1 INTRODUCTION TO EXOSOMES

1.1 Defining an Exosome

Numerous cells types have now been shown to release vesicles into the extracellular space *in vitro* and *in vivo* [1,2]. Generally, these vesicles have the following common characteristics: (i) they consist of a phospholipid bilayer enclosing a luminal space, (ii) the phospholipid bilayer contains peripheral and transmembrane proteins, and (iii) the intraluminal space contains soluble proteins and possibly RNA or DNA [3]. However, examination of extracellular vesicles (EVs) reveals divergent sizes and protein compositions. Based on this, vesicle size and protein composition were used to place the EVs into three broad categories. The largest vesicles were apoptotic blebs, 1–5 μm in diameter. The next smaller vesicles were called microvesicles ranging from 1 μm down to 100 nm in diameter. Finally, the smallest EVs were called exosomes with 30–100 nm diameters. As the EV field has grown, however, we now know that size and presence of one or two marker proteins cannot be used to categorize EVs. The current consensus is that EVs must be categorized based on their origin. To be called an apoptotic bleb (or body), the vesicle must come from the blebbing of the plasma membrane of a cell undergoing apoptosis [4]. A microvesicle (or ectosome) is generated from direct shedding of the plasma membrane in nonapoptotic cells [5]. For an EV to be called an exosome, it must have been formed by the invagination of the limiting membrane of a specialized endosome. Then this specialized endosome, called a multivesicular body or endosome (MVB or MVE), must fuse with the plasma membrane and release the contained vesicles into the extracellular space. Prior to the release of these vesicles, they are called intraluminal vesicles (ILVs) and only when they are released to the extracellular space are they referred to as exosomes [6]. The purification and characterization of these exosomes will be the focus of this chapter.

The degree to which the exosomes must be purified is dependent on the specifics of the experiment and the hypothesis in question. Some experiments will require highly purified exosomes, nearly devoid of contaminating proteins, lipoproteins, other extracellular vesicles, or cell debris. Other experiments may not require this degree of purification to address the hypothesis. In this chapter, we will present the various methods to purify and characterize exosomes from cell culture media and biological fluids. With each protocol presented, we will provide a best-use case, a rationale for each step, and a discussion of the advantages and disadvantages. The goal of this chapter is to provide the reader with: (i) easy-to-follow protocols for exosomes preparation, but more importantly (ii) an understanding of how best to choose a purification technique or characterization method based on the current consensus within the exosome research community.

1.2 Organization of this Chapter

Exosomes are released by cells into biological fluids *in vivo* and cell culture conditioned media *in vitro*. Currently, the "gold standard" for the purification of exosomes from these fluids is serial ultracentrifugation [7]. This method relies on the relatively low sedimentation coefficient of exosomes versus other biological materials commonly found in biofluids and cell culture conditioned media. For instance, dead cells, cell debris, and apoptotic bodies will pellet at low centrifugation forces while exosomes will not. The specifics of this technique are discussed in detail in Section 2.2. This, Basic Ultracentrifugation Protocol technique is broadly applicable, relies on common laboratory equipment, and will be suitable for the majority of situations where exosomes need to be purified. While this technique will be suitable for the preparation of exosomes in the majority of cases, special cases exist, such as purifying exosomes from viscous biological fluids, where extra steps are needed to ensure efficient exosomes isolation. These steps will be covered in Section 2.4. For convenience or necessity, we will present several techniques that can be substituted for some centrifugation steps in Section 2.3. These substitutions should not significantly affect the final exosomes preparation. Depending on the experiment, exosomes may need to be prepared to a higher degree of purity than possible via our Section 2.2. For example, some protein aggregates may have sedimentation coefficients similar to exosomes and therefore pellet with the exosomes. For these cases we have included Section 2.5, detailing the use of sucrose cushions that can separate exosomes from contaminating protein aggregates. As an alternative to the centrifugation-based methods for exosomes preparations, we have included Section 2.6. This method does not involve centrifugation and instead relies on antibody capture of exosome–associated proteins. While this method offers several advantages to Section 2.2, there are of course several drawbacks and it is up to the researcher to weigh these advantages/disadvantages when choosing a protocol.

Once a sample is prepared it needs to be characterized to demonstrate that the vesicles have the size, density, and protein composition consistent with exosomes. Most likely due to their biogenesis, exosomes are produced in a very narrow size range, and this size is below the resolution limits of optical microscopy. Because of this, exosomes can only be imaged using electron microscopy (EM). We have included Section 3.1 detailing the materials and methods needed to image exosomes. While determining the size of an individual exosome by EM is straightforward, determining the size distribution of an entire exosome preparation using EM can be laborious and subjective since it is essentially done by hand. This makes rapid determination of the size distribution and concentration of exosome preparation impossible. However, indirect methods are available to determine the size distribution and concentration of an exosome preparation. The most common method to do this is via a technique called nanoparticle tracking analysis (NTA). This technique has numerous advantages over EM including minimum sample preparation and rapid analysis of large numbers of exosomes, and it is mostly automated such that there is minimum human influence in the final measurements. Despite the advantages of NTA, EM and NTA should be viewed as complementary. We have included Section 3.6 detailing this technique.

Besides the characteristic size of exosomes, they have a density that differentiates them from other vesicles and membrane artifacts. Exosome density can be determined by its buoyancy in a continuous sucrose gradient. In Section 3.4, we detail the methods and protocols to separate exosome preparations on continuous sucrose gradients for analysis. Finally, the protein composition of exosomes likely reflects the cell from which it originates, yet across a wide variety of studied exosomes there appear to be proteins common to exosomes independent of their cellular origin [1,2]. The presence of these exosome marker proteins is a strong indication of the vesicle's biogenesis from a MVE, and thus indicates that the vesicle is truly an exosome. Section 3.5 focuses on analyzing the exosome protein content. Complementary to sodium dodecyl sulfate-polyacrylamide gel electrophoresis (SDS-PAGE) analysis of exosomes, we present a protocol for analyzing surface exposed exosome proteins using flow cytometry in Section 3.3.

Finally, we end this chapter with a general commentary to guide the reader in their exosome studies.

2 ISOLATION OF EXOSOMES

2.1 Preparing Exosome Collection Media

The procedures described in this chapter are intended to produce a concentrated exosome solution free of all possible external contaminants. To achieve this goal, it is advisable to start with cell culture media that is free of contaminants. For *in vitro* exosome collection, a major source of potential contamination is the often necessary, fetal bovine serum (FBS). The FBS commonly used in tissue culture applications contains exosomes, which will contaminate the cell-derived exosomes and make interpretation of experimental findings difficult. To eliminate this possible source of contamination, three major options are available: (i) if the particular cells of interest can tolerate serum-free conditions for the necessary time to collect an adequate amount of exosomes, collecting exosomes under serum-free conditions is advised. However, a number of cell types will not tolerate extended periods in the absence of serum. (ii) In this case, conditions permitting, substitution of FBS with a commercially available "serum supplement" or bovine serum albumin (BSA) is advisable. Cells in culture tend to need external sources of protein and by using protein sources with defined ingredients, as opposed to FBS, which is undefined, an experiment can be designed to account for or subtract any effect the known proteins may have. While this may not be ideal in every situation, the known proteins will allow for more repeatable results as the proteins and growth factors/cytokines in FBS may vary from lot to lot. Finally, if the exosome-producing cells require the presence of FBS, (iii) the FBS-containing media can be depleted of serum-derived exosomes via ultracentrifugation. However, due to the viscosity of FBS, which inhibits the sedimentation of the exosomes, the FBS must be diluted to obtain a practical viscosity for exosome sedimentation (similar methods are described for viscous biofluids in Section 2.4). The preparation of exosome-depleted, FBS-containing cell culture medium is described here.

2.1.1 Materials

Ultracentrifuge with fixed-angle or swinging-bucket rotor
- Polyallomer ultracentrifuge tubes or polycarbonate ultracentrifuge bottles
- 0.22 μm vacuum filter sterilization apparatus

- Sterile glass bottle (100 mL to 1 L)
- Serum-free cell culture media with nutrients and antibiotics (the exact media the cells require *minus* FBS)
- Cell culture grade FBS

2.1.2 Protocol

1. Combine serum-free culture media (containing nutrients and antibiotics) with the FBS to obtain a final 20% FBS solution (v:v).
2. Centrifuge this media overnight at 100,000g, 4°C.

   Overnight centrifugation is recommended to remove as many FBS-derived exosomes as possible. However, shorter centrifugation times can be equally effective, depending on the concentration of exosomes in the FBS, the specifics of the rotor used (i.e., k-factor), and the desired outcome of the experiment. It is up to the individual to determine their experimental needs and how to most efficiently meet those needs. To reduce possible cross-contamination with other exosome preparations, it is advisable to dedicate bottles or tubes solely for the preparation of exosome-depleted media and not use them for any other exosome preparation purposes.

   As indicated in Section 2.2, care must be taken when using fixed angle rotors to ensure accurate localization of the resulting pellet. This is best done by marking the bottle/tube with a water-proof marker and orienting the bottle/tube with the mark facing up.

3. Attach the 0.22 μm vacuum filter apparatus to the sterile glass bottle, connect it to a vacuum and filter the carefully collected supernatant resulting from step 2. This will result in sterile media containing 20% exosome-depleted FBS.
4. This 20% FBS media can be stored up to 1 month at 4°C.
5. To make the exosome collection media, add the exosome-depleted 20% FBS media to serum-free culture media in the correct proportions to achieve the desired final FBS concentration.

   For example, to make media with a 10% FBS final concentration, mix exosome-depleted 20% FBS media to serum-free culture media in a one to one ratio.

6. As in step 4, this final media can be stored up to 1 month at 4°C.

2.2 Basic Ultracentrifugation Protocol

The standard method to purify exosomes from conditioned media or biofluids is by several centrifugation steps, each with an increasing force. The first low speed centrifugation steps are intended to remove nonadherent cells, dead cells, and cellular debris. Once these potential contaminants are removed, the remaining supernatant is centrifuged at 100,000g to pellet exosomes. To reduce contamination of this exosome pellet, the pellet is resuspended and centrifuged a second time at 100,000g. This final pellet should primarily be exosomes. This protocol is optimized to purify exosomes from low viscosity fluids such as cell culture conditioned media or urine. To prepare exosomes from higher viscosity fluids such as serum, see Section 2.4.

2.2.1 Materials
- Phosphate buffered saline (PBS)
- Refrigerated centrifuge
- 50 mL or 15 mL polypropylene centrifuge tubes
- Ultracentrifuge with fixed-angle or swinging-bucket rotor
- Polyallomer ultracentrifuge tubes or polycarbonate ultracentrifuge bottles
- Micropipettor
- −80°C freezer

General notes: All centrifugations should be performed in refrigerated centrifuges at 4°C. Cleanliness will determine the final outcome of this procedure, so tubes should be suitably clean and free of residual proteins, detergents, or other contaminants. If the experiment requires the final exosome pellet to be sterile, centrifuge tubes and rotor buckets/lids must be sterilized and all reagents must be sterile. This can be achieved by cleaning the tubes and rotor components with 70% ethanol. Centrifuge tubes must then be rinsed with sterile PBS. Additionally, to ensure sterility, all handling steps should be performed in a tissue culture hood using proper sterile technique.

Centrifuge notes: Both fixed angle and swinging bucket can be used to prepare exosomes; however, swinging bucket rotors offer several advantages and therefore are the preferred option. That said, fixed angle rotors, such as a Beckman 45 Ti rotor, offer superior volume capacity (>550 mL with the 45 Ti) enabling the processing of larger volumes of conditioned media or biofluid at once. If a fixed-angle rotor is to be used, it is advisable to mark each tube and always place the marked side facing up in the rotor. This allows for an estimate of the pellet location, as an exosome pellet is often not visible to the naked eye.

2.2.2 Protocol
2.2.2.1 Removal of Cells, Dead Cells, and Cell Debris
1. Collect cell culture media or biofluids and transfer to 15 mL or 50 mL centrifuge tube.

2. Centrifuge at 2000*g* for 20 min, 4°C.
3. Using a pipet (do not pour), carefully transfer the supernatant to the polyallomer tubes (swinging-bucket rotor) or polycarbonate bottles (fixed-angle rotor).
   To ensure that the resulting pellet is not disturbed, leave behind enough supernatant so there is 0.5–1 cm of liquid remaining above the pellet. Disturbing the pellet can result in contamination of the supernatant and the resulting exosome preparation.

2.2.2.2 Purify Exosome Fraction
For the following section, all centrifugation times and forces are given for a Beckman SW28 rotor. This rotor offers several advantages for exosome preparation; (i) swinging bucket, which allows for easy identification of the pellet location (bottom of the tube) versus a fixed-angle rotor (pellet on the side of the tube), which aids in supernatant removal, (ii) large capacity, which allows for exosome preparation from up to ~230 mL of conditioned media or biofluid, and (iii) availability, this is a common, multiuse rotor found in many labs. If another rotor is to be used, centrifugation times should be adjusted based on the chosen rotor's k–factor.

4. Centrifuge at 100,000*g* for 70 min, 4°C.
   One hour of centrifugation at 100,000g is sufficient to pellet exosomes; how-ever, 70 min of total time is recommended to account for the rotor reaching full speed/max g-force. Centrifugation at 100,000g can be longer (up to several hours) without damaging the exosomes and it is up to the individual to determine what is necessary for their particular application.
5. Remove supernatant.
   Use a pipet to remove the supernatant while leaving a few millimeters of liquid above the exosome pellet. The pellet is most likely not visible.

2.2.2.3 Wash Exosomes
6. Resuspend the pellet from each tube with 1 mL of PBS. If multiple tubes were used, combine the resuspended pellets into one tube. Adjust the final volume by filling the tube with PBS.
7. Centrifuge at 100,000*g* for 70 min, 4°C.
8. Remove most of the supernatant with a pipet (do not pour) leaving ~5 mm of liquid above the pellet. With a pipet held at the meniscus, carefully aspirate the remaining liquid and as the final amounts of liquid are being removed rotate the tube, such that it is facing downward. Continue to hold the tube facing downward and carefully wipe the inner walls of the tube with a cleaning tissue. Do not touch the tissue to the bottom of the tube, which contains the exosome pellet. Once the tube walls are dried, the tube can be turned right–side up.

9. Resuspend the pellet (exosomes) with 25–100 μL of PBS or tris-buffered saline (TBS).

10. Exosome suspensions can be stored for up to 1 year at −80°C. Avoid multiple freezing and thawing.

2.3 Alternative Procedures

In Section 2.2, cells, dead cells, and cell debris were eliminated by centrifugation. However, in some special cases, it may be more desirable to achieve the same result without centrifugation. Steps 1–3 in our Section 2.2 can be replaced by a single filtration step whereby the collected cell culture media is passed through a 0.22 μm filter to eliminate living/dead cells and cell debris. If this procedure is more desirable than centrifugation, it is advisable to compare the final exosome preparations obtained with this procedure (using the characterization methods outlined later) against an exosome preparation using the Section 2.2

2.3.1 Materials
- 0.22 μm vacuum filter sterilization apparatus
- Two sterile glass bottles
- Exosome collection media
- PBS

2.3.2 Protocol
1. Attach the vacuum filter apparatus to one of the sterile glass bottles. Add PBS to completely cover the filter surface and apply vacuum.

In some instances, filters are coated in a preservative, such as glycerol, to ensure longevity; however, this could negatively impact the final exosome preparation. This washing step eliminates any coatings on the filter.

2. Add unused exosome collection media (~100 mL should be sufficient) to the vacuum filter apparatus. Allow this media to be vacuum filtered.

For reasons not fully understood, exosomes tend to "stick" to certain materials, including the materials used to manufacture the filter in the vacuum filter apparatus. This could result in loss of exosomes due to nonspecific binding to the filter materials. Passing the exosome collection media over the filter is intended to "block" the nonspecific binding sites in the filter and reduce exosome "sticking" therefore increasing exosome yield in the final preparation.

3. Disconnect the vacuum and carefully remove the vacuum filter apparatus from the sterile glass bottle now containing PBS and unused exosome collection media. Reattach the now washed and blocked vacuum filter apparatus to the second, unused, sterile glass bottle.

4. Add the conditioned media containing exosomes to the vacuum filter apparatus and apply vacuum to filter.

5. The resulting filtrate containing exosomes can be stored up to 1 week at 4°C, before proceeding to steps 4–10 in Section 2.2.

   If this procedure results in lower exosome yields when compared to the centrifugation steps listed in Section 2.2, steps 1–3 or if the starting conditioned media contains a large amount of living/dead cells or cell debris, this alternative procedure can be modified to include an initial filtration through a larger, 0.45 μm filter and a second filtration through a 0.22 μm filter. For best results, both filters should be washed and blocked using the listed steps 1–3.

2.4 Special Case Ultracentrifugation Protocol

Purifying exosomes from certain bodily fluids (biofluids) offers an increased challenge when compared to isolating exosomes from cell culture conditioned media. More specifically, collected biofluids can have widely ranging viscosities, high protein concentrations, and possibly living or dead cells, all of which make exosome isolation using our Section 2.2 impractical. The general strategy to deal with these obstacles is to substantially dilute the biofluids prior to ultracentrifugation and to increase the duration of centrifugation. While diluting precious biofluid samples may have an undesirable effect on a number of assays, exosome purification is a concentrating step and diluting the biofluid has a positive effect here. It should be noted, however, that if exosomes and soluble components of the biofluid are of equal interest, it is up to the individual to determine whether analysis of the soluble components is possible once diluted or whether the initial undiluted biofluid should be partitioned; one portion for exosome purification and one for separate analysis (i.e., an enzyme-linked immunosorbent assay or western blot).

2.4.1 Materials

- Viscous biofluid
- PBS
- Refrigerated centrifuge
- 50 mL or 15 mL polypropylene centrifuge tubes
- Ultracentrifuge with fixed-angle or swinging-bucket rotor
- Polyallomer ultracentrifuge tubes or polycarbonate ultracentrifuge bottles

2.4.2 Protocol

1. Dilute viscous biofluid 1:1 with PBS and transfer to a 50 mL centrifuge tube. Centrifuge 30 min, 2000*g*, 4°C.

2. Carefully transfer resulting supernatant, without disturbing the pellet, to an ultracentrifuge tube or bottle. Centrifuge 45 min, 10,000*g*, 4°C.

3. Again, carefully transfer the resulting supernatant to a clean ultracentrifuge tube or bottle. Centrifuge 2 h, 100,000*g*, 4°C.

4. Carefully remove the supernatant and resuspend the pellet in 1 mL of PBS. Combine all resuspended pellets into a single ultracentrifuge tube or bottle and fill the container to an appropriate volume using PBS.
   As noted in Section 2.2, the exosome pellet is most likely not visible.
   When compared to conditioned cell culture media, biofluids may have a more diverse composition and therefore more potential contaminants. If analysis of the exosomes from a biofluid shows considerable contamination, an additional filtration of the resuspended exosome pellet through a 0.22 μm filter is advisable prior to the next step.

5. Centrifuge 70 min at 100,000*g*, 4°C.

6. Remove the supernatant and resuspend the pellet in 1 mL of PBS then fill the tube with the appropriate amount of PBS. Centrifuge 70 min at 100,000*g*, 4°C.

7. Carefully remove the supernatant as described in Section 2.2, step 8.

8. Resuspend the pellet (exosomes) with 25–100 μL of PBS or TBS.

9. Exosome suspensions can be stored for up to 1 year at −80°C. Avoid multiple freezing and thawing.

2.5 High Purity Ultracentrifugation Protocol

The exosome purification protocol describe thus far will yield reasonably pure exosomes, free of contaminating proteins. However, some experiments will require exosomes to be prepared in a way that virtually eliminates all possible contaminants. For example, an investigator wants to identify all exosomal proteins from a biofluid by mass spectrometry (MS). In exosomes prepared with Section 2.2, MS identified several unexpected, high molecular weight proteins. Upon further investigation, it is discovered that these proteins have high sedimentation coefficients and thus may pellet independent of any exosomal association. The strategy to separate the exosomes from the potential protein contaminant relies on a sucrose cushion and the inherent low density of exosomes versus higher density proteins. In this protocol, exosomes will be floated on a sucrose cushion while proteins not bound to or associated with the exosomes will sediment below the cushion.

2.5.1 Materials

- Ultracentrifuge with SW28 rotor (for a smaller SW41 rotor volumes will have to be adjusted)

- Exosome pellet obtained using Section 2.2
- Tris/sucrose/deuterium oxide (D_2O) solution
- Polyallomer tubes for appropriate rotor
- 5 mL syringe with 18-G needle

2.5.2 Protocol

1. Add 4 mL of the Tris/sucrose/D_2O solution to the bottom of the SW28 tube.
2. Resuspend a concentrated exosome solution (obtained using Section 2.2) in 31 mL of PBS.
3. Carefully layer the resuspended exosome solution over the sucrose cushion. Avoid mixing.
4. Centrifuge for 90 min at 100,000g, 4°C.
5. Using the syringe with attached 18-G needle, puncture the side of the tube just below the interface between the PBS and sucrose cushion. Collect ~3.5 mL of the Tris/sucrose/D_2O cushion without disturbing the remaining bottom portion of the cushion.
   This 3.5 mL of the Tris/sucrose/D_2O cushion contains the exosomes while the remaining bottom portion of the cushion will contain large proteins, protein aggregates, and contaminants not bound to the exosomes.
6. Dilute the collected Tris/sucrose/D_2O solution with PBS up to 35 mL and transfer it to a fresh SW28 tube.
7. Centrifuge for 90 min at 100,000g, 4°C.
8. Remove the supernatant without disturbing the exosome pellet using the same procedure described in Section 2.2, step 8.
9. Resuspend the exosome pellet in 25–100 μL of PBS or TBS.
10. Exosome suspensions can be stored for up to 1 year at −80°C. Avoid multiple freezing and thawing.

2.6 Immunoisolation Protocol

Thus far the described exosome isolation techniques have relied on ultracentrifugation exclusively. Ultracentrifugation has a number of advantages as described earlier, but in certain situations it can be a cumbersome or impractical option. It requires expensive equipment (i.e., an ultracentrifuge with appropriate rotors), provides no immediate information about the exosomes, and can be less ideal for isolation of small amounts of exosomes or isolation of exosomes from small sample volumes. The most appealing alternative to the ultracentrifugation procedure is immunoisolation of exosomes using immobilized antibodies. This technique relies on recognition of epitopes exposed on the surface of an exosome by antibodies covalently bound to magnetic microspheres.

Immunoisolation has a number of advantages but several disadvantages when compared to isolation by ultracentrifugation. As an advantage, (i) this technique uses common, less expensive laboratory reagents and equipment, (ii) in a single step, exosomes are simultaneously isolated and demonstrated to have a known exosome protein, and (iii) this technique yields a sample that can be directly analyzed by flow cytometry (later described in Section 3.3) allowing rapid exosome isolation and characterization. However, there are a few potential disadvantages that must be considered before choosing this technique. (i) The major difficulty with this technique is deciding which antibody/epitope is best for isolating the exosome of interest. For example, the presence of CD63 is a characteristic of exosomes released from a wide variety of cells [1,2]. However, the abundance of CD63 in an exosome preparation can vary considerably, depending on the exosome source (unpublished observation). Additionally, through personal communications between the authors and fellow exosome investigators, it is now understood that exosomes from certain cell types or biofluids may completely lack some "canonical" exosomes marker proteins or only a portion of the total exosomes population contains a given marker protein. Therefore, (ii) this technique may only isolate a subset of the total exosome population, potentially skewing experimental results. (iii) Antibody-bound exosomes may be difficult, if not impossible to elute from the beads. This prohibits further use of the exosomes in most functional assays. Even surface enzymatic activity on the exosome may be negatively affected due to steric hindrance/reduced substrate access to catalytic sites. (iv) For isolation of small volumes/amounts of exosomes this technique is ideal; however, if this technique is scaled up to isolate large amounts of exosomes, the cost/resource uses outweigh the benefits. And finally, (v) to be confident that this technique is isolating the vast majority of total exosomes, it is best to prepare exosomes using the presented ultracentrifugation protocols and characterize them using EM, immunogold labeling, western blot analysis, MS, NTA and sucrose density gradients. If a laboratory chooses to use the presented immunoisolation protocol prior to fully characterizing the exosomes as described above, there is a risk of isolating only a subset of exosomes. It is for this reason that the authors strongly recommend using this technique only after a careful characterization of the exosomes of interest has been completed. As an example of an ideal case for this technique can be used (see our commentary in Section 3.3).

2.6.1 Materials

- Conditioned media from cell cultures or dilute biofluid (1:1 dilution with PBS)
- 3 mg/mL BSA in PBS, sterile filtered (0.22 μm filter)
- 50 mL centrifuge tubes, sterile
- Refrigerated centrifuge
- Tube roller for 50 mL centrifuge tubes
- Dynabeads and DynaMag-50 magnet (Life Technologies)
- Antibody against chosen exosome protein

Antibody and Dynabead notes: Immunoisolation of exosomes relies on antibody recognition of exposed epitopes on the surface of the exosomes. Therefore, proper selection of this antibody is key to achieving the desired results. It is up to the individual to determine which antibody will yield the best results for their specific application. Additionally, the proper Dynabead must be selected to match the chosen antibody. Take, for example, these two possible combinations: (i) anti-CD9 mouse monoclonal antibody + Protein G Dynabeads or (ii) biotinylated anti-CD81 mouse monoclonal antibody + streptavidin Dynabeads. In both cases, the antibody should be bound to the Dynabead prior to incubation with the exosome containing samples (the protocol below assumes the Dynabeads are coated with the chosen antibody). We will not go into the specific protocols to achieve this here and refer the reader to the instructions provided with the Dynabeads. If the experiment requires the exosomes to be eluted from the antibody/bead combination and the Protein G Dynabeads are being used, it is recommended that the antibody/bead combination be cross-linked (we suggest the cross-linker BS3) so that the elution separates the exosome from the antibody and not the exosome-bound antibody from the Protein G Dynabead. The high affinity of streptavidin for biotin eliminates the need for cross-linking if streptavidin Dynabeads are used. Finally, this section will not include a protocol to elute the exosomes from the Dynabeads. Ideal elution protocols are usually specific to the antibody used and can be based on extreme pHs and salt concentrations, which may affect the exosomes themselves or the proteins they possess. If a particular experiment requires exosomes to be free floating (i.e., not bound to a bead) then the authors prefer using an exosome isolation protocol based on ultracentrifugation. However, if immunoisolation of exosomes is required, the authors strongly urge that the efficiency of the exosome elution be determined as a partial elution may skew experimental results.

Maximum capture efficiency per Dynabead is determined by (i) the number of individual antibodies bound to each bead, (ii) the ratio of exosomes in the solution to the Dynabeads, and (iii) the incubation time. Ideally, the antibody binding sites on the bead are saturated. To achieve this, the user must carefully follow the manufacturer's protocol for absorbing the antibody to the Dynabead and experimentally determine if the bead has been saturated. Once this has been completed the user must then experimentally determine how many beads to add to a given exosome-containing solution to fully saturate the beads with exosomes. Due to the high variability in these two steps, we will not include protocols to address the conditions.

2.6.2 Protocol
2.6.2.1 Immunoprecipitate Exosomes
1. Transfer cell culture conditioned media to 50 mL centrifuge tubes. Centrifuge at 2000*g*, 10 min, 4°C.
2. Transfer supernatant to a new 50 mL tube and centrifuge again at 2000*g*, 10 min, 4°C.
3. Wash the antibody-coated Dynabeads with the 3 mg/mL BSA/PBS solution in a clean 50 mL tube using the DynaMag-50 magnet and the manufacturer's protocol.
4. Use the magnet to gather the Dynabeads onto the side of the tube and completely remove the remaining liquid. Add the supernatant resulting from step 2 to the beads.

   Do not allow the beads to dry at any point. Allowing the beads to dry may result in reduced binding efficiency and nonspecific binding.
5. Incubate for 18–24 h at 4°C on the 50 mL tube roller.

   Take care to not introduce bubbles into the exosome-containing solution through vortexing, pipeting, or shaking the tube. The bubbles tend to trap the Dynabeads, decreasing the isolation efficiency.

   In the authors' experience, overnight incubation (18 h) at 4°C is sufficient to bind the vast majority of exosomes to the beads. However, depending on the antibody, the concentration of exosomes and the number of beads used, incubation times may need to be extended in order to saturate the beads with exosomes. It is up to the user to determine the incubation time necessary to achieve the desired result.
6. Gather the beads with the magnet and remove the supernatant.
7. Wash the beads three times with the 3 mg/mL BSA/PBS solution to remove unbound material.
8. Wash the beads three additional times with PBS only to remove the BSA.
9. Resuspend the beads in 100 μL of PBS for subsequent analysis.

This solution of exosome-bound Dynabeads should not be frozen as this could negatively affect the beads such that exosomes and the beads are no longer bound together. This solution can be briefly stored at 4°C (for a few hours or possibly overnight) but it is advisable to continue directly to the analysis of the exosome by flow cytometry or western blot.

3 CHARACTERIZATION OF EXOSOMES

3.1 Imaging of Exosomes by Electron Microscopy

One characteristic that separates exosomes from other EVs is the small vesicle diameter. Determining the size distribution of vesicles in conditioned media or biofluids can suggest the origin of the vesicles (i.e., small exosomes versus larger microvesicles). However, exosomes are of such a small size that they fall below the limit of detection by optical microscopy. Therefore, the only available technique to visualize exosomes is EM. When using EM to visualize tissues or cells, the sample can be fixed, dried, embedded in resin, and sectioned to better reveal areas of interest. However, these standard-processing steps may not yield satisfactory images of the small exosomes. Instead, here we describe a protocol to mount exosomes, whole on Formvar/carbon coated EM grids, and treat them with contrast agents allowing the exosomes to be visualized under an electron microscope.

3.1.1 Materials
- Concentrated exosome suspension (i.e., product of the Section 2.2)
- PBS
- 4% paraformaldehyde (w/v)
- 1% glutaraldehyde
- Uranyl–oxalate, pH 7
- Methyl cellulose–uranyl acetate, pH 4
  Mix nine parts 2% methyl cellulose with one part 4% uranyl acetate just prior to use.
- Formvar/carbon coated EM grids
- Parafilm
- Glass dish
- Forceps
- Small stainless steel loop
  This loop can be made from stainless steel wire and a P1000 pipet tip. The inner diameter of the loop should be slightly larger than the outer diameter of the EM grid.
- Whatman filter paper, #1
- Transmission electron microscope (TEM)

3.1.2 Protocol
3.1.2.1 Fixing Exosomes on EM Grids
1. Mix the concentrated exosome suspension with an equal volume of the 4% paraformaldehyde.

   The resulting exosome/2% paraformaldehyde solution can be stored at 4°C for up to 1 week.
2. Pipet 5 μL of the exosome/2% paraformaldehyde solution onto the Formvar-coated EM grids. It is best to prepare several grids for each exosome sample. Cover the grid and allow the membrane to absorb the solution for 20 min in a dry area.
3. Pipet 100 μL of PBS on a sheet of parafilm. Carefully wash the EM grid by transferring it, membrane side down, onto the PBS with clean forceps.

   It is important to keep the membrane side of the grid wet at all times while the opposite side should remain dry.
4. Pipet 50 μL of the 1% glutaraldehyde solution onto a sheet of parafilm and transfer the EM grid onto the glutaraldehyde solution. Incubate for 5 min.
5. Pipet 100 μL of distilled water onto a parafilm sheet. Transfer the EM grid onto the water for 2 min. Repeat seven more times, each time using new, clean distilled water to wash the grids.

3.1.2.2 Contrast and Embed Exosome Samples
6. Pipet 50 μL of uranyl-oxalate solution onto a parafilm sheet. Place the grid onto the solution for 5 min.
7. Place a glass on ice. Place a sheet of parafilm onto the now, cold glass surface. Add 50 μL of the methyl cellulose-uranyl acetate solution onto the parafilm. Transfer the EM grid onto the cold methyl cellulose-uranyl acetate solution and incubate for 10 min.
8. Using the stainless steel loop, remove the EM grid. Blot excess liquid by touching the loop to the Whatman filter paper at a 90° angle and slowly drag the loop across the filter paper. Stop and lift the loop before the remaining liquid is completely absorbed by the Whatman filter paper.

   The thickness of the methyl cellulose film affects the final contrast and structural preservation of the sample. The film thickness is controlled by the rate of absorption of the fluid onto the filter paper. If the suggested #1 Whatman filter paper does not yield acceptable methyl cellulose thickness (i.e., contrast is not optimal), the authors suggest trying a different grade Whatman filter paper with a different absorbance.

9. Air dry the grid, still in the stainless steel loop, at room temperature for 5–10 min. This can be accomplished by placing the pipet (used to make the stainless steel loop) back onto a pipet tip rack.

   To assess the methyl cellulose film thickness, view the prepared EM grids under a stereomicroscope. A blue–gold color is indicative of an optimum methyl cellulose thickness.

10. The prepared grid can now be stored for many years in an appropriate EM grid storage box.

11. Examine the prepared grids with an electron microscope at 80 kV.

    Using this technique, exosomes will appear as cup-shaped vesicles with a diameter of ∼ 40–100 nm. Small lipid particles may also be present but should appear as smaller, 10–20 nm particles lacking the cup shape. It is important to note that the cup-shape morphology of the exosomes is widely considered to be an artifact of the fixation process. Using the more complicated technique of cryoEM (not covered in this chapter), exosomes have a spherical appearance, which is thought to be representative of their actual morphology. Analysis of the prepared exosomes may reveal contamination, seen as large and/or irregular debris, which may represent incomplete or inefficient separation of the exosomes from dead cells or cell membrane debris. If this is the case, the authors suggest reviewing the exosome preparation protocol used, paying careful attention to each step or altering the Section 2.2 to include the steps outlined in Section 2.4.

    Images obtained by EM can be used to measure exosome diameter. If this technique is going to be used to characterize the size distribution of the entire exosomes population present in the sample, a large number of exosomes must be measured. It is up to the individual to determine how many exosomes must be measured in order to have predictive power for the entire exosome population. This mostly likely will require a large human effort, which may introduce bias into the measurement. Due to these reasons, the authors recommend using EM analysis of the exosome preparation solely to determine the quality of the preparation (or presence of certain proteins, see Section 3.2). For a more rapid, quantitative measure of an exosome population's size and concentration, see Section 3.6.

3.2 Identifying Exosome Proteins by Immunogold Labeling

Characterization of prepared exosomes is critical and size alone is not fully indicative of a vesicle originating from a MVE (i.e., accepted exosome biogenesis). As the exosome research field has developed and exosomes from a variety of sources were characterized, a number of proteins were commonly seen in these diverse exosome samples [1,2]. These proteins came to be thought of as "canonical" exosome marker proteins. In this section, we present a protocol whereby exosome proteins can be labeled with gold conjugated antibodies and imaged by EM. This technique allows the simultaneous characterization of exosome size *and* protein composition.

3.2.1 Materials

- Concentrated exosome suspension (i.e., product of the Section 2.2)
- PBS
- 50 mM glycine in PBS
- Blocking buffer: 5% BSA in PBS (w/v) or 10% fetal calf serum in PBS (v/v)
  As mentioned earlier, it is up to the individual to determine which blocking buffer is ideal for their exosomes and antibodies.
- Primary antibody and (optional) secondary antibody
- Antibody diluent: 1% BSA in PBS (w/v) or 5% fetal calf serum in PBS (v/v)
- Wash buffer: 0.1% BSA in PBS (w/v) or 0.1% fetal calf serum in PBS (v/v)
- 0.5% BSA in PBS (w/v)
- Protein A-gold conjugate
- 4% paraformaldehyde (w/v)
- 1% glutaraldehyde
- Uranyl oxalate, pH 7
- Methyl cellulose-uranyl acetate, pH 4
  Mix nine parts 2% methyl cellulose with one part 4% uranyl acetate just prior to use.
- Formvar-carbon coated EM grids
- Parafilm
- Glass dish
- Forceps
- Small stainless steel loop
  This loop can be made from stainless steel wire and a P1000 pipet tip. The inner diameter of the loop should be slightly larger than the outer diameter of the EM grid.
- Whatman filter paper, #1
- TEM

3.2.2 Protocol

3.2.2.1 Fixing Exosomes on EM Grids

1. Mix the concentrated exosome suspension with an equal volume of the 4% paraformaldehyde
   The resulting exosome/2% paraformaldehyde solution can be stored at 4°C for up to 1 week.

2. Pipet 5 μL of the exosome/2% paraformaldehyde solution onto the Formvar-coated EM grids. It is best to prepare several grids for each exosome sample. Cover the grid and allow the membrane to absorb the solution for 20 min in a dry area.

3. Pipet 100 μL of PBS on a sheet of parafilm. Carefully wash the EM grid by transferring it, membrane side down, onto the PBS with clean forceps.

   It is important to keep the membrane side of the grid wet at all times while the opposite side should remain dry.

3.2.2.2 Wash and Block Exosomes

4. Pipet 100 μL of PBS onto a sheet of parafilm. Transfer EM grid to PBS. Wash for 3 min. Repeat.

5. Pipet 100 μL of the 50 mM glycine/PBS onto a sheet of parafilm. Transfer EM grid to the 50 mM glycine/PBS, incubate for 3 min. Repeat with fresh 50 mM glycine/PBS for a total of four washes.

   The glycine solution is used to quench free aldehyde groups. If needed, an alternative to 50 mM glycine/PBS is a 50 mM solution of ammonium chloride (NH_4Cl) in PBS.

6. Pipet 100 μL of the blocking buffer onto a sheet of parafilm. Transfer EM grid onto the blocking buffer, incubate for 10 min.

   For all remaining steps, use the same proteins in the diluents and wash buffers as was chosen for the blocking buffer.

3.2.2.3 Incubate Exosomes with Antibodies

7. Pipet 5 μL of the primary antibody (in the appropriate diluent) onto a sheet of parafilm. Transfer the EM grid onto this solution and incubate for 30 min.

   Each antibody has different characteristics and it is up to the individual to determine the correct antibody dilution for this application. Generally, an acceptable antibody concentration will fall between 2 and 20 μg/mL.

8. Pipet 100 μL of the appropriate wash buffer onto a sheet of parafilm. Transfer EM grid onto the wash buffer and wash for 3 min. Repeat this step with fresh wash buffer for a total of six washes.

   It is important to keep the membrane side of the grid wet at all times while the opposite side should remain dry. If the samples were allowed to dry during these antibody incubations/washes, high background or nonspecific staining may result.

9. If a secondary (bridging) antibody is to be used, repeat steps 7 and 8 using the secondary antibody.

3.2.2.4 Label Exosomes with Protein A-Gold

10. Pipet 5 μL of the protein A-gold conjugate (in the appropriate diluent) onto a sheet of parafilm. Transfer the EM grid onto this solution and incubate for 20 min.

11. Pipet 100 μL of PBS onto a sheet of parafilm. Transfer EM grid to PBS. Wash for 3 min. Repeat using fresh PBS for a total of eight washes.

12. To stabilize the antibodies, pipet 50 μL of the 1% glutaraldehyde onto a sheet of parafilm. Transfer the EM grid onto this solution and incubate for 5 min.

13. Pipet 100 μL of water onto a sheet of parafilm. Transfer the EM grid to the water. Wash for 3 min. Repeat using fresh clean water for a total of eight washes.

These washes are critical to ensure proper contrast enhancement of the exosomes. To label the exosomes with another primary antibody, repeat steps 3 through 13. Double or triple labeling can be accomplished using this protocol. With each primary antibody, a different size gold particle must be used (i.e., 10 nm, then 15 nm, then 25 nm) in order to distinguish the stained proteins from one another.

At this point, proceed to steps 6–11 described in Section 3.1.

3.3 Exosome Protein Analysis by Flow Cytometry

As discussed in Section 3.1, exosomes are so small that they fall below the resolution limits of imaging techniques relying on visible or near visible wavelengths. This means that exosomes themselves cannot be directly analyzed by flow cytometry since the device cannot discriminate light scatter from one exosome versus several. To overcome this hurdle, exosomes isolated by ultracentrifugation can be absorbed onto larger beads that are detectable by a flow cytometer and subsequent fluorescent antibody labeling can be used to detect exosome proteins. In the protocol below, we describe a known technique to absorb isolated exosomes onto aldehyde/sulfate latex beads. This technique has been used in a variety of laboratories to isolate exosomes from a wide range of sources [8–10] and should be applicable to exosomes in general. Additionally, this technique is complementary to Section 2.6 as that technique yields exosomes absorbed onto Dynabeads, which are large enough to be detected by flow cytometry.

3.3.1 Materials

- Purified exosomes
- 4 μm aldehyde/sulfate latex beads, 4% (w/v) (Life Sciences)

- PBS
- 1 M glycine in PBS
- 0.5% BSA in PBS (w/v)
- Antibodies
- Tube rotator for 1.5 mL microcentrifuge tubes
- Microcentrifuge
- Flow cytometer

Latex bead notes: Prepare the aldehyde/sulfate latex beads by washing according to the manufacturer's protocol.

Antibody notes: This technique requires either a primary antibody conjugated to a fluorophore or an appropriate secondary antibody conjugated to a fluorophore. It is up to the individual to determine what combination of antibodies and fluorophore are best for their application. As the proteins of interest on the exosomes are in their natural conformations and not denatured, it is important to choose a primary antibody that recognizes the folded protein.

Immunoisolated sample notes: If the sample has been isolated using Section 2.6 Immunoisolation Protocol, check to ensure that the flow cytometer is compatible with the magnetic Dynabeads. If it is compatible, proceed directly to step 6.

3.3.2 Protocol

1. Add 5 μg of exosomes (isolated using the ultracentrifugation) to 10 μL of washed latex beads in a 1.5 mL microcentrifuge tube. Incubate for 15 min at room temperature.
2. Add PBS for a 1 mL final volume and incubate on the tube rotator for 2 h at room temperature or overnight at 4°C.
3. Add 110 μL of 1 M glycine to achieve a 100 mM final concentration. Gently mix then let stand at room temperature for 30 min.
   The glycine will block all remaining free binding sites on the beads, reducing nonspecific binding by the antibodies.
4. Centrifuge beads/exosomes at 4000 rpm for 3 min at room temperature. Collect the supernatant.
   The collected supernatant can be used to determine the amount of exosomes absorbed onto the aldehyde/sulfate latex beads. To do this, determine the protein content of this solution and subtract that number from the original 5 μg of exosomes. If exosome absorption onto the beads is low, consult the manufacturer's protocol to optimize the binding conditions.
5. Resuspend the beads/exosomes in 1 mL of the 0.5% BSA/PBS solution.

6. Centrifuge beads/exosomes at 4000 rpm for 3 min at room temperature. Remove the supernatant and discard it. Repeat steps 5 and 6 for a total of three washes.

7. Resuspend the beads/exosomes in 250 μL 0.5% BSA/PBS solution.

8. Incubate 10 μL of beads with 50 μL of the primary antibody (diluted in the 0.5% BSA/PBS solution) for 30 min at 4°C on the tube rotator.

9. Centrifuge beads at 4000 rpm for 3 min at room temperature. Remove supernatant and discard.

10. Add 150 μL of 0.5% BSA/PBS solution to wash beads. Repeat step 8 for a total of two washes.

11. If a secondary antibody is to be used, repeat steps 7 through 9 with the secondary antibody.

12. Resuspend pelleted beads/exosomes in 200 μL of the 0.5% BSA/PBS solution.

13. Load exosome/bead solution into the flow cytometer. Adjust the forward and side scatter so that both single beads and bead doublets can be seen. Adjust the gating on both the single and double beads to measure fluorescence.

   If the primary antibody is conjugated to a fluorophore, it is advisable to prepare a separate tube of exosomes/beads incubated with an irrelevant primary antibody conjugated to the same fluorophore as a control. If a secondary antibody conjugated to a fluorophore is used, prepare a separate tube of exosomes/beads using only the fluorophore conjugated secondary antibody as a control.

3.4 Determination of Exosome Density

One of the characteristics of exosomes that separates them from other extracellular vesicles, intracellular vesicles, and membrane artifacts is their density. Exosomes have a density between 1.14 and 1.19 g/mL. This density allows for exosomes to be separated from the other vesicles and contaminants like apoptotic bodies (1.24–1.28 g/mL) [11], ectosome (1.03–1.08 g/mL) [12], and proteins (1.2–1.35 g/mL). In order to determine the density of the exosomes in a preparation, we will float the exosomes up into a continuous sucrose gradient. Fractions of this gradient will be collected and analyzed for the presence of known exosomal proteins by MS or western blot.

3.4.1 Materials

- Prepared exosomes
- 2.5 M sucrose/hydroxyethyl piperazineethanesulfonic acid (HEPES) solution, pH 7.4

- HEPES solution, pH 7.4
- 4× Laemmli buffer
- Gradient maker
- Magnetic stir plate, small stir bar
- Ultracentrifuge with SW41 rotor
- Polyallomer ultracentrifuge tubes
- P1000 pipet
- 96-well plate
- Refractometer

Gradient notes: Using a gradient marker to pour a continuous sucrose gradient can be difficult and may require some practice. The authors strongly advise the user to pour several test gradients and analyze them by refractometry to confirm they are near linear before pouring a gradient over the exosomes. If during the gradient making process there is a problem, stop pouring the sucrose/HEPES. The exosomes can be recovered by diluting the exosome containing sucrose/HEPES solution with PBS, one part exosome sucrose/HEPES solution to two parts PBS. This should yield a density low enough that exosomes can be recovered using steps 3 through 8 of Section 2.4.

3.4.2 Protocol
3.4.2.1 Pour Sucrose Gradient over Exosomes

1. Prepare 5 mL of a 2 M sucrose/HEPES solution by mixing 4 mL of the 2.5 M sucrose/HEPES solution with 1 mL of the HEPES solution.
2. Prepare 5 mL of a 0.25 M sucrose/HEPES solution by mixing 0.5 mL of the 2.5 M sucrose/HEPES solution with 4.5 mL of the HEPES solution.
3. Resuspend exosomes in 2 mL of the 2.5 M sucrose/HEPES solution.
4. Pipet the exosome suspension into the bottom of an SW41 ultracentrifuge tube.
5. Close all valves in the gradient maker. Pipet the 5 mL of the 2 M sucrose/HEPES solution into the proximal chamber (closest to the outlet) and add the small magnetic stir bar. Pipet the 5 mL of the 0.25 M sucrose/HEPES solution into the distal chamber.
6. Elevate the magnetic stir plate above the work surface by placing it on a high shelf. Open the valve between the proximal and distal chambers and start the magnetic stir plate.
7. Put the tubing from the gradient maker against the inner wall of the SW41 tube just above the 2 mL of 2.5 M sucrose/HEPES/exosome solution.

8. Open the outer valve of the gradient maker and slowly pour a continuous sucrose gradient on top of the exosome containing solution. Lower the SW41 ultracentrifuge tube as the gradient is being poured, keeping the gradient maker tubing against the inner wall of the SW41 tube and just above the surface of the solution.

9. Balance the ultracentrifuge tubes by slowly and carefully pipeting the 0.25 M sucrose/HEPES solution near the inner wall of the SW41 tube. Take care not to mix the solutions as it may disturb the linearity of the gradient.

3.4.2.2 Centrifuge Gradients and Collect Fractions

10. Centrifuge gradients overnight at 210,000g, 4°C with the centrifuge set to no braking.

    A minimum centrifugation time for the exosomes to reach their equilibrium density is 14 h. It is best to set the ultracentrifuge to "hold" overnight and not set a centrifugation time. It is advisable to take great care in handling the tubes/centrifuge buckets once the centrifugation is complete. The linear sucrose gradient (and the exosomes) may be disturbed if the tubes are handled roughly.

11. Collect 11 (1mL) fractions from the gradient starting at the top by placing the pipet tip at the meniscus of the gradient and slowly drawing the solution into the pipet tip. Put each fraction into a clean SW41 ultracentrifuge tube.

12. Transfer 50 μL of each fraction into individual wells of a 96-well plate. Cover the plate to prevent evaporation.

3.4.2.3 Prepare Gradient Fractions for Analysis

13. Dilute the collected fraction in the SW41 tubes with the HEPES solution and balance the tubes.

    Take care to mix the sucrose fractions completely with the HEPES solution. Incomplete mixing may inhibit proper exosome pelleting in the next steps and therefore skew results.

14. Centrifuge six of the fractions with the SW41 rotor at 100,000g for 70 min at 4°C. Leave the remaining six fractions on ice.

15. While the ultracentrifuge is running, measure the sucrose concentration of each fraction saved in the 96 well plates using the refractometer.

16. Carefully remove the supernatant from the tubes using the technique described in Section 2.2, step 8.

17. Resuspend the pelleted exosomes from each tube using 40 μL of PBS and transfer the suspension to a microcentrifuge tube. Keep these tubes on ice.

18. Repeat steps 14, 16, and 17 for the remaining six fractions.

19. Store 22 μL of each fraction at −80°C for up to a year.

20. Add 6 μL of the 4× Laemmli buffer to the remaining 18 μL of each fraction to yield a final 1× concentration.

3.5 SDS-PAGE Analysis of Exosomes

As previously mentioned, exosomes isolated from a variety of sources are consistently enriched with a number of proteins that differ from the originating cell and other types of vesicles. Identifying these proteins in a sample is indicative of the isolated vesicles being exosomes. The easiest method to demonstrate the presence of these exosomes marker proteins is by SDS-PAGE and western blotting. Here we give a brief protocol to prepare exosome samples for analysis by western blotting.

3.5.1 Materials

- Exosome sample
- Cell lysate – from the cells that produced the exosomes
- PBS
- 4× Laemmli buffer – reducing or nonreducing
- Standard equipment and reagents for western blotting and quantification of protein concentration

3.5.2 Protocol

1. Measure the protein concentration of all exosome samples and cell lysates.

2. For initial experiments, prepare two tubes from each sample with different amounts of protein (2 μg and 10 μg).

3. Adjust the final volume of each tube to 18 μL using the PBS.

4. Add 6 μL of 4× Laemmli buffer to each tube.

For western blotting, some proteins may require separation under reducing or nonreducing conditions for proper detection with the corresponding antibody. Consult the product information provided with the antibody to determine whether your samples require reducing or nonreducing conditions for proper detection.

5. Heat each tube at 95°C for 8 min.

6. Load samples onto a 10% SDS–PAGE gel.

7. Run gel, transfer proteins, and blot for exosome marker proteins using standard western blotting techniques.

3.6 Characterization of Exosomes by Nanoparticle Tracking Analysis

As mentioned in Section 3.1, the size distribution of exosomes is a characteristic that separates them from other intra- and extracellular vesicles.

Determining the size distribution of the vesicles in your preparation can be one way to demonstrate that your isolated vesicles are indeed exosomes. We have already covered one method to determine the size of prepared exosomes, EM. However, as mentioned in that section, measuring enough exosomes manually can be laborious and possibly subjective. To overcome these problems, we will utilize a technique called NTA to quantify the size distribution and concentration of exosomes in a sample. This technique is based on measuring the Brownian motion of each particle. The amount of Brownian motion correlates with diameter of the particle (assuming it is perfectly spherical) [13]. Therefore, this technique can use relatively basic equipment (visible solid state laser, microscope objective, high-sensitivity camera) combined with specialized particle tracking software to yield a histogram of particle size and concentration. This information can be valuable for experiments requiring incubation with a known number of exosomes, potentially reducing the variability that may come from using alternative methods to quantify exosomes, such as absolute protein amounts.

3.6.1 Materials

- Exosomes
- PBS
- Culture media (if samples were derived from cell culture conditioned media)
- Nanoparticle tracking device, such as NanoSight and ZetaView

  This technique requires a specialized device specifically designed to quantify nanoparticles via Brownian motion. Currently, a number of these devices are available from a couple manufacturers. We will present this protocol in a universal manner applicable to all of these devices. It is important to note that these devices may be built with components that limit the types of nanoparticles and buffers that can be analyzed/used. For example, a nanoparticle tracking device equipped to measure inorganic nanoparticles suspended in an organic solvent may not be compatible with biological samples suspended in PBS. Consult the manufacturer of the device to ensure it is equipped to measure biological samples.

  These devices are generally supplied with polystyrene nanospheres that can be used to check the calibration of the device. It is recommended that prior to measuring your samples, the supplied nanospheres are measured to ensure the device is working properly.

3.6.2 Protocol

1. Using the technique specific to your device, load an appropriate volume of PBS into the device for analysis.

This PBS should be the same solution used to resuspend the exosome pellet prepared by ultracentrifugation. It is critical that this PBS be relatively free of contaminating micro- or nanoparticles. If there are large microparticles detected in the PBS, filter the PBS through a 0.22 μm filter to clear the solution.

2. Repeat step 1 with the fresh culture media.

   If the culture media contains nanoparticles (ideally it should be free of nanoparticles) you can subtract this basal amount of vesicles from the amount of vesicles measured in your exosome sample.

3. Dilute a portion of the exosome sample (1/10th of the total exosome suspension, 10 μL from a 50 μL sample, is a good starting point) in the PBS free of nanoparticles.

4. Load the sample into the device and measure the particle concentration and size distribution three times (each time advancing the sample to analyze a new portion of it).

5. Based on the particle concentration, dilute a new portion of the original exosome suspension. If the first dilution was in the upper limit for an accurate particle count (nearly too many particles) dilute a smaller portion of the exosome solution in PBS. If the first dilution was close to the lower limit for an accurate particle count (nearly too few particles) dilute a larger portion of the exosome solution in PBS.

6. Load this newly diluted sample into the sample chamber of the NTA device and measure the particle concentration and size distribution three times (each time advancing the sample to analyze a new portion of it).

   The measured concentrations and size distributions between the two dilutions should be similar for a given sample. In cases of extremely, dilute, or concentrated samples these measures may diverge. Consult the manufacturer of the nanoparticle tracking device to determine the proper procedures to address this discrepancy. Most often, adjustments to the camera settings and the dilutions of the samples will correct the discrepancy.

4 CONCLUSIONS

The exosome research field is growing rapidly and information on exosome characteristics, biogenesis, and biological functions is expanding. This growth is so rapid that the definitions, nomenclature, and assumptions about exosomes are changing so quickly that it can be difficult to stay ahead of these topics. Since the field is evolving, the authors suggest that the term exosome is not used lightly. As we stated in the introduction, an exosome is an ILV, from a MVE, that has been released to the extracellular space. Specifically demonstrating that your vesicle of interest was generated in this manner can be difficult. For that reason, many publications will instead prepare vesicles and characterize them using the protocols presented here. In most cases, if the prepared vesicles have the size distribution, protein composition, and density characteristic of exosomes, the authors will call the vesicles exosomes. This is widely acceptable as it suggests with a high probability that the vesicle did derive from ILVs formed within a MVB. In fact, it is common in the literature to find only limited characterization of the vesicles of interest, such as size distribution, that the authors will use to call the vesicle an exosome.

Finally, we would like to make three recommendations for the readers to take forward into their exosome research. (i) Be cautious in your reading. A studied vesicle may be called an exosome by the author but how well was this vesicle characterized? Does the study have sufficient characterization of the isolated vesicles to satisfactorily test the hypothesis? Too often the term exosome is used loosely. (ii) Be cautious in your writing. The exosome field is changing rapidly. It is advisable to provide a solid foundation for the use of the term exosome in your studies. If the hypothesis in question does not require the extensive characterization described here, perhaps an alternative, looser term such as "exosome-like" or "having exosome characteristics" is more appropriate. (iii) We encourage all exosome researchers to stay up to date with exosome literature. This field is rapidly evolving with new techniques and methods being described and evaluated. If possible, a good strategy to keep up with the field is to become members of the American Society for Exosomes and Microvesicles (www.asemv.org) or the International Society for Extracellular Vesicles (www.isev.org) and attend their annual meetings.

ABBREVIATIONS

BSA	Bovine serum albumin
D₂O	Deuterium oxide
EM	Electron microscopy
EV	Extracellular vesicle
FBS	Fetal bovine serum
HEPES	Hydroxyethyl piperazineethanesulfonic acid
ILV	Intraluminal vesicles
MVB	Multivesicular body
MVE	Multivesicular endosome
NTA	Nanoparticle tracking analysis

PBS   Phosphate buffered saline
SDS-PAGE   Sodium dodecyl sulfate-polyacrylamide gel electrophoresis
TBS   Tris-buffered saline
TEM   Transmission electron microscope

REFERENCES

[1] Kalra H, Simpson RJ, Ji H, Aikawa E, Altevogt P, Askenase P, et al. Vesiclepedia: a compendium for extracellular vesicles with continuous community annotation. PLoS Biol 2012;10:e1001450.
[2] Mathivanan S, Fahner CJ, Reid GE, Simpson RJ. ExoCarta 2012: database of exosomal proteins, RNA and lipids. Nucleic Acids Res 2012;40:D1241–4.
[3] van der Pol E, Boing AN, Harrison P, Sturk A, Nieuwland R. Classification, functions, and clinical relevance of extracellular vesicles. Pharmacol Rev 2012;64:676–705.
[4] Mills JC, Stone NL, Pittman RN. Extranuclear apoptosis. The role of the cytoplasm in the execution phase. J Cell Biol 1999;146:703–8.
[5] Cocucci E, Racchetti G, Meldolesi J. Shedding microvesicles: artifacts no more. Trends Cell Biol 2009;19:43–51.
[6] Kowal J, Tkach M, Thery C. Biogenesis and secretion of exosomes. Curr Opin Cell Biol 2014;29:116–25.
[7] Raposo G, Nijman HW, Stoorvogel W, Liejendekker R, Harding CV, Melief CJ, et al. B lymphocytes secrete antigen-presenting vesicles. J Exp Med 1996;183:1161–72.
[8] Aboul Naga SH, Dithmer M, Chitadze G, Kabelitz D, Lucius R, Roider J, et al. Intracellular pathways following uptake of bevacizumab in RPE cells. Exp Eye Res 2014;131:29–44.
[9] Cai Z, Yang F, Yu L, Yu Z, Jiang L, Wang Q, et al. Activated T cell exosomes promote tumor invasion via Fas signaling pathway. J Immunol 2012;188:5954–61.
[10] Vargas A, Zhou S, Ethier-Chiasson M, Flipo D, Lafond J, Gilbert C, et al. Syncytin proteins incorporated in placenta exosomes are important for cell uptake and show variation in abundance in serum exosomes from patients with preeclampsia. FASEB J 2014;28:3703–19.
[11] Thery C, Boussac M, Veron P, Ricciardi-Castagnoli P, Raposo G, Garin J, et al. Proteomic analysis of dendritic cell-derived exosomes: a secreted subcellular compartment distinct from apoptotic vesicles. J Immunol 2001;166:7309–18.
[12] Ettelaie C, Collier ME, Maraveyas A, Ettelaie R. Characterization of physical properties of tissue factor-containing microvesicles and a comparison of ultracentrifuge-based recovery procedures. J Extracell Vesicles 2014;3.
[13] Filipe V, Hawe A, Jiskoot W. Critical evaluation of Nanoparticle Tracking Analysis (NTA) by NanoSight for the measurement of nanoparticles and protein aggregates. Pharm Res 2010;27:796–810.

CHAPTER 5

Stem Cell Extracellular Vesicles: A Novel Cell-Based Therapy for Cardiovascular Diseases

Ewa K. Zuba-Surma*, Marta Adamiak*, Buddhadeb Dawn**
*Department of Cell Biology, Faculty of Biochemistry, Biophysics and Biotechnology, Jagiellonian University, Krakow, Poland
**Division of Cardiovascular Diseases, Cardiovascular Research Institute, and the Midwest Stem Cell Therapy Center, University of Kansas Medical Center, Kansas City, MO, USA

Contents

1 INTRODUCTION

Cardiovascular diseases (CVDs) remain the leading cause of mortality and morbidity in western societies. Epidemiological data provided by the World Health Organization (WHO) indicate that nearly 17.3 million deaths were caused by CVDs in 2008, with a predicted increase to around 25 million CVD–related deaths per year by 2030 [1]. Ischemic heart disease (IHD), primarily representing occlusive atherosclerotic coronary artery disease and myocardial infarction (MI), accounts for a large number of deaths due to CVDs. In particular, ischemic cardiomyopathy resulting from progressive myocardial dysfunction after MI has poor prognosis, despite proper usage of the currently available therapies. Intense research efforts have therefore been focused on the development of newer therapeutic options to achieve myocardial repair and improve outcomes of CVDs.

Toward the beginning of this millennium, results reported in a number of papers from various laboratories indicated that injection of stem or progenitor cells into the infarcted heart could improve left ventricular (LV) function and remodeling in animal models of MI and heart failure. These preclinical studies incited profound enthusiasm among clinicians and were quickly followed by studies in humans, in which primarily bone marrow-derived cells were infused in patients with acute MI or ischemic heart failure [2]. Over the ensuing decade, numerous investigators have continued to test a large number of different cell types injected through various routes to induce cardiac repair in patients with IHD. Although the results from these trials have been divergent, several meta-analyses of pooled data from clinical trials indicate that injection of bone marrow cells improves heart structure and function as well as clinical outcomes in patients with IHD [2,3].

However, several studies of cell therapy have not shown any benefit, and cell therapy for IHD continues to be projected as an imperfect science, and questions continue to emerge regarding the efficacy vis-à-vis safety of this cell-based approach. Indeed, uncertainty about a number of variables integral to cell therapy has slowed the earlier rapid progress with clinical translation in this relatively new area. The therapeutic use of stem/progenitor cells is also mired in concerns regarding potential differentiation of highly proliferative and pluripotent cells to undesirable cell types that are not relevant for IHD and cardiovascular repair. These considerations have led to extensive investigations elucidating the molecular mechanisms involved in reparative processes induced by stem cell injection into damaged tissues. Some of these potential safety concerns have also resulted in attempts to use cellular substitutes that may render the regenerative benefits, and yet avoid any potential adverse effect inherent in injecting donor cells into unrelated human recipients. Thus, novel approaches in the future may focus on using stem cell-derived bioactive factors to induce myocardial repair, improve coronary perfusion, eliminate apoptotic cells, and regenerate tissue. In this context, stem cell-derived exosomes and microvesicles (MVs) are emerging as major regenerative tools that may be used to effectively repair ischemic tissues.

2 STEM CELL CONSIDERATIONS FOR CARDIAC REPAIR

Stem cells may be distinguished from other cell types based on two major characteristics. First, they are clonogenic cells capable of self-renewal through unlimited cell divisions, sometimes following long periods of quiescence.

Second, under specific physiological or experimental conditions, stem cells are able to give rise to diverse mature cell types with specialized functions, such as muscle cells, blood cells, neuronal cells, and many others [4–6]. The stem cell compartment of mammalian organisms may be organized in a hierarchy based on differentiation capacity and lineage commitment, from the most primitive (totipotent cells) through pluripotent and multipotent cells to tissue-resident committed progenitors (mono- or unipotent cells) [4–6]. In general, while the unipotent and tissue-committed stem cells have been described in adult tissues, stem cells with pluripotent and multipotent capacities were traditionally thought to be restricted to early embryonic stages [7]. However, the presence of cells that possess pluripotent or multipotent characters has now been confirmed in adult tissues. Indeed, reports from several laboratories have confirmed that cells that are related to multi- and pluripotent stem cells of embryonic origin do exist in adult tissues (including bone marrow and cord blood) [7–12]. Based on this knowledge, earlier efforts on cell-based cardiac repair were primarily focused on identifying and using cells with the capacity to differentiate into cells of cardiac lineage, including cardiomyocytes, endothelial cells, and smooth muscle cells.

Although the above ability of adult stem cells to differentiate into cells of unrelated lineages is a proven fact, a growing body of evidence suggests a lesser role played by this mechanism toward cardiac repair. In addition, although pluripotent or multipotent cells are able to differentiate into several different cell types, including cardiac cells, uncontrolled proliferation can be a problem for therapeutic transplantation. In particular, and despite their ability to give rise to all types of cells, embryonic stem cells and induced pluripotent stem cells (iPSCs) have been known to be tumorigenic and therefore may not be safe for clinical use [13–15]. Accordingly, clinical studies of cardiac repair thus far have almost exclusively been focused on a variety of adult stem and progenitor cells [3,16–18]. These adult stem cells comprise at least three different major populations depending on their origin: (i) bone marrow–derived stem/progenitor cells; (ii) circulating stem/progenitor cells (present in blood and may be derived from the bone marrow, at least partially); and (iii) other tissue-resident stem cells (e.g., cardiac stem cells). Human bone marrow harbors a complex reservoir of stem and progenitor cells, including hematopoietic stem cells, endothelial progenitor cells (EPCs), mesenchymal stem cells (MSCs), and very small embryonic-like stem cells [8,10,11,19–22]. Therefore, many preclinical studies as well as clinical trials in patients with IHD and cardiomyopathy have used relatively unpurified mononuclear cells derived from the bone marrow, while others have used more specialized stem cells, including MSCs from various tissues, and cardiac progenitor cells (CPCs) [3,17,20].

3 "PARACRINE" EFFECTS OF STEM CELLS

That stem cell transplantation may potentially lead to beneficial effects in recipients has been postulated since the 1960s [23,24]. These early discoveries paved the way for the field of regenerative medicine, which has significantly advanced over the last few decades. This increase in therapeutic use of stem cells has further confirmed that transplantation of stem/progenitor cells is associated with a number of beneficial effects in different organs, including the cardiovascular system [3,17,18,20]. Although these measurable beneficial outcomes of various cell-based therapies have been well documented, the experimental evidence for donor–recipient chimerism in the myocardium following cell injection has remained controversial [18,25–27]. Such observations indicate that mechanisms other than direct differentiation of injected stem cells into cells of damaged tissues may play an essential role in organ repair [28], giving credence to a constellation of alternative mechanisms, often collectively termed "paracrine effects," as the basis of positive effects after stem cell therapy [18,25–27]. This notion is entirely consistent with the fact that living cells in multicellular organisms communicate with other cells via mechanisms that involve direct cell–cell or cell–matrix interactions, and also by extracellular chemical molecules secreted by cells [29,30]. Such released factors may impact: (i) the same cell that produced the molecule (autocrine signaling); (ii) target cells in physical proximity (paracrine signaling); or (iii) target cells located at a distance (endocrine signaling) [29,30].

Growing evidence indicates that paracrine effects may indeed play a major role in cell-based therapies in regenerative medicine, where stem (or more differentiated progenitor) cells are utilized to repair damaged organs, such as the heart, kidney, or neural tissues [25,27,28]. Stem cells are known to secrete a variety of cytokines, chemokines, growth factors (GFs), and other small, bioactive molecules that regulate cellular biology and function in autocrine as well as paracrine manners, and thereby influence molecular interactions with the physiological microenvironment and cell niche [31,32]. Important for therapy, such factors may also be released from activated cells that are removed from their natural tissue setting (e.g., cells harvested from the bone marrow) or mobilized into circulation (e.g., progenitors in mobilized peripheral blood) [31–33]. Thus, different types of stem cells, including MSCs, EPCs, adipose tissue-derived stem cells (ASCs), and CPCs, which are currently used in cardiovascular regenerative medicine, represent potential sources of a broad array of paracrine factors [25,27,28]. A few key molecules known to be secreted by stem cells that have been proven to be responsible for beneficial effects in tissue regeneration include vascular endothelial growth factor (VEGF), stromal-derived factor-1, stem cell factor, hepatocyte growth factor (HGF), insulin-like growth factor-1 (IGF-1), insulin-like growth factor-2 (IGF-2), basic fibroblast growth factor (bFGF), matrix metalloproteinases, transforming growth factor β, and platelet-derived growth factor (PDGF) [31–35]. These chemical signals may inhibit apoptosis and increase survival of cells residing in the affected organs as well as stimulate cell proliferation or differentiation and promote vascularization in damaged tissues to improve oxygen delivery and metabolic turnover. Importantly, similar bioactive components may be either directly released to the environment or be contained in EVs shed by stem cells [25,27,28].

4 PARACRINE MECHANISMS OF STEM CELL-MEDIATED CARDIAC REPAIR

The cardiac tissue represents a complex mixture of several different cell types that collectively maintain adequate perfusion, preserve myocardial contractility, and ensure timely myocardial relaxation. The myocardium contains approximately 3 billion cardiac muscle cells, which comprise the majority of the volume of the heart. Cardiomyocytes, together with several other cell types, including endothelial cells, smooth muscle cells, fibroblasts, and a pool of cardiac progenitors, play interdependent and important roles in cardiac homeostasis, in both health and disease [36]. A well-controlled intra- and intercellular communication system is therefore essential to guarantee the balance between these different cell types residing within the heart.

Consistent with the above paracrine actions mediated by factors released from adult stem cells have been implicated in the reparative process following stem cell injection into the infarcted heart [25,27,28,37,38]. The protective and regenerative effects of different stem cell therapies have been attributed to specific paracrine actions in many cardiovascular studies [39]. It has also been shown that paracrine factors released by transplanted cells may significantly influence cytoprotection, inflammation, fibrosis, neovascularization, contractility, and new cell formation, thereby improving cardiac function after tissue injury [25,27,40]. However, many of these benefits may be stem cell specific, as the repertoire of released factors likely varies considerably depending on the type of stem cell used for treatment.

Observations from several studies indicate that adult stem cells, such as MSCs from various tissues, produce and secrete a broad variety of cytokines, chemokines, GFs, and proteases, which affect adjacent cells in a paracrine fashion [25,27,40]. Indeed, Gnecchi et al. have demonstrated that conditioned media from MSCs, particularly from genetically modified MSCs overexpressing Akt-1, can reduce apoptosis and necrosis of isolated rat cardiomyocytes exposed to low oxygen tension [40,41]. Another study has shown that MSCs and MSC-conditioned media enhance the excitation–contraction coupling in ventricular myocytes likely mediated by increase in sarco/endoplasmic reticulum calcium transport ATPase activity due to paracrine signals released from MSCs [42]. The authors proposed that Akt-induced change in calcium signaling also mediates the MSC-induced antiapoptotic effects and increased survival of ventricular myocytes [42].

MI is typically associated with a hypoxic environment, and hypoxic stress has been shown to upregulate cellular secretion of specific paracrine factors. Tissue concentrations of VEGF, bFGF, HGF, IGF-1, and adrenomedullin are notably increased in damaged hearts treated with MSCs or multipotent human bone marrow stem cells [43,44]. Findings of Takahashi et al. revealed that rat BM-derived mononuclear cells (BMMNCs) also produce and release a variety of cytoprotective factors, including VEGF, PDGF, interleukin 1β, and IGF-1, some of which are significantly upregulated in hypoxic conditions [45]. Most of these factors act through the activation of the prosurvival PI3K/Akt pathway. Additionally, Xiang et al. demonstrated that incubation of cardiomyocytes subjected to hypoxia and reoxygenation with conditioned media from MSCs results in a cardioprotective effect [46]. The treatment with MSC-conditioned medium reduced the injury by increasing the Bcl-2/Bax ratio and reducing the release of cytochrome C and apoptosis-induced factor from mitochondria into the cytosol [46]. Thus, it has been suggested that MSCs protect cardiomyocytes from hypoxia/reoxygenation-induced apoptosis through a mitochondrial pathway in a paracrine manner. Paracrine actions of BMMNCs also include promotion of angiogenesis confirmed by increased microvessel density in ischemic hearts injected with conditioned medium (CM) from BMMNCs [45]. Further, MSC-derived CM has been shown to inhibit cardiac fibroblast proliferation and attenuate type I and type III collagen expression by these cells, thereby supporting a role of paracrine effects toward improving myocardial remodeling after MI [47]. In addition, the authors demonstrated that collagen activity in MSC-CM was as high as the activity in the medium conditioned by cardiac fibroblasts and postulated that MSCs exerted paracrine antifibrotic effects possibly through inhibition of cardiac fibroblast proliferation and collagen synthesis [47]. Collectively, these studies support the notion that paracrine effects of stem/progenitor cells are responsible for many beneficial actions observed following cardiovascular cell therapy.

Although the above studies indicate a role of paracrine actions in stem cell–mediated cardiac reparative benefits, the precise mechanisms through which bioactive molecules are released from MSCs and other stem cells remain to be delineated. In addition to soluble factors, activated stem cells have recently been found to also secrete extracellular vesicles (EVs), which are nanosized, lipid bilayer-enclosed fragments shed from the cell surface or secreted from the endosomal compartment. Growing evidence indicates that by virtue of their ability to carry insulated bioactive molecules, EVs contribute significantly toward cell-to-cell communication in multicellular organisms, including humans. Indeed, a role of EVs in MSC-CM effects was suggested by findings in a study by Timmers et al. [48], in which infarct-sparing benefits were observed with therapy only with a specific fraction of MSC-CM containing products in the 100–220 nm size range. Therefore, it seems likely that EVs play an important and as yet under-appreciated role toward inducing effective repair of damaged hearts following stem cell therapy [28,49].

5 EXTRACELLULAR VESICLES AS IMPORTANT MEDIATORS OF CELL-TO-CELL COMMUNICATION

Individual cells in multicellular organisms are now known to communicate with other cells via complex methods that involve direct contact with neighboring cells with channels and receptors, interactions with the matrix, and through secretion of diverse factors and molecules that trigger complex signaling events [50,51]. In addition to various molecules that are directly secreted from cells, bioactive components that are enclosed in EVs have recently attracted attention as mediators of cellular functions. EVs represent heterogeneous populations of very small, naturally occurring membrane-enclosed fragments of cytoplasm and bioactive materials (mRNA, miRNA, protein, lipid, and other small molecules) that are produced by nearly all eukaryotic cells [50,51]. Although the existence of extracellular small particles has been known since the 1960s, these were considered to be unimportant cellular debris and referred to as procoagulant "dust" from blood platelets [52]. However, recent findings indicate that living cells actively release several different types of EVs carrying bioactive cargo and thereby influence their respective niche, and also biological function of neighboring as well as distant cells [50,51].

Three major fractions of EVs are commonly distinguished based on biogenesis and cell compartment of origin: (i) exosomes, (ii) MVs, and

(iii) apoptotic bodies [53–55]. EVs that are released from viable and healthy cells, such as MVs and exosomes, are usually smaller in size when compared with apoptotic bodies containing fragmented DNA derived from damaged cells [56]. Current research interests in the cardiovascular therapy field are focused primarily on two major types of EVs (exosomes and MVs, Table 5.1) that are produced universally by living cells through evolutionarily conserved cellular processes [50,51]. It is important to note that the vast majority of methods for EVs isolation (including the commonly employed ultracentrifugation strategy) from healthy nonapoptotic cells yield vesicle fractions that combine both MVs and exosomes [50,51]. Accordingly, in the present chapter, we will focus on MVs and exosomes as two major populations of intercellular messengers in stem cell-based therapies.

Figure 5.1 summarizes the biogenesis and major effects of EVs and identifies potential mechanisms for cardiovascular repair. MVs, also termed ectosomes or shedding vesicles, are heterogeneous objects with sizes ranging from 100 nm to 1000 nm [50,51]. They are shed into the extracellular

Table 5.1 Characteristics of nonapoptotic EVs

	Exosomes	MVs
Features	Homogeneous population, cup shape, size range 30–100 nm	Heterogeneous population, irregular shape, size range 100–1000 nm
Production mechanism	Release by exocytosis following fusion of MVBs with plasma membrane; biogenesis depends on cytoskeleton activation and Ca^{2+} independent [56]	Budding directly off plasma membrane; biogenesis depends on Ca^{2+}, calpain, and cytoskeleton reorganization [50]
Isolation	Differential centrifugation and sucrose gradient ultracentrifugation [57], 100,000–200,000g, vesicle density is 1.13–1.19 g/mL	Differential centrifugation 18,000–20,000g [58]
Markers	CD63, CD81, CD9, LAMP1, and TSG101	Cell-specific markers, PS, lipid raft-associated molecules (TF, flotillin)
Contents	Proteins, lipids, mRNA, miRNA, exosomal DNA	Proteins, lipids, mRNA, miRNA, rarely double-stranded DNA

Abbreviations: MVBs, multivesicular bodies; LAMP1, lysosome-associated membrane protein-1; TSG101, tumor susceptibility gene 101 protein; PS, phosphatidylserine; TF, tissue factor.

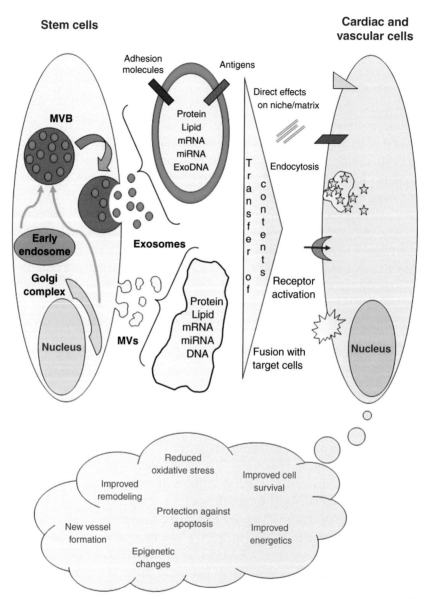

Figure 5.1 *Origin and Putative Modes of Action of EVs in the Cardiovascular System.* Stem cells release exosomes and membrane vesicles through distinct mechanisms. The released exosomes and MVs interact with the environment as well as target cells by exerting paracrine actions and delivering contents into target cells. These events lead to tissue protection, repair, and regeneration. MVs, membrane vesicles; MVB, multivesicular body.

environment by direct budding from the cellular plasma membrane. MVs may carry large amount of phosphatydylserine on their surface and are enriched in proteins associated with membrane lipid rafts, such as flotillin and tissue factor [50,59,60]. Additionally, shedding vesicles express surface receptors that vary according to the membrane composition of the parent cell [50,51]. In contrast, exosomes are typically smaller and more homogeneous compared with MVs, with sizes ranging from 30 nm to 100 nm, and are derived primarily from the endosomal compartment (Figure 5.1). Following production, exosomes may be stored as intraluminal vesicles within intracellular multivesicular bodies (MVBs) of the late endosome, and released following the fusion of MVBs with the cell membrane [54,57]. Exosomes carry and express cell type-specific proteins and molecules that are considered specific markers of exosomes of different cellular origin, such as CD63, CD81, and CD9 tetraspanin family members, TSG101 (tumor susceptibility gene 101 protein), and LAMP1 (lysosome-associated membrane protein-1) [57,61]. Thus, EVs differ in size and molecular composition depending on the cell of origin and the mechanism of biogenesis [50].

Following the release, EVs remain in extracellular space in the proximity of the cell of origin or enter diverse biological fluids (such as plasma, urine, amniotic fluid, and tumor-associated effusions). This widespread dissemination of EVs with biologically active substances ensures short- and long-range exchange of information among cells. Several reports have documented that the number of EVs released from cells increases depending on exposure to various cellular stressful conditions, including hypoxia, irradiation, oxidative injury, exposure to proteins from an activated complement cascade, and exposure to shear stress [51].

The cargo contained within EVs consists of a wide range of biologically active components, such as nucleic acids (e.g., mRNAs, miRNAs), proteins, and lipids similar to those present in the membranes, and cytoplasm of cells from which EVs originate [50,51]. Therefore, EVs originating from a given cell type may act as mediators of cell-to-cell communication by transferring surface determinants and genetic material from parent cells onto cells of other origins. Following fusion with target cells or endocytosis, EVs deliver diverse proteins, mRNAs, miRNAs, and lipids into target cells, influencing gene expression and numerous cellular functions (Figure 5.1). EVs may also stimulate other cells by surface-expressed ligands acting as a "signaling complex," and transfer surface receptors from one cell to another. EVs may also serve as a vehicle for the transfer of infectious particles (e.g., human immunodeficiency virus, prions) and may probably deliver even intact subcellular organelles, such as mitochondria, to the target cells under certain circumstances [51].

In the last decade EVs have gained increasing attention as mediators of signaling between cells, particularly in the field of immunology and oncology, where their roles in various processes have been already described in great detail. However, results from recent studies tend to indicate that EVs may also play an important role in regenerative biology [49,56,62]. The fact that EVs are integral parts of intercellular environment and important modulators of cell-to-cell cross-talk combined with their ability to deliver a large number of diverse bioactive contents to target cells raises the possibility of utilizing the innate therapeutic potential of EVs for organ repair [63–65]. This approach is particularly attractive as an alternative to cell-based therapy, because various stem and progenitor cells have been shown to serve as abundant sources of EVs that mediate transfer of stem cell-specific molecules and genetic components onto other cells and mature tissues [63–65]. Therefore, stem cell-derived EVs may potentially convey information that would affect the growth of target cells, influence their phenotype, contribute to cell-fate decision, and promote regeneration [33,50,51]. Thus, the observation that the phenotype of target cells may be modified by EV-mediated transfer of biologically active cargo provides a new perspective for the paracrine hypothesis of stem cell action [28,63,64].

6 EXTRACELLULAR VESICLES IN STEM CELL BIOLOGY AND CELL-BASED REPAIR

The results from several recent studies indicate that stem cells may utilize small vesicles as mediators of intercellular communication [28,63,64]. EVs released in physiological conditions may have innate therapeutic value and may be potentially employed for transfer of bioactive proteins, lipids, and nucleic acids into cells within injured tissues [50,63,64,66]. In particular, naturally shed vesicles selectively targeting certain cell types and delivering their bioactive contents may represent potential candidates for therapeutic applications in different areas of regenerative medicine. EVs derived from various types of stem cells (e.g., MSCs, EPCs, and such) have already been postulated to mimic the beneficial effects of the cell of origin, thus promoting repair and limiting injury in several organs, as they are able to trigger regenerative programs and coordinate self-repair in injured tissues [50,64,66,67].

The potential of iPSCs for directed differentiation and subsequent repair of multiple tissues has generated immense interest in these cells for regenerative purposes. In this regard, Bobis-Wozowicz et al. recently examined EVs derived from human iPSCs (hiPSC-EVs) and detected the presence of all tested transcripts related to pluripotency (Oct4, Nanog, Sox2, and Rex1), angio-, and cardiomyogenesis (GATA4, Tie2, Flk1, VE-Cadherin, vWF, CD105, and Nkx2.5) [63]. hiPSC-EVs were also shown to contain several miRNAs (e.g., miR-302d, miR-92a, miR-93, and miR-221) that are carried by their parental cell line [63]. Moreover, the authors found that hiPSC-EVs may transfer these contents in the form of mRNA to human cardiac MSCs [63]. Importantly, these data indicate that human iPSC-derived EVs may act as natural and effective nanocarriers of numerous proangiogenic, cardiomyogenic, and antiapoptotic agents to target mature heart cells [63]. Such hiPSC-EVs may be potentially utilized to both restore function and promote survival of mature heart cells when delivered into ischemic myocardium. Thus, findings from this study provide important rationale for future use of human iPSC-derived EVs in heart regeneration.

Because of the clinical success with bone marrow and other adult stem cell therapies for cardiac repair, EVs derived from other adult stem cells have also been subjects of great attention [67–69]. Adult human stem cells, such as bone marrow-derived multipotent stromal cells and human liver stem cells have been shown to secrete EVs carrying specific patterns of functional mRNA and miRNA associated with the mesenchymal phenotype and control transcription, proliferation, and proper regulation of the immune system [64,65,70]. By comparing the miRNA content of EVs with that of parental cells, an enrichment of certain subsets of miRNA within EVs was observed [64,65,70]. The selective differences in expression of mRNA in ASC-derived EVs and parental cells suggest the impact of a dynamic regulatory mechanism in the process of EV formation [71]. Compartmentalization of bioactive contents in EVs may be modulated by definite stimuli, as shown by EPCs, where hypoxic conditions were found to enhance the expression of miR-126 and miR-296 [71]. Furthermore, it was demonstrated that EVs released by EPCs activate angiogenesis in quiescent endothelial cells by transferring proangiogenic mRNA and miRNA from EPCs to endothelial cells [71,72]. In turn, observations by Professor Quesenberry's group led to a continuum theory of stem cell biology, in which MVs may be considered as a microenvironmental signal for phenotypic changes in stem cells [73]. Murine lung cell-derived MVs have been shown to enter into bone marrow cells and change their transcriptome profile by direct transfer of lung-specific mRNAs [73].

Thus, the exchange of biological information between the mature healthy or injured cells and stem cells has the potential to function in both directions. In the context of regenerative medicine, EVs may either transfer transcripts from injured cells to stem cells influencing their phenotype to acquire specific features of the tissue, or molecules, including transcripts, could be transferred from stem cells to injured cells restraining tissue injury, inducing developmental regulation of resident cells and leading to tissue repair. Since stem cell–derived EVs are proven to carry several biological properties of the cell of origin, the use of EVs may enable development of therapeutic strategies that avoid direct administration of multipotent cells.

Recent studies indicate that transplantation of stem or progenitor cells into injury sites induces regenerative processes and epigenetic changes, which are mediated, at least in part, by paracrine actions of biological factors and possibly EVs released by injected cells [48–50]. These observations are supported by several studies that have shown that injection of stem cell CM is often associated with beneficial actions similar to those imparted by those stem cells that conditioned the medium [40,42]. In addition to this exogenous administration, it is also very likely that located in physiological niches together, adult stem cells and differentiated cells communicate with each other to regulate the self-renewal and differentiation mechanisms leading to tissue repair and regeneration. In this context, stem cell–derived EVs could play an important role in the reparative process. Due to the expression of surface receptors derived from the stem cell of origin, EVs may effectively target the site of damage and deliver their bioactive contents to injured cells, inducing favorable events.

7 CARDIOVASCULAR THERAPEUTIC POTENTIAL OF EXOSOMES

Stem cell–derived EVs have been recently proposed as a new tool for tissue repair and regeneration in various diseased organs, including the heart and the vasculature [33,69,72,74–77]. In this regard, it has been shown that stem cell–derived EVs may effect functional changes in cells within target organs, such as cardiomyocytes and endothelial cells, influencing cell survival mechanisms, energetics, regeneration, and cellular differentiation [72,75]. Table 5.2 summarizes studies that have investigated the cardiac reparative and/or angiogenic potential of EV therapy in *in vivo* models or assays.

Table 5.2 Stem cell-derived EV therapy for cardiovascular repair and angiogenesis

Source of EVs	Species	Type of injury (or assay)	Therapy	Results	References
Studies of cardiac repair					
Human MSC–like cells	Mouse	Acute MI (30 min occlusion of LAD followed by reperfusion) *in vivo* and *ex vivo* (Langendorff)	Exosomes (3 μg or 0.4 μg [HPLC F1] in 200 μL) administered i.v. 5 min before reperfusion	Injection of exosomes as well as the purified F1 fraction reduced infarct size	[74]
Human MSC–like cells	Mouse	Acute MI (30 min occlusion of LAD followed by reperfusion) *in vivo* and *ex vivo* (Langendorff)	Purified exosomes (1–16 μg/kg) administered as i.v. bolus 5 min before reperfusion	Exoscme therapy reduced infarct size, and improved LV function and remodeling	[75]
Human BM MSCs	Rat	Acute MI (ligation of LAD)	MSC-EVs (20 μL containing 80 μg of protein) were injected i.m. at the infarct borderzone at 30 min after coronary ligation	MSC-EV injection reduced infarct size, enhanced angiogenesis, and improved LV function	[67]
Murine Sca-1+ CPCs	Mouse	Acute MI (45 min occlusion of LAD followed by reperfusion)	Exosomes from 5×10^5 CPCs in 25 μL injected i.m. immediately after coronary occlusion	Exosome injection reduced cardiomyocyte apoptosis after I/R injury	[68]

Human CPCs	Rat	Acute MI (ligation of LAD)	EV-CPC (300 μg in 150 μL) was injected i.m. at the infarct borderzone at 60 min after coronary ligation	EV-CPC injection reduced cardiomyocyte apoptosis, enhanced angiogenesis, and improved LV function	[69]
Studies of angiogenesis					
Human PB EPCs	SCID mouse	Matrigel plug	HMECs incubated with 30 mg/mL MVs	HMECs incubated with MVs resulted in formation of perfused vessels connected with host vasculature *in vivo*	[72]
Human CD34+ cells and MNCs	Nude mouse	Matrigel plug, corneal angiogenesis	Exosomes from 0.5×10^6 CD34+ cells used for both assays	Exosomes from CD34+ cells increased angiogenesis compared with MNC exosomes	[33]
Human PB EPCs	SCID mouse	Matrigel plug	Cells cultured in presence of MVs (10 μg/mL)	MVs induced enhanced vascularization of hIEC xenografts and Matrigel plugs	[76]

Abbreviations: BM, bone marrow; CPCs, cardiac progenitor cells; EPCs, endothelial progenitor cells; EV, extracellular vesicle; hEC, human islet-derived endothelial cell; I/R, ischemia/reperfusion; i.m., intramyocardial; i.v., intravenous; LAD, left anterior descending coronary artery; LV, left ventricular; MI, myocardial infarction; MNCs, mononuclear cells; MSC, mesenchymal stem cells; PB, peripheral blood; SCID, severe combined immunodeficiency.

7.1 MSC-Derived Exosomes

MSCs, also known as multipotent mesenchymal stromal cells, were identified more than 40 years ago as a population of bone marrow-derived adherent bone/cartilage-forming progenitor cells [78]. The International Society for Cellular Therapy has recently proposed a minimal set of standard criteria to define these cells, including (i) plastic adherence when maintained in standard culture conditions and (ii) expression of CD105, CD73, and CD90 antigens, along with lack of CD45, CD34, CD14, or CD11b, CD79α, or CD19, and HLA-DR surface molecules. Moreover, MSCs must differentiate into osteocytes, chondrocytes, and adipocytes *in vitro* [79]. However, several recent studies indicate that bone marrow MSCs may also give rise to other mature cell types, including endothelial cells and cardiomyocytes [4,20,80]. Thus, MSCs are currently one of the commonly used types of stem cells in regenerative medicine. They can be isolated from many sources, including bone marrow and adipose tissue, and expanded *ex vivo* for transplantation [4,20]. MSCs have been the focus of great interest in cellular therapy for their ability to migrate to damaged tissues, their potential to differentiate to lineages of mesodermal origin, and for their robust *in vitro* expansion. Transplanted MSCs have been shown to induce cardiac repair via stimulation of angiogenesis, cell survival and proliferation, and regulation of the inflammatory process in host tissues [4,20]. However, several recent studies have suggested that cell-to-cell communication network involving soluble factors and MVs is responsible for beneficial effects of MSC transplantation in injured tissue [48,75]. Indeed, the currently documented robust contribution of MSCs in cardiac repair *in vivo* could be explained by release of EVs from these cells, resulting in horizontal transfer of mRNA, miRNA, and proteins. The regenerative role of MSC-CM and EVs studied in the context of cardiac regeneration is discussed below.

Gnecchi et al. [40,41] were the first to demonstrate that CM from hypoxic, genetically modified MSCs overexpressing Akt-1 (Akt-MSCs) was sufficient to inhibit hypoxia-induced apoptosis and trigger spontaneous contraction of adult rat cardiomyocytes *in vitro* [40,41]. When injected into infarcted hearts, the Akt-MSC-CM limited infarct size and improved LV function [40,41]. Subsequently, investigating the different fractions of MSC-CM, Timmers et al. demonstrated a significant reduction in infarct size in pig and mouse models of ischemia/reperfusion injury when CM from human MSC-like cells was intravenously administrated prior to reperfusion [48]. Subsequent work from this group further determined that the cardioprotective effects were mediated by the release of exosomes from these

MSC-like cells [74,77]. In another study from these investigators, Arslan et al. established that MSC-derived exosomes may be potential candidates for adjunctive therapy for MI [75]. These authors demonstrated that injected exosomes reduced infarct size by 45% when compared with controls in a murine model of myocardial ischemia/reperfusion injury [75]. Moreover, the administration of vesicles resulted in long-term preservation of cardiac function and reduced adverse remodeling. More specifically, exosome-treated animals exhibited significant preservation of LV geometry and contractility after 28 days of follow-up [75]. Interestingly, exosome treatment increased levels of adenosine triphosphate and nicotinamide adenine dinucleotide, decreased oxidative stress, increased phosphorylated-Akt and phosphorylated-GSK-3β, and reduced phosphorylated-c-JNK in ischemic/reperfused hearts within an hour after reperfusion [75]. It was shown that both local and systemic inflammation was significantly reduced 24 h after reperfusion, which may be related to the above - mentioned molecular processes [75]. Using a rat model of acute MI, Bian et al. found that intramyocardial injection of MSC-EVs greatly enhanced blood flow recovery, consistent with reduced infarct size and preserved cardiac systolic and diastolic performance compared with those treated with saline [67]. Collectively, the above results indicate that EVs released from MSCs are able to induce beneficial effects in cardiac tissue *in vivo* in a cell-independent manner, suggesting that EVs may contribute to cardiac regeneration and protection conferred by MSCs.

7.2 Exosomes Derived from Other Stem Cells/Progenitor Cells

Recent reports indicate that EVs obtained from other stem and progenitor cells may also exert beneficial effects in injured cardiac and vascular tissues. In the study by Deregibus et al., MVs derived from human circulating EPCs were able to promote endothelial cell survival, proliferation and formation of capillary-like structures *in vitro*, and trigger angiogenesis by horizontal transfer of mRNA to human microvascular endothelial cells (HMECs) *in vivo* [72]. In the severe combined immunodeficiency (SCID) mouse model, MV-stimulated HMECs (cells of human origin) were capable of organizing vessels of variable size connected with the murine vasculature [72]. In another study [33], exosomes collected from human CD34+ cells increased endothelial cell viability, proliferation, and tube formation *in vitro*, and *in vivo*, stimulated angiogenesis in Matrigel plug and corneal angiogenesis assays. In the study by Cantalupi et al. [76], EPC-derived MVs were shown to induce an early vascularization of human pancreatic islets xenotransplanted in SCID mice. Furthermore, MVs accelerated and enhanced the formation of capillary-like structures by human islet-derived endothelial cells (hIECs) *in vivo*. Thus, these results suggest that MVs are able to activate an angiogenic program in islet endothelium that may sustain revascularization and β-cell function [76]. These findings suggest that benefits observed in numerous studies of EPC-based therapy were derived, at least in part, from neoangiogenesis stimulated by EPC-derived MVs and paracrine mechanisms.

CPCs derived from the adult heart have emerged as one of the promising stem cell types for cardioprotection and repair, and also source of exosomes. Recent results reported by Chen et al. suggest that the therapeutic effect of CPCs may be attributed to paracrine effects, and particularly to the actions of EVs [68]. In this study, injection of murine Sca-1$^+$ CPC-derived exosomes into the myocardium immediately after coronary occlusion resulted in a significant decrease in apoptotic cells in the ischemic/reperfused myocardium compared with vehicle-treated control infarcted hearts [68]. In another study [69], intramyocardial injection of human CPC-derived EVs in the infarct borderzone reduced infarct size, decreased apoptosis, and improved LV function in a rat model of acute MI. Together, these data indicate that EVs derived from CPCs are able to induce cardiac protection and repair.

8 EXTRACELLULAR VESICLE-BASED THERAPY FOR CARDIOVASCULAR REPAIR: ADVANTAGES AND CHALLENGES

The discovery that secreted factors and particulate vesicles from stem cells can recapitulate benefits conferred by the parent cells is changing the conventional therapeutic use of stem cells in regenerative medicine. Indeed, EVs from different cellular origins have been shown to harbor sets of molecules present in the parent cell, and therefore may be used instead of stem cells to induce effective cardiovascular repair. In fact, EV-based therapy would very likely preclude several of the safety concerns and practical limitations associated with the use of replicating stem/progenitor cells, such as the risk of maldifferentiation. The use of EVs offers certain advantages even over the use of soluble factors. EVs are naturally capable of carrying diverse cell-specific bioactive contents, and also protect that cargo from activities of degrading enzymes present in the extracellular matrix. Thus, the lipid bilayer of EVs may secure the contents allowing for careful maintenance of stability, integrity, and biological activity of these factors during manufacture, storage, and subsequent administration. Moreover, several surface receptors present on the membrane of EVs possess specific binding affinities to ligands on target cells and may be involved not only in EV homing to the sites of cardiovascular injuries, but also internalization of EVs into recipient cells. Thus, EVs may deliver a complex pattern of biologically active components including proteins and nucleic acids derived from stem cells (e.g., MSCs) to damaged tissues such as ischemic myocardium or vessel wall [28,63,67]. Important from a therapeutic standpoint, EVs express only a limited number of histocompatibility antigens and have been shown to be nonimmunogenic during repeated administration of allogenic MVs obtained from MSCs [56]. Consistent with the earlier considerations, a growing number of studies in preclinical models continue to confirm the safety of EV administration for cardiovascular repair [33,68,69,72,74–76].

Despite the above advantages, a few important issues need to be resolved before effective and routine clinical use of EVs. Therapeutic hurdles include establishing long-term safety of systemic EV administration, disease specificity, biodistribution, and persistency of EV biological effects. Moreover, the terminology used to describe EV fractions needs to be clarified and standardized for effective comparison of therapeutic effects. For instance, vesicles isolated from CM using the same or very similar methods are currently referred to as exosomes, MVs or microparticles by different authors. It is also important to recognize that EVs derived from the commonly used method of ultracentrifugation represent a highly heterogeneous fraction produced in different cellular compartments. Therefore, future efforts should be directed toward developing novel isolation methods to obtain more pure subpopulations of EVs and to define their specific role in cardiovascular repair and regeneration. Following the resolution of these preclinical uncertainties, large-scale production and quality control of the most efficacious fraction would need to be addressed.

9 CONCLUSIONS

It has been recently shown that stem and progenitor cells represent an abundant source of EVs. Stem cell-derived EVs are important mediators of signaling within cell niches influencing phenotype and function of target cells, and perhaps may be exploited as potential therapeutic tools for tissue regeneration. Several recent studies including *in vitro* and *in vivo* experimental models have indicated excellent potential in developing effective EV-based therapies for various cardiovascular disorders for which the current therapeutic options are inadequate and ineffective. Importantly, multiple recent studies using EVs derived from MSCs, EPCs, and CPCs have provided encouraging results particularly toward salvaging ischemic tissue, inducing angiogenesis, and reducing apoptosis. Upon therapeutic administration, EVs mimic the effects of parent stem/progenitor cells in various experimental models by exerting antiapoptotic action, enhancing cell proliferation, and promoting regeneration. Thus, EVs released from adult stem cells possess the potential to be exploited in novel therapeutic approaches in regenerative medicine to repair damaged cardiac tissue, which may be an alternative to stem cell-based therapy. However, further studies of the basic mechanisms responsible for EV-mediated myocardial repair and regeneration are required. In perspective, with the progress in molecular biology techniques, engineering of EV surface membrane or bioactive contents to influence their disease specificity and efficacy may be envisioned. Such endeavors may lead to formulation of novel and exciting approaches in cardiovascular regenerative medicine.

ACKNOWLEDGMENTS

Supported by the TEAM grant (TEAM/2012-9/6) from the Foundation for Polish Science (E.Z.S.), Ministry of Polish Science grant (N N302 177338) (E.Z.S.), EU grant No. POIG 01.02-00-109/09 "Innovative methods of stem cells applications in medicine" (E.Z.S.), and NIH grant R01 HL-117730 (B.D.).

ABBREVIATIONS

ASCs	Adipose tissue-derived stem cells
bFGF	Basic fibroblast growth factor
BMMNCs	BM-derived mononuclear cells
CM	Conditioned medium
CPCs	Cardiac progenitor cells
CVDs	Cardiovascular diseases
EPCs	Endothelial progenitor cells
EVs	Extracellular vesicles
GFs	Growth factors
HGF	Hepatocyte growth factor
hIECs	Human islet-derived endothelial cells
HMECs	Human microvascular endothelial cells
IGF-1	Insulin-like growth factor-1
IGF-2	Insulin-like growth factor-2
IHD	Ischemic heart disease
iPSCs	Induced pluripotent stem cells
LAD	Left anterior descending coronary artery

LAMP1	Lysosome-associated membrane protein-1
LV	Left ventricular
MI	Myocardial infarction
MMPs	Matrix metalloproteinases
MSCs	Mesenchymal stem cells
MVBs	Multivesicular bodies
MVs	Microvesicles
PDGF	Platelet-derived growth factor
SCF	Stem cell factor
SCID	Severe combined immunodeficiency
SDF-1	Stromal-derived factor-1
TGF-β	Transforming growth factor-β
TSG101	Tumor susceptibility gene 101 protein
VEGF	Vascular endothelial growth factor

REFERENCES

[1] Laslett LJ, Alagona P Jr, Clark BA 3rd, Drozda JP Jr, Saldivar F, Wilson SR, et al. The worldwide environment of cardiovascular disease: prevalence, diagnosis, therapy, and policy issues: a report from the American College of Cardiology. J Am Coll Cardiol 2012;60:S1–S49.

[2] Abdel-Latif A, Bolli R, Tleyjeh IM, Montori VM, Perin EC, Hornung CA, et al. Adult bone marrow-derived cells for cardiac repair: a systematic review and meta-analysis. Arch Intern Med 2007;167:989–97.

[3] Jeevanantham V, Butler M, Saad A, Abdel-Latif A, Zuba-Surma EK, Dawn B. Adult bone marrow cell therapy improves survival and induces long-term improvement in cardiac parameters: a systematic review and meta-analysis. Circulation 2012;126:551–68.

[4] Dawn B, Bolli R. Adult bone marrow-derived cells: regenerative potential, plasticity, and tissue commitment. Basic Res Cardiol 2005;100:494–503.

[5] Ratajczak MZ, Zuba-Surma EK, Machalinski B, Kucia M. Bone-marrow-derived stem cells – our key to longevity? J Appl Genet 2007;48:307–19.

[6] Zhang H, Wang ZZ. Mechanisms that mediate stem cell self-renewal and differentiation. J Cell Biochem 2008;103:709–18.

[7] Zuba-Surma EK, Kucia M, Ratajczak J, Ratajczak MZ. Small stem cells in adult tissues: very small embryonic-like stem cells stand up! Cytometry A 2009;75:4–13.

[8] Asahara T, Murohara T, Sullivan A, Silver M, van der Zee R, Li T, et al. Isolation of putative progenitor endothelial cells for angiogenesis. Science 1997;275:964–7.

[9] Kucia M, Dawn B, Hunt G, Guo Y, Wysoczynski M, Majka M, et al. Cells expressing early cardiac markers reside in the bone marrow and are mobilized into the peripheral blood after myocardial infarction. Circ Res 2004;95:1191–9.

[10] Kucia M, Reca R, Campbell FR, Zuba-Surma E, Majka M, Ratajczak J, et al. A population of very small embryonic-like (VSEL) CXCR4(+)SSEA-1(+)Oct-4+ stem cells identified in adult bone marrow. Leukemia 2006;20:857–69.

[11] Ratajczak MZ, Zuba-Surma EK, Wojakowski W, Ratajczak J, Kucia M. Bone marrow – home of versatile stem cells. Transfus Med Hemother 2008;35:248–59.

[12] Beltrami AP, Barlucchi L, Torella D, Baker M, Limana F, Chimenti S, et al. Adult cardiac stem cells are multipotent and support myocardial regeneration. Cell 2003;114:763–76.

[13] Ahmed RP, Ashraf M, Buccini S, Shujia J, Haider H. Cardiac tumorigenic potential of induced pluripotent stem cells in an immunocompetent host with myocardial infarction. Regen Med 2011;6:171–8.

[14] Ben-David U, Benvenisty N. The tumorigenicity of human embryonic and induced pluripotent stem cells. Nat Rev Cancer 2011;11:268–77.

[15] Moretti A, Bellin M, Jung CB, Thies TM, Takashima Y, Bernshausen A, et al. Mouse and human induced pluripotent stem cells as a source for multipotent ISL1+ cardiovascular progenitors. FASEB J 2010;24:700–11.

[16] Dawn B, Abdel-Latif A, Sanganalmath SK, Flaherty MP, Zuba-Surma EK. Cardiac repair with adult bone marrow-derived cells: the clinical evidence. Antioxid Redox Signal 2009;11:1865–82.

[17] Jeevanantham V, Afzal MR, Zuba-Surma EK, Dawn B. Clinical trials of cardiac repair with adult bone marrow-derived cells. Methods Mol Biol 2013;1036:179–205.

[18] Zuba-Surma EK, Guo Y, Taher H, Sanganalmath SK, Hunt G, Vincent RJ, et al. Transplantation of expanded bone marrow-derived very small embryonic-like stem cells (VSEL-SCS) improves left ventricular function and remodelling after myocardial infarction. J Cell Mol Med 2011;15:1319–28.

[19] Beltrami AP, Cesselli D, Bergamin N, Marcon P, Rigo S, Puppato E, et al. Multipotent cells can be generated in vitro from several adult human organs (heart, liver and bone marrow). Blood 2007;110:3438–46.

[20] Chugh AR, Zuba-Surma EK, Dawn B. Bone marrow-derived mesenchymal stems cells and cardiac repair. Minerva Cardioangiol 2009;57:185–202.

[21] D'Ippolito G, Diabira S, Howard GA, Menei P, Roos BA, Schiller PC. Marrow-isolated adult multilineage inducible (miami) cells, a unique population of postnatal young and old human cells with extensive expansion and differentiation potential. J Cell Sci 2004;117:2971–81.

[22] Kucia M, Halasa M, Wysoczynski M, Baskiewicz-Masiuk M, Moldenhawer S, Zuba-Surma E, et al. Morphological and molecular characterization of novel population of CXCR4(+)SSEA-4(+)Oct-4(+) very small embryonic-like cells purified from human cord blood – preliminary report. Leukemia 2007;21:297–303.

[23] Till JE, Mc CE. A direct measurement of the radiation sensitivity of normal mouse bone marrow cells. Radiat Res 1961;14:213–22.

[24] Becker AJ, McCulloch CE, Till JE. Cytological demonstration of the clonal nature of spleen colonies derived from transplanted mouse marrow cells. Nature 1963;197:452–4.

[25] Gnecchi M, Zhang Z, Ni A, Dzau VJ. Paracrine mechanisms in adult stem cell signaling and therapy. Circ Res 2008;103:1204–19.

[26] Dawn B, Tiwari S, Kucia MJ, Zuba-Surma EK, Guo Y, Sanganalmath SK, et al. Transplantation of bone marrow-derived very small embryonic-like stem cells attenuates left ventricular dysfunction and remodeling after myocardial infarction. Stem Cells 2008;26:1646–55.

[27] Duran JM, Makarewich CA, Sharp TE, Starosta T, Zhu F, Hoffman NE, et al. Bone-derived stem cells repair the heart after myocardial infarction through transdifferentiation and paracrine signaling mechanisms. Circ Res 2013;113:539–52.

[28] Ratajczak MZ, Kucia M, Jadczyk T, Greco NJ, Wojakowski W, Tendera M, et al. Pivotal role of paracrine effects in stem cell therapies in regenerative medicine: can we translate stem cell-secreted paracrine factors and microvesicles into better therapeutic strategies? Leukemia 2012;26:1166–73.

[29] Denef C. Paracrinicity: the story of 30 years of cellular pituitary crosstalk. J Neuroendocrinol 2008;20:1–70.

[30] Verkhratsky A, Rodriguez JJ, Parpura V. Neurotransmitters and integration in neuronal-astroglial networks. Neurochem Res 2012;37:2326–38.

[31] Janowska-Wieczorek A, Majka M, Ratajczak J, Ratajczak MZ. Autocrine/paracrine mechanisms in human hematopoiesis. Stem Cells 2001;19:99–107.

[32] Majka M, Janowska-Wieczorek A, Ratajczak J, Ehrenman K, Pietrzkowski Z, Kowalska MA, et al. Numerous growth factors, cytokines, and chemokines are secreted by human CD34(+) cells, myeloblasts, erythroblasts, and megakaryoblasts and regulate normal hematopoiesis in an autocrine/paracrine manner. Blood 2001;97:3075–85.

[33] Sahoo S, Klychko E, Thorne T, Misener S, Schultz KM, Millay M, et al. Exosomes from human CD34(+) stem cells mediate their proangiogenic paracrine activity. Circ Res 2011;109:724–8.

[34] Lataillade JJ, Clay D, Bourin P, Herodin F, Dupuy C, Jasmin C, et al. Stromal cell-derived factor 1 regulates primitive hematopoiesis by suppressing apoptosis and by promoting G(0)/G(1) transition in CD34(+) cells: evidence for an autocrine/paracrine mechanism. Blood 2002;99:1117–29.

[35] Ohnishi S, Yasuda T, Kitamura S, Nagaya N. Effect of hypoxia on gene expression of bone marrow-derived mesenchymal stem cells and mononuclear cells. Stem Cells 2007;25:1166–77.

[36] Tirziu D, Giordano FJ, Simons M. Cell communications in the heart. Circulation 2010;122:928–37.

[37] Wojakowski W, Tendera M, Kucia M, Zuba-Surma E, Milewski K, Wallace-Bradley D, et al. Cardiomyocyte differentiation of bone marrow-derived Oct-4+CXCR4+ SSEA-1+ very small embryonic-like stem cells. Int J Oncol 2010;37:237–47.

[38] Zuba-Surma EK, Wojakowski W, Ratajczak MZ, Dawn B. Very small embryonic-like stem cells: biology and therapeutic potential for heart repair. Antioxid Redox Signal 2011;15:1821–34.

[39] Burchfield JS, Dimmeler S. Role of paracrine factors in stem and progenitor cell mediated cardiac repair and tissue fibrosis. Fibrogen Tiss Rep 2008;1:4.

[40] Gnecchi M, He H, Liang OD, Melo LG, Morello F, Mu H, et al. Paracrine action accounts for marked protection of ischemic heart by akt-modified mesenchymal stem cells. Nat Med 2005;11:367–8.

[41] Gnecchi M, He H, Noiseux N, Liang OD, Zhang L, Morello F, et al. Evidence supporting paracrine hypothesis for Akt-modified mesenchymal stem cell-mediated cardiac protection and functional improvement. FASEB J 2006;20:661–9.

[42] DeSantiago J, Bare DJ, Semenov I, Minshall RD, Geenen DL, Wolska BM, et al. Excitation-contraction coupling in ventricular myocytes is enhanced by paracrine signaling from mesenchymal stem cells. J Mol Cell Cardiol 2012;52:1249–56.

[43] Yoon YS, Wecker A, Heyd L, Park JS, Tkebuchava T, Kusano K, et al. Clonally expanded novel multipotent stem cells from human bone marrow regenerate myocardium after myocardial infarction. J Clin Invest 2005;115:326–38.

[44] Nagaya N, Kangawa K, Itoh T, Iwase T, Murakami S, Miyahara Y, et al. Transplantation of mesenchymal stem cells improves cardiac function in a rat model of dilated cardiomyopathy. Circulation 2005;112:1128–35.

[45] Takahashi M, Li TS, Suzuki R, Kobayashi T, Ito H, Ikeda Y, et al. Cytokines produced by bone marrow cells can contribute to functional improvement of the infarcted heart by protecting cardiomyocytes from ischemic injury. Am J Physiol Heart Circ Physiol 2006;291:H886–93.

[46] Xiang MX, He AN, Wang JA, Gui C. Protective paracrine effect of mesenchymal stem cells on cardiomyocytes. J Zhejiang Univ Sci B 2009;10:619–24.

[47] Ohnishi S, Sumiyoshi H, Kitamura S, Nagaya N. Mesenchymal stem cells attenuate cardiac fibroblast proliferation and collagen synthesis through paracrine actions. FEBS Lett 2007;581:3961–6.

[48] Timmers L, Lim SK, Arslan F, Armstrong JS, Hoefer IE, Doevendans PA, et al. Reduction of myocardial infarct size by human mesenchymal stem cell conditioned medium. Stem Cell Res 2007;1:129–37.

[49] Sabin K, Kikyo N. Microvesicles as mediators of tissue regeneration. Transl Res 2013;163:286–95.

[50] Camussi G, Deregibus MC, Bruno S, Grange C, Fonsato V, Tetta C. Exosome/microvesicle-mediated epigenetic reprogramming of cells. Am J Cancer Res 2011;1:98–110.

[51] Ratajczak J, Wysoczynski M, Hayek F, Janowska-Wieczorek A, Ratajczak MZ. Membrane-derived microvesicles: important and underappreciated mediators of cell-to-cell communication. Leukemia 2006;20:1487–95.

[52] Wolf P. The nature and significance of platelet products in human plasma. Br J Haematol 1967;13:269–88.

[53] Gyorgy B, Szabo TG, Pasztoi M, Pal Z, Misjak P, Aradi B, et al. Membrane vesicles, current state-of-the-art: emerging role of extracellular vesicles. Cell Mol Life Sci 2011;68:2667–88.

[54] Akers JC, Gonda D, Kim R, Carter BS, Chen CC. Biogenesis of extracellular vesicles (ev): exosomes, microvesicles, retrovirus-like vesicles, and apoptotic bodies. J Neurooncol 2013;113:1–11.

[55] Loyer X, Vion AC, Tedgui A, Boulanger CM. Microvesicles as cell-cell messengers in cardiovascular diseases. Circ Res 2014;114:345–53.

[56] Biancone L, Bruno S, Deregibus MC, Tetta C, Camussi G. Therapeutic potential of mesenchymal stem cell-derived microvesicles. Nephrol Dial Transplant 2012;27:3037–42.

[57] Thery C, Amigorena S, Raposo G, Clayton A. Isolation and characterization of exosomes from cell culture supernatants and biological fluids. Curr Protoc Cell Biol 2006. Chapter 3: Unit 3.22.

[58] Yuana Y, Bertina RM, Osanto S. Pre-analytical and analytical issues in the analysis of blood microparticles. Thromb Haemost 2011;105:396–408.

[59] Heijnen HF, Schiel AE, Fijnheer R, Geuze HJ, Sixma JJ. Activated platelets release two types of membrane vesicles: microvesicles by surface shedding and exosomes derived from exocytosis of multivesicular bodies and alpha-granules. Blood 1999;94:3791–9.

[60] Del Conde I, Shrimpton CN, Thiagarajan P, Lopez JA. Tissue-factor-bearing microvesicles arise from lipid rafts and fuse with activated platelets to initiate coagulation. Blood 2005;106:1604–11.

[61] Mathivanan S, Ji H, Simpson RJ. Exosomes: extracellular organelles important in intercellular communication. J Proteomics 2010;73:1907–20.

[62] Camussi G, Quesenberry PJ. Perspectives on the potential therapeutic uses of vesicles. Exosomes Microvesicles 2013;1:1–9.

[63] Bobis-Wozowicz S, Adamiak M, Bik-Multanowski M, Madetko-Talowska A, Boruczkowski D, Kmiotek K, et al. IPS- and ESC-derived microvesicles as carriers of bioactive content to target mature cells: implications for tissue regeneration. World Conf Regenerat Med 2013.

[64] Bruno S, Grange C, Deregibus MC, Calogero RA, Saviozzi S, Collino F, et al. Mesenchymal stem cell-derived microvesicles protect against acute tubular injury. J Am Soc Nephrol 2009;20:1053–67.

[65] Herrera MB, Fonsato V, Gatti S, Deregibus MC, Sordi A, Cantarella D, et al. Human liver stem cell-derived microvesicles accelerate hepatic regeneration in hepatectomized rats. J Cell Mol Med 2010;14:1605–18.

[66] Bruno S, Grange C, Collino F, Deregibus MC, Cantaluppi V, Biancone L, et al. Microvesicles derived from mesenchymal stem cells enhance survival in a lethal model of acute kidney injury. PLoS One 2012;7:e33115.

[67] Bian S, Zhang L, Duan L, Wang X, Min Y, Yu H. Extracellular vesicles derived from human bone marrow mesenchymal stem cells promote angiogenesis in a rat myocardial infarction model. J Mol Med (Berl) 2014;92:387–97.

[68] Chen L, Wang Y, Pan Y, Zhang L, Shen C, Qin G, et al. Cardiac progenitor-derived exosomes protect ischemic myocardium from acute ischemia/reperfusion injury. Biochem Biophys Res Commun 2013;431:566–71.

[69] Barile L, Lionetti V, Cervio E, Matteucci M, Gherghiceanu M, Popescu LM, et al. Extracellular vesicles from human cardiac progenitor cells inhibit cardiomyocyte apoptosis and improve cardiac function after myocardial infarction. Cardiovasc Res 2014;103:530–41.

[70] Collino F, Deregibus MC, Bruno S, Sterpone L, Aghemo G, Viltono L, et al. Microvesicles derived from adult human bone marrow and tissue specific mesenchymal stem cells shuttle selected pattern of mirnas. PLoS One 2010;5:e11803.

[71] Cantaluppi V, Gatti S, Medica D, Figliolini F, Bruno S, Deregibus MC, et al. Microvesicles derived from endothelial progenitor cells protect the kidney from ischemia-reperfusion injury by microrna-dependent reprogramming of resident renal cells. Kidney Int 2012; 82:412–27.

[72] Deregibus MC, Cantaluppi V, Calogero R, Lo Iacono M, Tetta C, Biancone L, et al. Endothelial progenitor cell derived microvesicles activate an angiogenic program in endothelial cells by a horizontal transfer of mrna. Blood 2007;110:2440–8.

[73] Quesenberry PJ, Dooner MS, Aliotta JM. Stem cell plasticity revisited: the continuum marrow model and phenotypic changes mediated by microvesicles. Exp Hematol 2010; 38:581–92.

[74] Lai RC, Arslan F, Lee MM, Sze NS, Choo A, Chen TS, et al. Exosome secreted by msc reduces myocardial ischemia/reperfusion injury. Stem Cell Res 2010;4:214–22.

[75] Arslan F, Lai RC, Smeets MB, Akeroyd L, Choo A, Aguor EN, et al. Mesenchymal stem cell-derived exosomes increase ATP levels, decrease oxidative stress and activate PI3K/ Akt pathway to enhance myocardial viability and prevent adverse remodeling after myocardial ischemia/reperfusion injury. Stem Cell Res 2013;10:301–12.

[76] Cantaluppi V, Biancone L, Figliolini F, Beltramo S, Medica D, Deregibus MC, et al. Microvesicles derived from endothelial progenitor cells enhance neoangiogenesis of human pancreatic islets. Cell Transpl 2012;21:1305–20.

[77] Lai RC, Chen TS, Lim SK. Mesenchymal stem cell exosome: a novel stem cell-based therapy for cardiovascular disease. Regen Med 2011;6:481–92.

[78] Friedenstein AJ, Piatetzky S II, Petrakova KV. Osteogenesis in transplants of bone marrow cells. J Embryol Exp Morphol 1966;16:381–90.

[79] Dominici M, Le Blanc K, Mueller I, Slaper-Cortenbach I, Marini F, Krause D, et al. Minimal criteria for defining multipotent mesenchymal stromal cells. The international society for cellular therapy position statement. Cytotherapy 2006;8:315–17.

[80] Makino S, Fukuda K, Miyoshi S, Konishi F, Kodama H, Pan J, et al. Cardiomyocytes can be generated from marrow stromal cells *in vitro*. J Clin Invest 1999;103:697–705.

CHAPTER 6

Therapeutic Potential of Stem Cell-Derived Extracellular Vesicles in Cardioprotection and Myocardium Repair

Bin Yu*, Muhammad Ashraf*,**, Meifeng Xu*
*Department of Pathology and Laboratory Medicine, University of Cincinnati Medical Center, Cincinnati, OH, USA
**Department of Pharmacology, University of Illinois College of Medicine at Chicago, Chicago, IL, USA

Contents

1 GENERAL OUTLINE

Stem cell–mediated protection of ischemic myocardium is partially related to the secretion of paracrine factors, including extracellular vesicles (EVs), which can transfer a range of biological molecules and regulate signaling pathways in recipient cells.

2 INTRODUCTION

Heart disease is the leading cause of death with the highest economic cost of all major disease categories. Although several treatments can mitigate the initial cardiac damage during an acute myocardial ischemia, there is a need for a novel strategy to minimize subsequent cardiac remodeling. The discovery of stem cells and their potential in heart repair has evoked much excitement in treating cardiac diseases. It is found that stem cells secrete a variety of bioactive factors that inhibit fibrosis and apoptosis, enhance angiogenesis, and stimulate mitosis and differentiation of tissue-intrinsic reparative or stem cells [1]. EVs secreted from stem cells play a critical role in the transfer of these bioactive molecules. The role of EVs engenders novel approaches to the development of biologics for regeneration medicine. Despite being smaller than a cell, EVs are relatively complex biological entities that contain a range of biological molecules. EVs are internalized by recipient cells and allow for horizontal transfer of proteins, mRNA, and miRs, inducing epigenetic and functional changes in the recipient cells [2–5]. It is postulated that therapeutic efficacy of EVs is derived from the synergy of a select permutation of individual EV components [6,7].

3 STEM CELL THERAPY

Stem cells have the potential to reduce infarct size and enhance cardiac function in animal models. To date, stem cell therapy for the heart accounts for a third of publications in the regenerative medicine field [8]. Stem cell therapy in the acute phase of a myocardial infarction (MI) is geared towards prevention of cardiomyocytes (CMs) apoptosis, promotion of local neoangiogenesis, improvement of myocardial perfusion, and reduction of the local inflammatory response. In the late phase of MI, cell therapy can be used to replace the dead cells with viable CMs, smooth muscle cells, and/or endothelial cells, thereby reducing scarring and improving cardiac performance.

Many different types of stem cells and their derivatives have been the subjects of transplantation studies. Among the cells under investigation, mesenchymal stem cells (MSCs) are undergoing extensive clinical testing to evaluate therapeutic efficacy. These cells are easily available in accessible tissues, such as bone marrow aspirate [9,10] and fat tissue [11,12], and have a high capacity for *ex vivo* expansion [13]. Bone marrow-derived MSCs (BM-MSCs) are also known to have immunosuppressive properties [12] and, therefore, are appropriate for use in allogeneic transplantation.

MSC transplantation in most animal models of acute MI has generally resulted in reduced infarct size, improved left ventricular (LV) ejection fraction, and increased vascular density and myocardial perfusion [9,10,14]. Implantation of MSCs results in sustained engraftment, myogenic differentiation, and improved cardiac function in the left anterior descending (LAD) artery ligation model [9]. Contractile dysfunction was significantly attenuated and wall thinning was markedly reduced [9]. Cellular transplantation also resulted in long-term engraftment, profound reduction in scar formation, and near-normalization of cardiac function [10]. It has been indicated that not only BM-MSCs but also adipose-derived MSCs (AMSCs) improved cardiac function and perfusion after intracoronary cell transplantation. The thickness of the ventricular wall in the infarction area was significantly greater [14].

More importantly, it has been reported that the direct injection of BM-MSCs into damaged myocardium is safe and effective [10]. In Phase I clinical trials, single infusion of allogeneic MSCs in the patients with acute MI was documented to be safe with improved outcomes of left ventricular function [15]. Ambulatory electrocardiogram monitoring showed reduced ventricular tachycardia episodes. Global symptom score and ejection fraction in the anterior MI patients were both significantly better in human MSCs (hMSCs) compared to placebo subjects [15].

3.1 Transdifferentiation of Stem Cells

It was hypothesized that stem cells could differentiate into CMs and supporting cell types in the damaged myocardium. Stem cells have been demonstrated to differentiate into cardiac phenotypes in many experiments [16–18]. It has also been reported that the purified hMSCs engrafted in the myocardium and appeared to differentiate into CMs. Immunohistochemistry revealed the expression of desmin, β-myosin heavy chain, α-actinin, and cardiac troponin T; the sarcomeric organization of the contractile proteins was also observed [19]. Furthermore, these stem cells can also differentiate into endothelial cells, smooth muscle cells, and possibly cardiac fibroblasts [20–22]. However, transdifferentiation is extremely rare under physiological conditions, although extensive regeneration of myocardial infarcts was observed after direct stem cell injection [23]. Careful experimentation has also demonstrated that few of the transplanted stem cells engraft and survive, and even fewer cells differentiate into CMs or supporting cells [23,24]. It has been shown that MSC engraftment within infarcted myocardium was transient, MSC fusion with CMs was infrequent, and the differentiation rate of MSC into CMs was very low. Although new capillaries and CMs were formed around the infarcted area, only a small percentage of the trans-planted cells could be detected [24].

It has also been demonstrated that the functional improvement follow-ing intramyocardial injection of BM-MSCs overexpressing Akt (Akt-MSCs) occurs within 72 h [25]. This early remarkable effect cannot be attributed to myocardial regeneration from the donor cells. All of these results sug-gest that MSC therapy improved cardiac repair by enhancing the survival of existing myocytes and promoting cardiac regeneration by stimulating an endogenous regenerative capacity of the heart upon cell transplantation. The endogenous regeneration potentially resulted from release of growth factors or cytokines and other paracrine molecules by the stem cells – also referred to as paracrine hypothesis.

3.2 Paracrine Effect

It has been reported that human umbilical cord tissue-derived MSCs (UCX(R)) were delivered via intramyocardial injection to C57BL/6 fe-male mice that were subjected to permanent ligation of the LAD. UCX(R) preserved cardiac function and attenuated the cardiac remodeling through increasing capillary density and decreasing apoptosis in the injured tissue. The myocardial regenerative effects of UCX(R) were attributed to para-crine mechanisms that appear to enhance angiogenesis, limit the extent of the apoptosis in the heart, augment proliferation, and activate a pool of resi-dent cardiac progenitor cells (CPCs) [26]. Our previous studies have shown that stem cells transplantation increased LV ejection fraction and reduced infarct area, which is related to the paracrine effect [27].

MSC secretion could result in the reduced infarct size and improvement in cardiac function. Following lipopolysaccharide (LPS) or IL-1β treatment, neonatal mouse ventricular CMs manifested chaotic $[Ca^{2+}]_i$ handling, quantified as a three- to five-fold increase in spontaneous $[Ca^{2+}]_i$ tran-sients. Normal $[Ca^{2+}]_i$ signaling was preserved when CMs were cocultured with hMSCs in a transwell culture. Flow cytometry analyses revealed that hMSCs also block the LPS- and IL-1β-dependent activation of cardiac transcription factor, NF-κB [28]. It has been shown that conditioned me-dium (CdM) obtained from MSCs attenuates myocardial reperfusion injury when they were added at the onset of reperfusion [29]. CdM of hMSC (hMSC-CdM) restores normal Ca^{2+} signaling in LPS- and IL-1β-damaged CMs [28]. The CdM from hypoxic Akt-MSCs markedly inhibits hypoxia-induced CM apoptosis and triggers vigorous spontaneous contraction of adult rat CMs *in vitro*. Being injected into infarcted hearts, CdM of Akt-MSCs significantly limits infarct size and improves ventricular function [25].

CdM not only enhances CM and/or progenitor survival after hypoxia-induced injury, but also induces angiogenesis. The angiogenesis by secretome has been demonstrated in stem cells from human sources. CdM collected from human BM-MSCs of 15 patients with acute MI protected rat CMs from cell death induced by simulated ischemia or ischemia followed by reperfusion. CdM also stimulated coronary artery endothelial cell proliferation, migration, and tube formation, and induced cell sprouting in a mouse aortic ring assay [30]. In another study, pigs were subjected to left circumflex coronary artery ligation and randomized to intravenous hMSC-CdM treatment for 7 days. Three weeks post MI, myocardial capillary density was higher in the animals treated with hMSC-CdM [31].

By using a three-dimensional collagen assay, it was demonstrated that CdM obtained from MSCs enhanced mitogenic activity of cardiac Sca-1^+ progenitor cells, and mobilized and activated other endogenous cardiac stem cells. CdM upregulated the expression of CM-related genes in CPCs such as β-MHC and atrial natriuretic peptide (ANP). CdM also promoted proliferation of CPCs and inhibited apoptosis of CPCs induced by serum starvation and hypoxia [32].

The paracrine effect of MSCs has also been demonstrated to maintain mitochondrial membrane potential ($\Delta\Psi$m). Exposure of mouse CMs to an ischemic challenge depolarized their $\Delta\Psi$m, increased their diastolic Ca^{2+}, and significantly attenuated cell shortening. Reperfusion of CMs with CdM resulted in an increase of the Ca^{2+} transient amplitudes in all cells and prolonged survival [33].

The soluble factors secreted from stem cells include many growth factors, cytokines, and chemokines that orchestrate interactions within the tissue microenvironment to inhibit apoptosis, stimulate proliferation, and promote vascularization [1]. The expression level of cytokines has been measured in either MSC-CdM or CdM obtained from CMs (CM-CdM) in our previous study. Under normoxic conditions, vascular endothelial growth factor (VEGF), basic fibroblast growth factor (bFGF), insulin-like growth factor 1 (IGF-1), and stromal cell-derived factor (SDF) in MSC-CdM were significantly higher compared with CM-CdM. After cells were exposed to anoxia for 4 h, these proteins were further increased by 30–150% in MSC-CdM [27]. It has been demonstrated that VEGF, FGF-2, HGF, and IGF-1 are potential mediators of the effects exerted by the CdM of Akt-MSCs [25]. Direct injection of HGF or VEGF into the border zone of acute MI murine similarly improved the left ventricular ejection fraction and reduced scar size compared with MSC transplantation. HGF or VEGF overexpressing MSCs (MSC-HGF, MSC-VEGF) significantly increased *in vitro* and *in vivo* proliferation of MSCs, increased peri-infarct vessel densities, and better preserved left ventricular function. Moreover, fewer CMs were apoptotic in CMs maintained in CdM obtained from MSC-HGF or MSC-VEGF cultures [34]. These results indicated that soluble factors are cardioprotective by increasing the tolerance of CMs to ischemia and reducing CM apoptosis.

4 EXTRACELLULAR VESICLES

4.1 EVs Derived from Stem Cells

In addition to the paracrine factors, increasing attention has been focused on the release of extracellular membrane vesicles which may mediate ischemic tissue regeneration [7,35–37]. The vesicles released by cells have a lipid bilayer and contain a cell-specific cargo of proteins, lipids, and genetic material [38]. They can influence the cells they encounter via different mechanisms to alter their function and behavior [39,40]. Vesicle uptake takes place via direct binding with membrane surface molecules, endocytotic internalization, or fusion with the recipient plasma membrane [41,42]. EVs secreted from MSCs are small, spherical membrane fragments and can be divided into exosomes and shedding vesicles, which include microvesicles (MVs) and apoptotic bodies. These vesicles are capable of transferring proteins, mRNA, and miRNA between cells and represent a potential means of intercellular communication.

Exosomes are the most extensively characterized class of vesicles and represent a specific and unique subset. An exosome is generally classified as a secreted vesicle ranging in size from 30 nm to 100 nm. They are released from the cell following fusion of a multivesicular body (MVB) with the plasma membrane [43,44]. MSCs can produce copious amounts of exosomes than other cells, but there are no differences on morphological features, isolation, and storage conditions between exosomes derived from MSCs and other sources [35]. Exosomes are emerging as an attractive carrier of paracrine signals delivered by various cells. The proteins, lipids, and RNAs packaged into an exosome are highly dependent on the cell type of origin, the trigger or stimulus for release, and the lipid content of the surrounding membranes [42]. MVs are larger than exosomes, ranging from 100 nm to 1 μm in size, with most MV preparations comprised of a heterogeneous mixture of particles [45]. Unlike exosomes, MVs are formed from the outward bub-bliing of the plasma membrane upon an activation stimulus. Similarly, MVs carry functional protein, mRNA, and miRNA cargo, deliver these compo-nents to neighboring cells, and regulate the recipient cell function [46]. Here, we use the collective term EVs to encompass exosomes and MVs.

Proteins commonly found on EVs include CD9, CD63, and CD81, all members of the tetraspanin family. Additionally, heat shock proteins (HSPs) like HSP70 and other cytosolic proteins are detected in EVs. EVs also con-tain metabolic enzymes, ribosomal proteins, signal transduction molecules, adhesion molecules, ATPases, cytoskeletal and ubiquitin molecules, growth factors, cytokines, mRNA, and miR molecules [2,47].

Apoptotic bodies are generally much larger (~500–>2000 nm) and have a heterogeneous size distribution [48]. Apoptotic bodies are released at the early stages of apoptosis, and contain both a lipid bilayer and cyto-plasmic contents that originate from the parent cell. Zernecke et al. [49] showed that endothelial cells release miR-126 in apoptotic bodies to alter chemokine responses in neighboring cells.

4.2 EVs Mediate Cardioprotection

The physiological function of EVs is still a matter of debate, but increas-ing results obtained from various experimental systems suggest their in-volvement in multiple biological processes [50]. Timmers et al. [51] first reported that MSC secretion, after sterile filtration with a 0.2 μm filter, reduces infarct size in both murine and porcine models of myocardial ischemia/reperfusion (I/R) injury. Another study using a combination of *in vivo* and *ex vivo* techniques has demonstrated that MSC-derived EVs reduced myocardial damage on myocardial I/R injury [52]. Lim and col-leagues have identified a cardioprotective effect of EVs when EVs were injected into a rat model of I/R [37].

EVs secreted by human cardiosphere-derived cells (CDCs) might be pinpointed as critical agents of regeneration and cardioprotection. It has been indicated that EVs inhibit apoptosis and promote proliferation of CMs. Injection of EVs into injured mouse hearts recapitulates the regenerative and functional effects produced by CDC transplantation [53]. Moreover, mouse ESC-derived EVs (mES-EVs) enhanced neovascularization, CM survival, and reduced fibrosis post infarction consistent with resurgence of cardiac proliferative response [54]. CPCs have been shown to exert poten-tially antifibrotic effects by transferring EVs to fibroblasts, and by promoting angiogenesis and cardiac myocyte survival *in vitro* [55].

CPC-EVs have been shown to reduce myoblast apoptosis *in vitro* and decrease CM death in an animal MI model [56]. Importantly, mES- EVs augmented CPC survival, proliferation, and promoted formation of bona fide new CMs in the ischemic heart [54]. We have directly demonstrated that the purified EVs from BM-MSCs, in a concentration-dependent manner, directly increase the resistance of CMs against hypoxic injury [57]. High peripheral white blood cell (WBC) count is related to larger infarct size, worst cardiac performance, and poor clinical outcome. EV treatment reduced neutrophil and macrophage infiltration in the hearts at days 1 and 3 after reperfusion [52].

It is well known that mitochondrial dysfunction is one of the most important determinants of viability loss after ischemia. Mitochondrial dysfunction induces the loss of ATP and NADH during ischemia. Restoring ATP and NADH levels is therefore critical for the recovery of ischemic myocardium. It has been found that the ATP/ADP and NADH/NAD$^+$ ratios in exosome-treated animals were significantly increased [52]. $\Delta\Psi$m plays a pivotal role in maintaining mitochondrial integrity. We have observed that $\Delta\Psi$m was reduced in CMs exposed to hypoxia and was restored to near normal value in CMs treated with EVs [57].

EVs have also been found to serve as protective agent against other oxidative stress-induced cell injury. For example, pretreatment with CPC-EVs significantly decreased the levels of caspase-3/7 activity in H_2O_2-treated H9C2, indicating that CPC-EVs attenuated H_2O_2-induced apoptosis [56].

4.3 Internalization of EVs

The homing potential to target tissues is important for an effective use of EVs as a therapeutic agent to treat different disorders. The internalization of EVs has been shown to modulate cellular activities consistent with the transfer of biologically functional proteins or RNAs [7]. We investigated the internalization of EVs by CMs using a state-of-the-art time-lapse imaging system, which showed direct evidence of the dynamic uptake of EVs into recipient cells. The internalization was assayed by adding EVs prelabeled with PKH26 (PKH26-EVs) into cultured CMs. Time-lapse images of CMs at one hour intervals during 24 h following the addition of PKH26-EVs were taken. No clear red EVs could be seen in CMs during the first 2 h. Very negligible red fluorescent spots appeared in CMs following EV treatment for 3 h. PKH26-EVs were clearly visible in many CMs after culture with EVs for 8 h. To further confirm the internalization, cultured cells were fixed and labeled with anti-α-actinin. Many PKH26-EVs were found inside α-actinin-positive CMs [57]. Cellular uptake of EVs has been demonstrated to occur through endocytosis, phagocytosis, and membrane fusion [58–60].

Our *in vivo* study further demonstrated that the EVs were more frequently inside damaged cells in the ischemic area than in normal myocardial cells [57]. It has been indicated that the efficiency of EV uptake is correlated directly with intracellular and microenvironmental acidity [58]. This may be a mechanism by which EVs exert their cardioprotective effects on ischemic CMs that have a low intracellular pH [61]. The membrane proteins may provide a potential target for the homing of EVs to a specific tissue or microenvironment. For example, integrins on EVs can home EVs to CMs that express intercellular adhesion molecule-1 (ICAM1), a ligand of integrins [62] after I/R injury or to VCAM-1 on endothelial cells [63].

4.4 Proteomic Complementation

The proteomes of I/R mouse hearts were first profiled using mass spectrometry and antibody array [64]. Of the 509 quantified proteins identified, 121 proteins exhibited significant quantitative changes after 30 min of ischemia, and persisted for up to 120 min post reperfusion. These proteins participated in a wide array of biochemical and cellular processes such as communication, structure and mechanics, inflammation, and metabolism. EVs could compensate for these proteomic alterations to alleviate I/R injury and promote cellular recovery [64]. EVs express not only common surface markers such as CD9, CD81, and Alix [37], but also proteins that are characteristic of MSC self-renewal and differentiation. Kim et al. [65] performed proteomic analysis using liquid chromatography–mass spectrometry and identified 730 proteins in EVs. The proteome includes (1) surface receptors that promote therapeutic potential of MSCs; (2) signaling molecules downstream of the surface receptors that promote recruitment and proliferation; and (3) cell adhesion molecules that are mostly involved in cell adhesion. The appearance of these proteins indicates that several biological processes involved in the therapeutic effects of MSCs are represented in EVs, including vesicle-mediated transport, cell cycle and proliferation, cell migration, morphogenesis, and developmental processes. The MSC-EV proteome provides a comprehensive basis for understanding the potential of MSC-MVs to affect tissue repair and regeneration [65].

4.5 Genetic Materials (mRNA/miRs) Delivery

Rapidly emerging evidence indicates that EVs represent an ideal vehicle to deliver genetic materials such as mRNA and miRs from one cell to another and to alter the fate of recipient cells [6,57,66–69]. A recent breakthrough in EV biology is that RNA molecules, in particular, miRs, are present in these vesicles.

mRNA microarray analyses confirmed that mRNA profiles in EVs reflect their parent cell phenotypes [67,70]. ES-EVs are highly enriched selectively in mRNA for several pluripotent transcription factors as compared to parental embryonic stem cells. These mRNAs could be delivered to target cells and translated into the corresponding proteins. The biological effects of ES-EVs were inhibited after heat inactivation or pretreatment with RNase [71]. The direct evidence for EVs shuttling mRNA to recipient cells is that addition of EVs enriched human IGF-1R mRNA into mouse fibroblast cell line. After culturing for 3 h, IGF-1R mRNA transcript was detected in the cells incubated with human EVs [69]. Moreover, horizontal transfer of IGF-1R mRNA was time dependent suggesting that IGF-1R expression was due to the mRNA transfer instead of residual mRNA [69]. More recently, it has also been directly demonstrated that EVs derived from human stem cells may also deliver mRNA to rat cells, resulting in protein translation [70]. EVs derived from human liver stem cells induced proliferation and increased apoptosis resistance of human and rat hepatocytes. However, these effects were abrogated if EVs were pretreated with RNase. Using human *AGO2* as a reporter gene present in EVs, the expression of human *AGO2* mRNA and protein was observed in the liver of hepatectomized rats treated with EVs [70]. To further study the transduction of mRNA directly, endothelial cells were cultured with EVs carrying green fluorescent protein (GFP) mRNA. mRNA transfer was shown by transduction of GFP into endothelial cells [68]. Deregibus et al. [68] systematically investigated EVs acting as a vehicle for mRNA transport among cells using microarrays. A total of 298 transcripts were found with this procedure, of which 183 were associated to RefSeq identifiers and the remaining were Unigene expressed sequence tags. This observation indicates that EVs are not shuttling a random sample of cellular mRNA but represents a specific subset. It has been demonstrated that RNAs in EVs are more stable and are reportedly resistant to degradation during prolonged storage and freeze/thaw cycles compared to cellular RNAs [72].

Besides mRNA, miRs are also transferred from donor cell into a recipient cell by EVs and microparticles with intact functionality to regulate protein expression in recipient cells [6,54,73–77]. miRs are 18–25-nucleotide small noncoding regulatory RNAs that play a pivotal regulatory role in diverse biological processes by decreasing protein translation resulting in fine-tune cellular function. The change in miR expression levels in response to cell stressors may affect entire cellular regulatory pathways and fundamentally alter cellular biology. EV-mediated transfer of miRs regulates the target proteins in acceptor cells and alters cellular phenotypes.

The paracrine transfer of miRs via EVs is a new, dynamic area of research. The original idea of EV transfer of miRs began in 2007 when Valadi et al. [2] grew mast cell cultures and collected their secreted EVs. Chen and coworkers [56] demonstrated that CPC-derived EVs contained a high level of miR-451 compared to CPCs. miR-451 is an miR already known to improve the clinical outcome when transfected into ischemic rodent hearts. Ibrahim et al. [53] identified EVs as critical agents of cardiac regeneration triggered by cardiosphere-mediated cell therapies and was enriched with miR-146a. EVs obtained from embryonic stem cells can transfer a subset of miRs to mouse embryonic fibroblasts *in vitro* by incubating ES-EVs with γ-irradiated mouse embryonic fibroblasts. The abundance of several miRs (miR-290, miR-291-3p, miR-292-3p, miR-294, and miR-295) increased in recipient cells [76]. Using state-of-the-art time-lapse confocal imaging, the dynamic course of miR-22-loaded EVs released from the MSCs were successfully captured. A coculture system of MSCs with neonatal CMs was employed, in which the cells were separated by a membrane of 0.3 μm pore size. The expression of miR-22 was dramatically upregulated in CMs. EVs enriched with miR-22 were secreted from MSCs and mobilized to CMs [6].

It was observed that 45 of the 60 miRs identified in EVs were also presented in MSCs, suggesting that released EVs echo the contents of their parent cells [75]. Some miRs were found only in EVs, suggesting that the miR profiles in EVs do not completely reflect the miR profiles observed in the parental cells and that miRs were selectively enriched in EVs [2,56,78,79]. For example, miR-124, one of the miRs present only in EVs, is a well-characterized brain-specific miR, which is involved in neurogenesis [79]. Squadrito et al. [80] suggested that one mechanism of miR selection for EVs is not needed for local cellular regulation. Transferring miRs contained in EVs opens the possibility that stem cells can alter the expression of genes in recipient cells.

4.6 EVs Derived from Preconditioned and Genetically Modified Stem Cells

Accumulating evidence suggests that the EV secretion profile is influenced by preconditioning or genetic manipulation of the parent cells. It has been well established that ischemic preconditioning (IPC) is a potent approach to enhance survival and regeneration of MSCs in an ischemic environment. During preconditioning, BM-MSCs upregulate several cytokines, growth factors, and survival proteins, which serve as antiapoptotic and myoangiogenic differentiation stimulants [27]. The number and expression level of miRs were also different in EVs secreted from MSCs subjected to IPC (EV^{IPC}). It has been reported that miR-22 was highly upregulated in EV^{IPC}. Treatment of infarcted hearts with EV^{IPC} resulted in a better cardiac outcome [6].

It is found that EVs from hypoxic CPCs improved cardiac function and reduced fibrosis, enhanced tube formation of endothelial cells, and decreased profibrotic gene expression in TGF-β stimulated fibroblasts [81]. Microarray analysis of EVs secreted by hypoxic CPCs identified 11 miR-NAs that were upregulated compared with exosomes secreted by CPCs grown under normoxic conditions [81].

Genetically modified MSCs have also been widely tested in various studies. It has been hypothesized that EVs derived from pretreated MSCs could be used as ideal vehicles for gene delivery to facilitate gene and cell therapy. Treatment of adipose-derived stem cells with platelet-derived growth factor (PDGF) stimulated the secretion of EVs, which carried c-kit and SCF, which played a role in vessel-like structure formation [82].

The CD34$^+$ hematopoietic stem cells have shown significant promise in addressing myocardial ischemia by promoting angiogenesis [5]. Unfortunately, the viability and angiogenic effect of autologous CD34$^+$ cells decreases with advanced age and diminished cardiovascular health. To offset age- and health-related angiogenic declines in CD34$^+$ cells, Dr. Losordo's group [5] explored whether the therapeutic efficacy of human CD34$^+$ cells could be enhanced by augmenting secretion of the known angiogenic factors, sonic hedgehog (Shh). Shh-modified CD34$^+$ cells (CD34-Shh) protected against ventricular dilation and acute MI-induced cardiac functional decline. Treatment with CD34-Shh reduced infarct size and increased border zone capillary density. CD34-Shh primarily store and secrete Shh protein in EVs, which transfer functional Shh to elicit induction of the canonical Shh signaling pathway in recipient cells [5]. CXCR4-enriched EVs acquired from CXCR4 overexpressing MSCs can also enhance the protection of CMs from ischemic injury [83].

We have demonstrated previously that MSCs overexpressing GATA-4 released more growth factors and promoted angiogenesis [84].

Overexpression of GATA-4 increases MSC survival in the ischemic environment and regulates miR expression in MSCs [85]. EVs derived from MSC^{GATA-4} (EV^{GATA-4}) showed a higher potential in maintaining $\Delta\Psi$m, reducing CM apoptosis and ischemic myocardium infarct size, and improving cardiac function compared to the EVs released from control empty-vector transduced MSCs. Recently, we directly and systematically investigated whether EV^{GATA-4} enriched antiapoptotic miRs and delivered these miRs into cultured CMs and ischemic myocardium [57]. miR-19a was selected to test the hypothesis that antiapoptotic miRs play an important role in EV^{GATA-4}-mediated cardioprotection. Knockdown of miR-19a by transfecting miR-19a inhibitor partially abolished EV^{GATA-4}-mediated cardioprotection [57].

EV-delivered miR can regulate target protein expression in recipient cells, which is an important transfer mechanism among neighboring cells [74]. To confirm the transferred miRs regulated their target proteins in CMs, we have reported a downregulation of the level of phosphatase and tensin homolog (PTEN), one of miR-19 target proteins, in CMs, which were treated with EVs enriched with miR-19 [57]. It has also been demonstrated that EVs produced by human umbilical cord MSCs inhibited signal transducer and activator of transcription 3 (STAT3) signaling in isolated human pulmonary artery endothelial cells via increasing lung levels of miR-204 [86]. These results indicate that miRs transferred from EVs can regulate the expression of their target proteins in recipient cells.

5 ADVANTAGES OF EV THERAPY

EV therapy may reduce cardiac injury and delay the loss of cardiomyocytes, although EVs do not replace the lost cardiomyocytes. The refinement of MSC therapy from a cell to secretion-based therapy opens a novel cell-free therapeutic perspective and offers several advantages.

In adults, MSCs are scarce with a frequency of 0.01% of nucleated cells in bone marrow, which generally needs further culture expansion to generate sufficient cells for therapeutic application [87]. In addition, impaired cell functionality in aged people and in those with advanced cardiovascular disease limits autologous cell transplantation [88]. Moreover, some potential side effects of cell therapy also limit clinical application. For example, the use of myogenic cells has been demonstrated to increase the risk of arrhythmias [89,90]. The amplification and delivery of beneficial paracrine signals (e.g., EVs) generated by these cells could overcome obstacles associated with cell injection-based approaches to repair damaged myocardium [91].

EVs have gained considerable attention due to their experimental and therapeutic significance. Compared with cells, EVs are more stable and reservable. It is reported that high centrifugal forces (110,000g for up to 22 h) have no impact on the size and integrity of the EVs [92]. Multiple deep-freezing and thawing did not affect their size and integrity [92]. EVs have no risk of aneuploidy and may provide an alternative therapy for various diseases. EVs have the potential for avoiding many of the limitations of viable cells for therapeutic applications in regenerative medicine [93]. Cellular secretions are more amenable to development as an "off-the-shelf" therapeutics that can be delivered to patients in a timely manner [94]. As a bilipid membrane vesicle, EVs not only have the capacity to carry a large cargo load, but can also protect the contents from degradative enzymes and chemicals. In addition, EVs have been shown to cross the plasma membrane to deliver their cargo into target cells. More importantly, EVs are amenable to membrane modifications that enhance cell-specific targeting. In perspective, with the advances of cellular techniques, the engineering of the EV surface or content may be envisaged in order to enhance disease-specific targeting.

In addition to mitigating the risks associated with cell transplantation, EVs can also circumvent some of the challenges associated with the use of small soluble biological factors such as growth factors, chemokines, cytokines, and transcription factors [95]. The administration of a single factor cannot effectively mimic the therapeutic effect of MSCs that concomitantly release a number of different factors. EVs may deliver a complex array of biologically active molecules to injured cells, which may favor tissue regeneration.

6 CONCLUSIONS

EVs present a complex composition and strongly contribute to the paracrine effects of stem cells. In regenerative therapies EVs mimic the beneficial effects of the cells from which they originate, and could represent an important potential therapeutic tool. Therefore, the use of MSC-derived EVs represents an interesting alternative for heart repair, which may overcome the limitations and risks commonly associated with cell-therapy approaches. Engineering or modification of the EV surface antigen and internal content through preconditioning or engineering parent cells will enable EVs to target other more complex and specific diseases.

ABBREVIATIONS

$\Delta\Psi m$	Mitochondrial membrane potential
AMSCs	Adipose-derived MSCs
ANP	Atrial natriuretic peptide
β-MHC	Beta-myosin heavy chain
bFGF	Basic fibroblast growth factor
MI	Myocardial infarction
CdM	Conditioned medium
CMs	Cardiomyocytes
CPCs	Cardiac progenitor cells
CSCs	Cardiac stem cells
ES-EVs	EVs derived from ESCs
EVs	Extracellular vesicles
EV^{GATA-4}	EVs derived from MSC^{GATA-4}
h-EVs	EVs released from human BM-MSCs
hESCs	Human embryonic stem cells
HSPs	Heat shock proteins
IGF-1	Insulin-like growth factor 1
ICAM1	Intercellular adhesion molecule-1
IPC	Ischemic preconditioning
I/R	Ischemia/reperfusion
LAD	Left anterior descending artery
LPS	Lipopolysaccharide
LV	Left ventricular
MSCs	Mesenchymal stem cells
MVB	Multivesicular body
MVs	Microvesicles
PDGF	Platelet-derived growth factor
SDF	Stromal cell-derived factor
Shh	Sonic hedgehog
STAT3	Signal transducer and activator of transcription 3
UCX(R)	Human umbilical cord tissue-derived MSCs
VEGF	Vascular endothelial growth factor
WBC	White blood cell

ACKNOWLEDGMENT

This work was supported by National Institutes of Health grants HL105176 and HL114654 (M. Xu), and R37-HL-074272, HL-095375, and HL-087246 (M. Ashraf). The authors thank Christian Paul for technical assistance.

REFERENCES

[1] Caplan AI, Dennis JE. Mesenchymal stem cells as trophic mediators. J Cell Biochem 2006;98(5):1076–84.

[2] Valadi H, Ekstrom K, Bossios A, Sjostrand M, Lee JJ, Lotvall JO. Exosome-mediated transfer of mRNAs and microRNAs is a novel mechanism of genetic exchange between cells. Nat Cell Biol 2007;9(6):654–9.

[3] Hergenreider E, Heydt S, Treguer K, et al. Atheroprotective communication between endothelial cells and smooth muscle cells through miRNAs. Nat Cell Biol 2012;14(3):249–56.

[4] Montecalvo A, Larregina AT, Shufesky WJ, et al. Mechanism of transfer of functional microRNAs between mouse dendritic cells via exosomes. Blood 2012;119(3):756–66.

[5] Mackie AR, Klyachko E, Thorne T, et al. Sonic hedgehog-modified human CD34+ cells preserve cardiac function after acute myocardial infarction. Circ Res 2012;111(3):312–21.

[6] Feng Y, Huang W, Wani M, Yu X, Ashraf M. Ischemic preconditioning potentiates the protective effect of stem cells through secretion of exosomes by targeting Mecp2 via miR-22. PLoS One 2014;9(2):e88685.

[7] Camussi G, Deregibus MC, Bruno S, Cantaluppi V, Biancone L. Exosomes/microvesicles as a mechanism of cell-to-cell communication. Kidney Int 2010;78(9):838–48.

[8] Mummery CL, Davis RP, Krieger JE. Challenges in using stem cells for cardiac repair. Sci Transl Med 2010;2(27):27ps17.

[9] Shake JG, Gruber PJ, Baumgartner WA, et al. Mesenchymal stem cell implantation in a swine myocardial infarct model: engraftment and functional effects. Ann Thorac Surg 2002;73(6):1919–25. discussion 26.

[10] Amado LC, Saliaris AP, Schuleri KH, et al. Cardiac repair with intramyocardial injection of allogeneic mesenchymal stem cells after myocardial infarction. Proc Natl Acad Sci USA 2005;102(32):11474–9.

[11] Lee RH, Kim B, Choi I, et al. Characterization and expression analysis of mesenchymal stem cells from human bone marrow and adipose tissue. Cell Physiol Biochem 2004;14(4–6):311–24.

[12] Le Blanc K, Pittenger M. Mesenchymal stem cells: progress toward promise. Cytotherapy 2005;7(1):36–45.

[13] Giordano A, Galderisi U, Marino IR. From the laboratory bench to the patient's bedside: an update on clinical trials with mesenchymal stem cells. J Cell Physiol 2007;211(1):27–35.

[14] Valina C, Pinkernell K, Song YH, et al. Intracoronary administration of autologous adipose tissue-derived stem cells improves left ventricular function, perfusion, and remodelling after acute myocardial infarction. Eur Heart J 2007;28(21):2667–77.

[15] Hare JM, Traverse JH, Henry TD, et al. A randomized, double-blind, placebo-controlled, dose-escalation study of intravenous adult human mesenchymal stem cells (prochymal) after acute myocardial infarction. J Am Coll Cardiol 2009;54(24):2277–86.

[16] Goumans MJ, de Boer TP, Smits AM, et al. TGF-beta1 induces efficient differentiation of human cardiomyocyte progenitor cells into functional cardiomyocytes in vitro. Stem Cell Res 2007;1(2):138–49.

[17] Laflamme MA, Chen KY, Naumova AV, et al. Cardiomyocytes derived from human embryonic stem cells in pro-survival factors enhance function of infarcted rat hearts. Nat Biotechnol 2007;25(9):1015–24.

[18] van Laake LW, Passier R, Monshouwer-Kloots J, et al. Human embryonic stem cell-derived cardiomyocytes survive and mature in the mouse heart and transiently improve function after myocardial infarction. Stem Cell Res 2007;1(1):9–24.

[19] Toma C, Pittenger MF, Cahill KS, Byrne BJ, Kessler PD. Human mesenchymal stem cells differentiate to a cardiomyocyte phenotype in the adult murine heart. Circulation 2002;105(1):93–8.

[20] Oswald J, Boxberger S, Jorgensen B, et al. Mesenchymal stem cells can be differentiated into endothelial cells in vitro. Stem Cells 2004;22(3):377–84.

[21] Gong Z, Niklason LE. Use of human mesenchymal stem cells as alternative source of smooth muscle cells in vessel engineering. Methods Mol Biol 2011;698:279–94.

[22] Lee CH, Shah B, Moioli EK, Mao JJ. CTGF directs fibroblast differentiation from human mesenchymal stem/stromal cells and defines connective tissue healing in a rodent injury model. J Clin Invest 2010;120(9):3340–9.

[23] Murry CE, Soonpaa MH, Reinecke H, et al. Haematopoietic stem cells do not transdifferentiate into cardiac myocytes in myocardial infarcts. Nature 2004;428(6983):664–8.

[24] Noiseux N, Gnecchi M, Lopez-Ilasaca M, et al. Mesenchymal stem cells overexpressing Akt dramatically repair infarcted myocardium and improve cardiac function despite infrequent cellular fusion or differentiation. Mol Ther 2006;14(6):840–50.

[25] Gnecchi M, He H, Noiseux N, et al. Evidence supporting paracrine hypothesis for Akt-modified mesenchymal stem cell-mediated cardiac protection and functional improvement. FASEB J 2006;20(6):661–9.

[26] Nascimento DS, Mosqueira D, Sousa LM, et al. Human umbilical cord tissue-derived mesenchymal stromal cells attenuate remodeling following myocardial infarction by pro-angiogenic, anti-apoptotic and endogenous cell activation mechanisms. Stem Cell Res Ther 2014;5(1):5.

[27] Uemura R, Xu M, Ahmad N, Ashraf M. Bone marrow stem cells prevent left ventricular remodeling of ischemic heart through paracrine signaling. Circ Res 2006;98(11):1414–21.

[28] Rogers TB, Pati S, Gaa S, et al. Mesenchymal stem cells stimulate protective genetic reprogramming of injured cardiac ventricular myocytes. J Mol Cell Cardiol 2011;50(2):346–56.

[29] Angoulvant D, Ivanes F, Ferrera R, Matthews PG, Nataf S, Ovize M. Mesenchymal stem cell conditioned media attenuates *in vitro* and *ex vivo* myocardial reperfusion injury. J Heart Lung Transplant 2011;30(1):95–102.

[30] Korf-Klingebiel M, Kempf T, Sauer T, et al. Bone marrow cells are a rich source of growth factors and cytokines: implications for cell therapy trials after myocardial infarction. Eur Heart J 2008;29(23):2851–8.

[31] Timmers L, Lim SK, Hoefer IE, et al. Human mesenchymal stem cell-conditioned medium improves cardiac function following myocardial infarction. Stem Cell Res 2011;6(3):206–14.

[32] Windmolders S, De Boeck A, Koninckx R, et al. Mesenchymal stem cell secreted platelet derived growth factor exerts a pro-migratory effect on resident cardiac atrial appendage stem cells. J Mol Cell Cardiol 2014;66:177–88.

[33] DeSantiago J, Bare DJ, Banach K. Ischemia/reperfusion injury protection by mesenchymal stem cell derived antioxidant capacity. Stem Cells Dev 2013;22(18):2497–507.

[34] Deuse T, Peter C, Fedak PW, et al. Hepatocyte growth factor or vascular endothelial growth factor gene transfer maximizes mesenchymal stem cell-based myocardial salvage after acute myocardial infarction. Circulation 2009;120(11 Suppl.):S247–54.

[35] Yeo RW, Lai RC, Zhang B, et al. Mesenchymal stem cell: an efficient mass producer of exosomes for drug delivery. Adv Drug Deliv Rev 2013;65(3):336–41.

[36] Bruno S, Grange C, Collino F, et al. Microvesicles derived from mesenchymal stem cells enhance survival in a lethal model of acute kidney injury. PLoS One 2012;7(3):e33115.

[37] Lai RC, Arslan F, Lee MM, et al. Exosome secreted by MSC reduces myocardial ischemia/reperfusion injury. Stem Cell Res 2010;4(3):214–22.

[38] Laulagnier K, Motta C, Hamdi S, et al. Mast cell- and dendritic cell-derived exosomes display a specific lipid composition and an unusual membrane organization. Biochem J 2004;380(Pt 1):161–71.

[39] Vlassov AV, Magdaleno S, Setterquist R, Conrad R. Exosomes: current knowledge of their composition, biological functions, and diagnostic and therapeutic potentials. Biochim Biophys Acta 2012;1820(7):940–8.

[40] Skog J, Wurdinger T, van Rijn S, et al. Glioblastoma microvesicles transport RNA and proteins that promote tumour growth and provide diagnostic biomarkers. Nat Cell Biol 2008;10(12):1470–6.

[41] Chaput N, Thery C. Exosomes: immune properties and potential clinical implementations. Semin Immunopathol 2011;33(5):419–40.

[42] Record M, Subra C, Silvente-Poirot S, Poirot M. Exosomes as intercellular signalosomes and pharmacological effectors. Biochem Pharmacol 2011;81(10):1171–82.

[43] Mori Y, Koike M, Moriishi E, et al. Human herpesvirus-6 induces MVB formation, and virus egress occurs by an exosomal release pathway. Traffic 2008;9(10):1728–42.

[44] Buning J, von Smolinski D, Tafazzoli K, et al. Multivesicular bodies in intestinal epithelial cells: responsible for MHC class II-restricted antigen processing and origin of exosomes. Immunology 2008;125(4):510–21.

[45] Muralidharan-Chari V, Clancy JW, Sedgwick A, D'Souza-Schorey C. Microvesicles: mediators of extracellular communication during cancer progression. J Cell Sci 2010;123(Pt 10):1603–11.

[46] Lee Y, El Andaloussi S, Wood MJ. Exosomes and microvesicles: extracellular vesicles for genetic information transfer and gene therapy. Hum Mol Genet 2012;21(R1): R125–34.

[47] Mathivanan S, Simpson RJ. ExoCarta: a compendium of exosomal proteins and RNA. Proteomics 2009;9(21):4997–5000.

[48] Huber J, Vales A, Mitulovic G, et al. Oxidized membrane vesicles and blebs from apoptotic cells contain biologically active oxidized phospholipids that induce monocyte-endothelial interactions. Arterioscler Thromb Vasc Biol 2002;22(1):101–7.

[49] Zernecke A, Bidzhekov K, Noels H, et al. Delivery of microRNA-126 by apoptotic bodies induces CXCL12-dependent vascular protection. Sci Signal 2009;2(100):ra81.

[50] Thery C, Amigorena S, Raposo G, Clayton A. Isolation and characterization of exosomes from cell culture supernatants and biological fluids. Curr Protoc Cell Biol 2006;22.

[51] Timmers L, Lim SK, Arslan F, et al. Reduction of myocardial infarct size by human mesenchymal stem cell conditioned medium. Stem Cell Res 2007;1(2):129–37.

[52] Arslan F, Lai RC, Smeets MB, et al. Mesenchymal stem cell-derived exosomes increase ATP levels, decrease oxidative stress and activate PI3K/Akt pathway to enhance myocardial viability and prevent adverse remodeling after myocardial ischemia/reperfusion injury. Stem Cell Res 2013;10(3):301–12.

[53] Ibrahim AG, Cheng K, Marban E. Exosomes as critical agents of cardiac regeneration triggered by cell therapy. Stem Cell Rep 2014;2(5):606–19.

[54] Khan M, Nickoloff E, Abramova T, et al. Embryonic stem cell-derived exosomes promote endogenous repair mechanisms and enhance cardiac function following myocardial infarction. Circ Res 2015;117(1):52–64.

[55] Gray WD, French KM, Ghosh-Choudhary S, et al. Identification of therapeutic covariant microRNA clusters in hypoxia-treated cardiac progenitor cell exosomes using systems biology. Circ Res 2015;116(2):255–63.

[56] Chen L, Wang Y, Pan Y, et al. Cardiac progenitor-derived exosomes protect ischemic myocardium from acute ischemia/reperfusion injury. Biochem Biophys Res Commun 2013;431(3):566–71.

[57] Yu B, Kim HW, Gong M, et al. Exosomes secreted from GATA-4 overexpressing mesenchymal stem cells serve as a reservoir of anti-apoptotic microRNAs for cardioprotection. Int J Cardiol 2015;182:349–60.

[58] Parolini I, Federici C, Raggi C, et al. Microenvironmental pH is a key factor for exosome traffic in tumor cells. J Biol Chem 2009;284(49):34211–22.

[59] Tian T, Wang Y, Wang H, Zhu Z, Xiao Z. Visualizing of the cellular uptake and intracellular trafficking of exosomes by live-cell microscopy. J Cell Biochem 2010;111(2):488–96.

[60] Feng D, Zhao WL, Ye YY, et al. Cellular internalization of exosomes occurs through phagocytosis. Traffic 2010;11(5):675–87.

[61] Schrader J. Mechanisms of ischemic injury in the heart. Basic Res Cardiol 1985;80(Suppl.2): 135–9.

[62] Kukielka GL, Hawkins HK, Michael L, et al. Regulation of intercellular adhesion molecule-1 (ICAM-1) in ischemic and reperfused canine myocardium. J Clin Invest 1993;92(3):1504–16.

[63] Rieu S, Geminard C, Rabesandratana H, Sainte-Marie J, Vidal M. Exosomes released during reticulocyte maturation bind to fibronectin via integrin alpha4beta1. Eur J Biochem 2000;267(2):583–90.

[64] Li X, Arslan F, Ren Y, et al. Metabolic adaptation to a disruption in oxygen supply during myocardial ischemia and reperfusion is underpinned by temporal and quantitative changes in the cardiac proteome. J Proteome Res 2012;11(4):2331–46.

[65] Kim HS, Choi DY, Yun SJ, et al. Proteomic analysis of microvesicles derived from human mesenchymal stem cells. J Proteome Res 2012;11(2):839–49.

[66] Yellon DM, Davidson SM. Exosomes: nanoparticles involved in cardioprotection? Circ Res 2014;114(2):325–32.

[67] Bruno S, Grange C, Deregibus MC, et al. Mesenchymal stem cell-derived microvesicles protect against acute tubular injury. J Am Soc Nephrol 2009;20(5):1053–67.

[68] Deregibus MC, Cantaluppi V, Calogero R, et al. Endothelial progenitor cell derived microvesicles activate an angiogenic program in endothelial cells by a horizontal transfer of mRNA. Blood 2007;110(7):2440–8.

[69] Tomasoni S, Longaretti L, Rota C, et al. Transfer of growth factor receptor mRNA via exosomes unravels the regenerative effect of mesenchymal stem cells. Stem Cells Devel 2013;22(5):772–80.

[70] Herrera MB, Fonsato V, Gatti S, et al. Human liver stem cell-derived microvesicles accelerate hepatic regeneration in hepatectomized rats. J Cell Mol Med 2010;14(6B): 1605–18.

[71] Ratajczak J, Miekus K, Kucia M, et al. Embryonic stem cell-derived microvesicles reprogram hematopoietic progenitors: evidence for horizontal transfer of mRNA and protein delivery. Leukemia 2006;20(5):847–56.

[72] Reid G, Kirschner MB, van Zandwijk N. Circulating microRNAs: association with disease and potential use as biomarkers. Crit Rev Oncol Hematol 2011;80(2):193–208.

[73] Stoorvogel W. Functional transfer of microRNA by exosomes. Blood 2012;119(3): 646–8.

[74] Xin H, Li Y, Buller B, et al. Exosome-mediated transfer of miR-133b from multipotent mesenchymal stromal cells to neural cells contributes to neurite outgrowth. Stem Cells 2012;30(7):1556–64.

[75] Chen TS, Lai RC, Lee MM, Choo AB, Lee CN, Lim SK. Mesenchymal stem cell secretes microparticles enriched in pre-microRNAs. Nucleic Acids Res 2010;38(1):215–24.

[76] Yuan A, Farber EL, Rapoport AL, et al. Transfer of microRNAs by embryonic stem cell microvesicles. PLoS One 2009;4(3):e4722.

[77] Das S, Halushka MK. Extracellular vesicle microRNA transfer in cardiovascular disease. Cardiovasc Pathol 2015;24(4):199–206.

[78] Collino F, Deregibus MC, Bruno S, et al. Microvesicles derived from adult human bone marrow and tissue specific mesenchymal stem cells shuttle selected pattern of miRNAs. PLoS One 2010;5(7):e11803.

[79] Akerblom M, Sachdeva R, Jakobsson J. Functional studies of microRNAs in neural stem cells: problems and perspectives. Front Neurosci 2012;6:14.

[80] Squadrito ML, Baer C, Burdet F, et al. Endogenous RNAs modulate microRNA sorting to exosomes and transfer to acceptor cells. Cell Rep 2014;8(5):1432–46.

[81] Tang YL, Zhu W, Cheng M, et al. Hypoxic preconditioning enhances the benefit of cardiac progenitor cell therapy for treatment of myocardial infarction by inducing CXCR4 expression. Circ Res 2009;104(10):1209–16.

[82] Lopatina T, Bruno S, Tetta C, Kalinina N, Porta M, Camussi G. Platelet-derived growth factor regulates the secretion of extracellular vesicles by adipose mesenchymal stem cells and enhances their angiogenic potential. Cell Commun Signal 2014;12:26.

[83] Kang K, Ma R, Cai W, et al. Exosomes secreted from CXCR4 overexpressing mesenchymal stem cells promote cardioprotection via Akt signaling pathway following myocardial infarction. Stem Cells Int 2015;2015:659890.

[84] Li H, Zuo S, He Z, et al. Paracrine factors released by GATA-4 overexpressed mesenchymal stem cells increase angiogenesis and cell survival. Am J Physiol Heart Circ Physiol 2010;299(6):H1772–81.

[85] Yu B, Gong M, He Z, et al. Enhanced mesenchymal stem cell survival induced by GATA-4 overexpression is partially mediated by regulation of the miR-15 family. Int J Biochem Cell Biol 2013;45(12):2724–35.

[86] Lee C, Mitsialis SA, Aslam M, et al. Exosomes mediate the cytoprotective action of mesenchymal stromal cells on hypoxia-induced pulmonary hypertension. Circulation 2012;126(22):2601–11.

[87] Takashima S, Tempel D, Duckers HJ. Current outlook of cardiac stem cell therapy towards a clinical application. Heart 2013;99(23):1772–84.

[88] Dimmeler S, Leri A. Aging and disease as modifiers of efficacy of cell therapy. Circ Res 2008;102(11):1319–30.

[89] Chang MG, Tung L, Sekar RB, et al. Proarrhythmic potential of mesenchymal stem cell transplantation revealed in an *in vitro* coculture model. Circulation 2006;113(15):1832–41.

[90] Price MJ, Chou CC, Frantzen M, et al. Intravenous mesenchymal stem cell therapy early after reperfused acute myocardial infarction improves left ventricular function and alters electrophysiologic properties. Int J Cardiol 2006;111(2):231–9.

[91] Sahoo S, Losordo DW. Exosomes and cardiac repair after myocardial infarction. Circ Res 2014;114(2):333–44.

[92] Sokolova V, Ludwig AK, Hornung S, et al. Characterisation of exosomes derived from human cells by nanoparticle tracking analysis and scanning electron microscopy. Colloids Surf B 2011;87(1):146–50.

[93] Barile L, Gherghiceanu M, Popescu LM, Moccetti T, Vassalli G. Ultrastructural evidence of exosome secretion by progenitor cells in adult mouse myocardium and adult human cardiospheres. J Biomed Biotechnol 2012;2012:354605.

[94] Lai RC, Chen TS, Lim SK. Mesenchymal stem cell exosome: a novel stem cell-based therapy for cardiovascular disease. Regen Med 2011;6(4):481–92.

[95] Mirotsou M, Jayawardena TM, Schmeckper J, Gnecchi M, Dzau VJ. Paracrine mechanisms of stem cell reparative and regenerative actions in the heart. J Mol Cell Cardiol 2011;50(2):280–9.

CHAPTER 7

Engineered/Hypoxia-Preconditioned MSC-Derived Exosome: Its Potential Therapeutic Applications

Wei Zhu, Han Chen, Jian'an Wang
Department of Cardiology, Provincial Key Cardiovascular Research Laboratory, Second Affiliated Hospital, Zhejiang University School of Medicine, Hangzhou, Zhejiang, China

Contents

1 PARACRINE EFFECTS OF MESENCHYMAL STEM CELLS

Stem cell therapy holds great promise for treating various cardiovascular diseases and has become a hot topic in the area of regenerative medicine. Among various stem cells, mesenchymal stem or stromal cells (MSCs) are the most widely used in clinical trials for treating heart diseases, due to the following unique properties of MSCs [1]. First, MSCs are easily available in accessible tissues, such as bone marrow aspirate and fat tissue [2] and have the large capacity for *ex vivo* expansion [3]. Second, MSCs are also known to have immunomodulatory effects [4,5] for which they could be used in allogeneic transplantation. Third, MSCs are multipotent adult stem cells that are capable of self-renewal and have great differentiation potentials for adipogenesis, osteogenesis, and chondrogenesis [6], but also for endothelial, cardiovascular [7,8], and neurogenic [9] differentiation. Most earlier experimental studies have demonstrated that MSC transplantation can significantly improve cardiac function [7,10,11]. The beneficial effects of MSC transplantation were initially mainly ascribed to the hypothesis that MSCs differentiate into cardiomyocytes and supporting cell types to repair cardiac tissues. However, careful study showed that most transplanted MSCs are entrapped in the lungs and the capillary beds of tissues other than the heart [12]. Furthermore, depending on the method of infusion, 6% or less of the transplanted MSCs persist in the heart 2 weeks after delivery [13]. In fact, the differentiation of MSCs into cardiomyocytes is reported to be a rare event [7]. In addition, data are also available showing that ventricular function can be rapidly restored less than 72 h after MSCs transplantation [14]. It is unreasonable for MSCs to differentiate into enough cardiomyocytes to support an increase in heart function in such a short period. Therefore, MSCs differentiation-mediated tissue repair contributes minimally to the beneficial effects of MSCs therapy. Now it is generally agreed that it is the secreted paracrine factors through which MSCs transplantation mediates their therapeutic effects.

More than 15 years ago, it was reported that MSCs synthesize and secrete a broad spectrum of growth factors and cytokines [15] such as vascular endothelial growth factor, stromal cell-derived factor-1, fibroblast growth factor, transforming growth factor β, and interleukin 1 receptor antagonist [16–20], all of which exert their effects on cells in their vicinity. The beneficial effects offered by these factors are observed in the cardiovascular system, including neovascularization [21], improved ventricular thinning [22], and increased angiogenesis [23]. In fact, these paracrine factors secreted by MSCs are transported and delivered in the form of cargos, and are contained within a membrane vesicle, which confers protection from degradation and facilitates targeted cell delivery through membrane receptors. Recent studies have shown that exosomes, membrane-bound vesicles, are one form of the vesicles produced by and released from almost all cell types, and can carry critical messengers for cell–cell communications [24].

2 DISCOVERY OF EXOSOMES

Exosomes are one of several groups of cell-secreted vesicles, including microvesicles (MVs) and apoptotic bodies. The discovery of cell-derived vesicles dates back more than a half century. In 1946, Chargaff and West [25] tested the clotting times of plasma at different centrifugation speeds, and observed a significantly prolonged clotting time with a supernatant obtained from high-speed centrifugation at 31,000g for 150 min, which, however, can be reversed when the pellet of centrifugation that contains "the clotting factor of which the plasma is deprived" was added back to the plasma. Twenty years later, this fraction was identified by electron microscopy and shown to have a diameter between 20 nm and 50 nm and a density of 1.020–1.025 g/mL [26]. Pan et al. [27] observed that maturing sheep reticulocytes can disclose an intracellular sac filled with small uniform-sized membrane-enclosed structures; these sacs can be fused with the plasma membrane and be released through exocytosis. The same phenomenon was observed by another research group in rat reticulocytes [28]. Later on, Johnstone's research group interpreted this process as a way of releasing internal vesicular content; these cell-extruded membrane-bound structures were named exosomes [29]. It is now known that exosomes can be released by many cell types, including B and T lymphocytes, dentritic cells, mast cells, intestinal epithelial cells, neurons, tumor cells, and MSCs [30–36]. Exosomes are found in physiological fluid such as urine, plasma, cerebrospinal fluid, and human milk, and also in exudates [37–39].

3 BIOGENESIS OF EXOSOMES

Exosomes originate from internal budding of the plasma membrane during endocytic internalization, a mature process from an early endosome, through an interaction with the Golgi complex, to a late endosome at which stage the bilayer membrane gives rise to intraluminal vesicles, i.e., exosomes. At completion of biogenesis, exosomes are contained within multivesicular bodies that have incorporated recycled proteins from coated pits in the cellular membrane; proteins directly from the Golgi complex; and mRNA, microRNA, and DNA. The multivesicular bodies can either fuse with the plasma membrane to release exosomes through exocytosis or can be sent to lysosomes for degradation [40]. Therefore, exosomes differ from MVs or apoptotic bodies that are released from the cell as a result of a direct budding process of the plasma membrane. The structure of exosomes by transmission electron microscopy appears cup shaped; however, this could be a result of the processing and fixation that cause a collapse of the circular molecules. A cryo-electron microscopic examination using quickly frozen exosomes actually demonstrates perfectly round-shaped exosomes [41]. Typically, exosomes have a diameter of 30–100 nm, a density of 1.13–1.19 g/mL, and are isolated through sucrose cushion or density gradient by ultracentrifugation at 100,000g [42]. With fusion of the multivesicular bodies and the cell membrane, secretion of exosomes occurs. It has been shown that several Rab GTPase proteins are involved in this process [43].

Exosomes released from cells can act in a paracrine or even an endocrine manner to modify the behavior of adjacent cells or distant cells. Intercellular communication is essential for multicellular organisms to maintain vital functions. Direct cell-to-cell contact or transfer of secreted molecules can accomplish this communication. The release of extracellular vesicles such as exosomes is another mode of intercellular contract, with direct contact between the exosomes and the cell membrane, either by cell surface receptors, by fusion of the two membranes, or by endocytosis. A combination of specific cell surface molecules on exosomes is critical for cell targeting and cell adhesion. Exosomes will certainly display net negative surface charges that facilitate the solubility and integrity of exosomes in body fluids such as blood plasma. A message by exosomes can be transferred to distant cells through three possible mechanisms: (i) direct contact between the exosomal membrane and the plasma membrane of the target cell; (ii) fusion of the two membranes; and (iii) internalization of the exosomes by a target cell [44]. Thus, a parental cell is able to communicate with a target cell in its vicinity or at a distance through an amplification process. The number of exosomes present in the plasma of the healthy individual is extraordinary ($\approx 10^{10}$ mL^{-1}) in the plasma of healthy individuals [45], which is seemingly unreasonable; however, exosomes are so minute that the volume contained within 10^{10} spherical exosomes of 50 nm diameter is only ≈ 5 nL, i.e., 0.0005% of 1 mL. Thus, even though exosomes constitute a limited space, a significant quantity of proteins can be delivered to effect changes in recipient cells. Of note, it is reasonable to speculate that the interactions between cells that are mediated by exosomes will be more complicated than those mediated by a single ligand with a single receptor.

4 MOLECULAR COMPOSITIONS OF EXOSOMES

Exosomes carry a unique cargo of proteins, lipids, and RNAs that can be distinct, and reflect the cell of origin. However, exosomes from different cellular origins share some common characteristics, such as lipid bilayer, which has an exceptionally high cholesterol/phospholipid ratio, their size and density, and basic compositions of lipid and protein. The different protein compositions of exosomes have their specific roles: annexins and flotillin are important for transport and fusion; tetraspanins are involved in cell targeting; and other proteins, such as Alix and TSG101, are involved in their biogenesis from multivesicular bodies. In addition, some protein kinases and heterotrimeric G-proteins are implicated in signaling transduction and lipid metabolism [46]. Exosome proteins also include those derived from the cytoplasm or membrane-bound proteins. Proteomic analyses have revealed the presence of structural components: (i) extracellular matrix and cell surface proteins, such as collagens, integrins, and galectin; (ii) cell surface receptors such as platelet-derived growth factor receptor B and epidermal growth factor receptor; and (iii) intracellular cytoskeletal components, including signaling molecules, metabolic enzymes and G-proteins [47,48].

Importantly, the lipid compositions of exosomes are enriched in cholesterol, ceramide, phosphoglycerides, and long and saturated fatty-acyl chains, because all of these compositions could provide structural stability for exosomes [49]. In addition, prostaglandins that are bound to the exosomal membrane for delivery to target cells can potentially enhance their biological activities [50]. It is increasingly recognized that RNA, especially microRNAs, contained in exosomes are the important signaling messengers that can be transferred between cells [51]. MicroRNAs released within exosomes, as a subset of the cellular RNA, can be unique or tissue specific [52]. In addition, many exosomes contain major histocompatibility complex class I and class II molecules that are involved in antigen binding and presentation [53,54].

It is important to note that proteins like integrins and annexins as compositions of exosomes play important roles in cell adhesion to recipient cells, as do tetraspanins, which can direct targeting to specific cells such as endothelial cells to promote angiogenesis and vasculogenesis [55]. The long-range targeting and tissue uptake of exosomes and their stability in the circulation or in other biological fluids make exosomes attractive as a biomarker as well as a therapeutic vehicle in the treatment of various diseases.

5 THERAPEUTIC EFFECTS OF EXOSOMES IN MYOCARDIAL INFARCTION

Over the past decades, great progresses have been made in interventional cardiology, leading to a significant increase in the survival rate for patients with acute myocardial infarction. A timely and efficient reperfusion of the ischemic myocardium can rescue the endangered myocardium. Ironically, however, reperfusion in turn could also cause damage to the ischemic myocardium, which is called ischemia/reperfusion (I/R) injury. This paradox is attracting much attention in the world of cardiovascular research. It has been shown that ischemic preconditioning can protect the myocardium from I/R injury and its underlying mechanism has been a hot topic in this area. It has been shown that MSCs-released exosomes contain heat shock proteins, including αB-crystallin, HSP60, and HSP70, that have been well established as being cardioprotective [56,57], and can potentially be transferred to adjacent cells to confer protection against oxidative stress [58].

Study has shown that culture medium conditioned by human embryonic stem cell-derived MSCs can significantly reduce infarct size by approximately 50% in both pig and mouse models of myocardial I/R injury when administered intravenously in a single bolus just before reperfusion. Furthermore, using electron microscopy, ultracentrifugation, mass spectrometry, and biochemical assays, this complex was eventually identified as an exosome, a secreted bilipid membrane vesicle. Interestingly, when exosomes derived from conditioned medium of human embryonic stem cell-derived MSCs were further purified by size exclusion using high-performance liquid chromatography, only one-tenth of the protein dosage from conditioned medium was enough to achieve a reduction in infarct size, indicating that exosomes constitute about 10% of the conditioned medium in terms of protein amount [35,59]. Therefore, the favorable effects offered by conditioned medium of human embryonic stem cell-derived MSCs could be attributed primarily to the exosomes. The cardioprotective biological activity by exosomes is not only observed in human embryonic stem cell-derived MSCs, but is also found to be present in nonhypoxic cultured conditions of aborted fetal tissue-derived MSCs [60]. In addition, in hypoxia-induced lung injury and pulmonary hypertension, using exosomes from murine marrow-derived MSCs isolated by using size exclusion chromatography, intravenous injection of MSC-derived exosomes suppressed hypoxia-induced lung inflammation, pulmonary hypertension, and hence right ventricular hypertrophy compared with phosphate-buffered saline therapy. In contrast, fibroblast-derived exosomes had no such inflammatory-inhibitive effects [36].

Heart failure is still a major burden to public health. During the early pathological process of heart failure, there is an adaptive hypertrophic change in the myocardium, which could maintain cardiac function for the time being; however, if the cause of heart failure persists, i.e., the loss of myocytes (myocardial infarction), cardiac overload (valve calcification, hypertension, or volume overload, etc.), maladaptive hypertrophy eventually occurs, resulting in a cardiac remodeling and hence a decrease in cardiac performance. It is becoming well recognized that insufficient blood flow to the hypertrophied myocardium is the main cause of cardiac remodeling [61]. Therefore, an enhanced angiogenesis to maintain a sufficient oxygen and nutrient supply to the hypertrophied myocardium would reverse this adverse remodeling and improve cardiac function.

There is increasing evidence for exosomes to have an important role in angiogenesis. The role of exosomes in angiogenesis has previously been recognized in cancer [62]; recent studies showed that the MVs formed in the blood also possess a therapeutic potential regarding angiogenesis, through interaction between endothelial cells and endothelial progenitor cells [63]. The role of $CD34^+$ stem cells in growing vasculature has been documented in multiple studies; however, the magnitude of transplanted cells seemed modest compared with significant overall physiological impact. To test whether it is the exosomes secreted from $CD34^+$ cells that are responsible for angiogenesis, Sahoo et al. [64] have performed a series of experiments. They confirm that conditioned media contain exosomes, which are characterized by their size being between 40 nm and 90 nm as determined by dynamic light scattering and with a cup-shaped morphology (by electron microscopy). In addition, the exosomes express exosome markers, including CD63, phosphatidylserine, and TSG101. Importantly, the $CD34^+$ cell-derived exosomes can replicate the angiogenic activity by increasing endothelial cell proliferation and tube formation on Matrigel. Moreover, the exosomes obtained from the medium of the $CD34^+$ cell, but not those from mononuclear cells, exhibit proangiogenic activity as shown in the *in vivo* Matrigel plug and corneal assays. As expected, the proangiogenic activities are abolished when the medium is exosome depleted, further confirming that it is the exosomes that mediate the proangiogenic activity.

Recently, Salmon et al. tested the angiogenic activity of placental MSCs [65] when exposed to hypoxic mediates. They isolated the placental MSCs from placental villi of 8–12 weeks' gestation and cultured under an atmosphere of 1, 3, or 8% O_2, respectively. Placental MSC-derived exosomes were isolated from cell-conditioned media by differential and buoyant density centrifugation, and further identified by electron microscopy as being spherical vesicles, with a typical cup shape and around 100 nm in diameter. Exosomes released by placental MSCs are positive for exosome markers, such as CD63, CD9, and CD81. The results showed that oxygen tension could modify exosomal contents from placenta MSCs, which significantly affected the migration and angiogenic tube formation by placental microvascular endothelial cells. Under hypoxic conditions (1% and 3% O_2) released exosomes by placental MSCs increased by 3.3- and 6.7-fold, respectively, compared to the controls (8% O_2; $P < 0.01$). Using a real-time live-cell imaging system, the dose-dependent effects (ranging from 5 µg to

20 μg exosomal protein/mL) of exosomes by placental MSCs on human microvascular endothelial cell migration and tube formation were tested. Exosomes released from placental MSCs increased human microvascular endothelial cell migration by 1.6-fold compared to the control ($P < 0.05$) and increased human microvascular endothelial cell tube formation by 7.2-fold ($P < 0.05$). The data suggest that exosomes released by placental MSCs can be secreted and adapted to the cell environment such as low oxygen and play important roles in both physiological and pathological conditions.

6 HYPOXIA-PRECONDITIONING-MODIFIED MSC-DERIVED EXOSOMES

In myocardial infarction, most MSCs that were directly injected into the ischemic myocardium were lost several days after transplantation [1,13]. Typically, MSCs are cultured under ambient, or normoxic, conditions (21% oxygen). However, the physiological niches for MSCs in the bone marrow and other sites have much lower oxygen tension. When used as a therapeutic tool to repair tissue injuries, MSCs cultured in standard conditions must adapt from 21% oxygen in culture to less than 1% oxygen in the ischemic tissue. It is thought that hypoxia-preconditioning (HPC) pretreatment would mimic HPC and render MSCs more resistant to ischemic stress, resulting in much improved survival. Our previous study showed that HPC conferred a protective effect against MSC apoptosis induced by hypoxia/reoxygenation via stabilizing mitochondrial membrane potential, upregulating B-cell lymphoma 2 (Bcl-2) and vascular endothelial growth factor (VEGF), and promoting ERK and Akt phosphorylation. Interestingly, cyclosporine A showed similar effects to HPC. Thus, HPC has been used as a novel strategy to protect MSCs against ischemic condition after implantation into the ischemic myocardium [66]. Our further studies confirmed that HPC-enhanced capacity of MSCs to treat infarcted myocardium against the remodeling process is mainly attributable to increased MSC survival, which is closely associated with enhanced angiogenesis through hypoxia-inducible factor 1α [67,68]. The same results were also observed by another research group [69]. Nolta's research group observed that injecting HPC-treated MSCs can result in a much earlier restoration of blood flow in a murine hind limb ischemia model, which was associated with activation of hepatocyte growth factor signaling in MSCs [70].

It has been well established that pharmacological preconditioning can equally protect against ischemic stress and enhance cell survival. Using trimetazidine (1-[2,3,4-trimethoxybenzyl]piperazine, TMZ), a drug that was widely used as an anti-ischemic agent, Wisel et al. [71] pharmacologically preconditioned MSCs with 10 μM TMZ for 6 h and then evaluated the cell viability and metabolic activity when MSCs were exposed to culture medium under hypoxia conditions (2% O_2) combined with 100 μM H_2O_2 for 1 h. The preconditioned MSCs showed a significantly strong resistance against oxidative stress, higher cellular viability, less membrane damage, and well-preserved oxygen metabolism. Protection by TMZ was associated with a significant increase in HIF-1α, survivin, phosphorylated Akt and Bcl-2 protein levels, and Bcl-2 gene expression. When TMZ-preconditioned MSCs were implanted into rat hearts in an *in vivo* myocardial infarction model, a significant increase in the recovery of cardiac function was achieved, which was again associated with upregulation of pAkt and Bcl-2 levels [71]. Similarly, TGF-α stimulates VEGF production and decreases inflammatory responses and apoptosis activities via a p38 mitogen-activated protein kinase (MAPK)-dependent mechanism in MSCs; the beneficial effects were further enhanced when combined with tumor necrosis factor-α or hypoxia intervention. And these effects were responsible for augmented MSCs' protective abilities for myocardium against I/R injury. In addition, application of SB202190, a p38 MAPK inhibitor, abolished transforming growth factor (TGF)-α-mediated cardioprotection [72]. These data imply that various preconditioning approaches can efficiently improve MSCs survival; however, it remains to be tested whether it is the increase in the number of surviving MSCs or the different messengers induced and contained in the exosomes that are secreted by preconditioned MSCs, or both that account for the improved therapeutic effects of MSCs.

7 ENGINEERING-MODIFIED MSC-DERIVED EXOSOMES

The most important part of the functions for exosomes is transferring the signaling as a way of cell cell communications, for which the molecules contained in the exosomes take major responsibility to exert their biological effects. Increasing amounts of data demonstrate that engineering-modified exosomes exhibit enhanced therapeutic effects. Studies showed that the condition medium obtained from Akt-overexpressing MSCs exhibited protection within 72 h in an *in vivo* myocardial infarction model, which was associated with upregulation of VEGF, FGF-2, HGF, IGF-I, and TB4, as potential mediators [14,73]. Interestingly, gene chip analysis that aimed to quantify the changes in transcripts driven by Akt in MSCs showed that 650 transcripts are differentially regulated compared with green fluorescent protein-expressing control MSCs. Among those upregulated genes, Sfrp2, but not the other Sfrp family members, turned out to be the most dramatically upregulated one. Coincidentally, Sfrp2 activation is actually Akt pathway dependent. On the other hand, when Sfrp2 was silenced by siRNA, the salutary effects offered by Akt-overexpressing MSCs were abolished. In fact, Sfrp2 can bind to frizzled receptor to suppress Wnt signaling and simultaneously upregulate antiapoptotic genes such as Birc1b possibly via mediating β-catenin [74]. In addition, Sfrp2 can directly inhibit cardiac fibrosis and hence modulate the remodeling process [75]. Thus, the data indicate that genetically modified Akt-overexpressing MSCs offered superior beneficial effects mainly through paracrine effects. Even though direct delivery of protective factor of Sfrp2 can also offer favorable effects [75], we assume that the exosomes secreted by modified MSCs may contain other multiple factors; for recognizing the targeted cells, for protein molecule protection while being transported, etc., thus leading to much more stable and consistent results. Certainly it would be interesting to test whether exosomes directly obtained from Sfrp2-overexpressing MSCs would offer the same effects. These data suggest that specific molecule-targeted, engineering-modified MSC-derived exosomes potentially can offer more efficient and consistent cardiac protection.

Another example of engineering-modified MSC-secreted exosomes for achieving better therapeutic effects is a Notch signaling-targeted molecule. Notch signaling plays key roles in several biologic processes, such as differentiation, cell fate determination, as well as angiogenesis. Inhibition of Notch signaling reduces the levels of stalk cell markers and confers a tip cell phenotype, resulting in vessel branching and sprouting. Interaction of a Notch receptor is traditionally considered to be a membrane-tethered, receptor–ligand interaction. Human umbilical vein endothelial cells naturally produce exosomes that contain Delta-like 4. The Delta-like 4-containing exosomes can be transferred to the membrane of the recipient cells such as endothelial cells and tumor cells, reduce the Notch 1 receptor levels, and are eventually endocytosed by the recipient cells. Thus, Delta-like 4-containing exosomes increase capillary-like structure formation both *in vitro* and *in vivo* by a mechanism that implicates the transfer of Delta-like 4 into the endothelium. These data not only suggest that the Delta-like/Notch pathway does not require direct cell–cell contact to expand its signaling potential on angiogenesis but also indicate that genetically modified Delta-like 4 overexpressing MSCs could be used to promote angiogenesis [76]. Exosomes in this case serve as a tool of cargo for transferring molecules. Other proteins also could be carried by endothelial-derived exosomes and implicated in their proangiogenic potency. Taraboletti et al. [77] have shown that matrix metalloproteinases harbored by exosomes from endothelial cells are functionally active and lead to endothelial cell invasion and capillary-like formation. It remains unknown if simultaneous overexpressing of these molecules that are produced within exosomes and secreted by MSCs would result in much improved effects.

Engineering-manipulated, or -modified MVs or exosomes that contain the protein Sonic hedgehog (Shh) play key roles in nitric oxide generation in endothelial cells, leading to endothelium-dependent relaxation, angiogenesis, and new vessel formation – all of these effects have well been demonstrated in experimental studies using an ischemic hind limb model [78–80]. In contrast, sonic hedgehog-depleted MVs inhibit angiogenesis [78]. The interrupted balance for the angiogenesis process also indicates that antiangiogenic molecules may exist naturally in MVs or exosomes secreted by endothelial cells. It is, therefore, reasonable that (Shh)-overexpressing exosomes derived from engineering modified MSCs could be another practical approach for MSCs therapy, in this case to stimulate neovascularization; whereas Shh-depleted exosomes secreted by MSCs could be used to reduce tumorigenesis via inhibiting tumor vascularization and progression [81]. The therapeutic effects of Shh-containing exosomes released by CD34$^+$ stem cells have been proved in a myocardial infarction model [64]. However, the CD34$^+$ stem cells showed an age-related decline in their function, indicating that Shh contents can decrease in exosomes released by aging CD34$^+$ stem cells. Thus, it is more desirable to obtain engineering-modified Shh-containing exosomes. The feasibility of obtaining Shh overexpressing exosomes from CD34$^+$ stem cells via a genetic engineering approach has been tested experimentally. The data showed that injection of engineering-modified Shh overexpressing CD34 cells to infracted heart did result in a reduction in infarct size, an increase in capillary density, and much improved long-term functional recovery [82]. Interestingly, however, simultaneous transfer of Shh protein with regular CD34$^+$ stem cells failed to confer significant benefits in an *in vivo* study.

8 MicroRNAs AS MAJOR COMPOSITIONS OF EXOSOMES FOR MOLECULE TARGETING

Even though exosomes contain important compositions of proteins as information for cell–cell communication [76], considering the size of the molecules, it is easier to transfer minute microRNAs compared with the delivery of proteins, which might have more dramatic effects. The discovery that mRNA and microRNAs are localized within exosomes has attracted much attention. It has been shown that exosomes contain the majority of plasma microRNAs [83], and exosomes (and MVs) can transfer micro-RNAs to recipient cells [51,84]. This is important for RNAs as they are protected from RNAse or trypsin [51]. On the other hand, microRNAs or mRNA contents of exosomes might reflect the cell they originate from, or even the status of cell conditions [85]. For example, exosomes from cells that are exposed to oxidative stress can contain stress-induced microRNAs or mRNAs, which conferred resistance against oxidative stress to recipient cells [86]. The different profiles of microRNA expression, such as miR-210 induced by HPC [69,87], pharmacological preconditioning-induced miR-146a with diazoxide [88], or miR-155 with hydrogen peroxide [89], is closely related to the different cell types and the different activities of their specific upstream transcriptional factors. Manipulation of this microRNA expression by MSCs might be an interesting approach to confer MSC protection against apoptosis.

Exosomal microRNA contents are closely related to cardiovascular diseases. Patients with acute myocardial infarction exhibit increased serum levels of miR-1 and miR-133a, and *in vitro* experiments also suggest that exosomes from cardiac cells can release miR-133a and transfer it to recipient cells for modulating their gene expressions [90]. The change in miR-133a could reflect their potentially compensatory roles in the pathological process after myocardial infarction, as miR-133 can function to suppress hypertrophy by restraining the expression of the inositol 1,4,5′-triphosphate receptor II calcium channel. Cellular levels of miR-133a have been found to decrease during hypertrophic response to pressure overload [91].

Human cardiomyocyte progenitor cells have also been shown to release exosomes, which can stimulate the migration of microvascular endothelial cells [92]. Interestingly, the exosomes or MVs that have been shown to protect the kidney from I/R injury [93] could also induce neovascularization and stimulate angiogenesis in a murine model of hindlimb ischemia [94,95], through transferring microRNAs or mRNA. RNAse-treated MVs lost their protective effects; the same is true after proangiogenic miR-126 and miR-296 are depleted [93,94]. Thus, either MVs or exosomes exert their proangiogenic effects through transferring common components, microRNAs, to the endothelial cells.

The microRNAs contained in exosomes are also involved in vascular homeostasis. Human umbilical vein endothelial cells that are subjected to shear stress can release exosomes or MVs enriched in miR-143/145, which control target gene expression in co-cultured smooth muscle cells. Importantly, the released vesicles are able to reduce atherosclerotic lesion formation in a mouse model of atherosclerosis [84].

9 OPEN QUESTIONS FOR PRODUCING ENGINEERING-MODIFIED EXOSOMES

MSCs are increasingly recognized as a safe source for cell therapy, due to their unique characteristics [1]. The exosomes obtained from MSCs, especially those from engineered MSCs that have the capacity for signal communication, can be isolated and used for unique therapeutics. While this generates great enthusiasm in the research area, there are still a lot of issues that need to be cleared and questions that need to be answered to further facilitate their potential therapeutic applications.

First, it is known that in addition to exosomes, cells can also shed other types of extracellular-membrane vesicles, namely, MVs and apoptotic bodies [96,97], after various biological stimuli, including induction of programmed cell death. Exosomes differ from the larger MVs in more than just their size. MVs are formed by budding off or shedding directly from the plasma membrane with distinct properties from those of exosomes [24,98,99]. In the cardiovascular system, MVs are claimed to mainly exert detrimental effects [100]. Numerous studies have identified a strong association between elevated numbers of circulating MVs and cardiovascular diseases [101,102]. Increases in procoagulant MVs in the blood can be detected in patients with events of acute coronary syndrome [103,104]. Increases in circulating MVs can be used as independent markers to predict the occurrence of cardiovascular events [105]. However, these data only provide an association between existence of MVs and cardiovascular pathological process and do not necessarily suggest the causative effects of MVs on cardiovascular diseases. Meanwhile, a significant amount of data is now available demonstrating the roles of MVs in mediating the pathological process, with beneficial effects observed in experimental studies [63,106]. On the other hand, MVs are more straightforward to isolate and easier to quantify using flow cytometry, and contain significant amounts of protein, all of which make them a favorable therapeutic tool. In contrast, exosomes are below the detection limit of flow cytometry and require specialized equipment, such as nanoparticle tracking or light scattering, for their quantification. However, exosomes represent a totally distinct population of vesicles from that of MVs and contain unique important information for signaling communication. Of note, even though new technology has been developed to isolate exosomes from MVs, contamination still exists. A routine check-up for their specific markers should always be performed. Up to now, it is still too early to make comments on their differential roles for these membrane-derived vesicles; much work needs to be done to further elucidate their unique roles in cardiovascular diseases.

Second, even though exosomes are produced by MSCs, the detailed process of their generation and secretion is still not fully understood. Further elucidation of this process would allow us to better design engineering-modified exosomes that are secreted from MSCs for much improved specificity and efficacy. The specifically designed exosomes should be better for targeting local or remote cells for which exosome components produced by MSCs will be recognized by the membrane protein or receptor of the recipient cells. To achieve this purpose, great efforts should be taken to design the exosomes. An interesting result has been reported where purified exosomes are used as an *in vivo* transfection reagent for delivering microRNAs into the brains of mice; in this case microRNAs are loaded into the exosomes by electroporation [107]. It would be desirable to obtain specific miRNA expressing exosomes that can be recognized by recipient cells.

Third, exosomes also have other unique functions. Exosomes are regarded as a way for waste management to protect the cells against stress. For example, a substantial amount of active caspase 3-enriched exosomes is detected, which cannot be detected in the releasing cells. This process indicates that exosomes can function as waste management for self-protection against stress [108]. Thus, quality control is an essential step when producing the targeted exosomes; any unwanted molecules should be avoided.

Fourth, once MSCs become the major tool cells for producing specific molecule-targeted exosomes, finite expansion capacity will be one of the major technical hurdles. And stability and reproducibility of the exosomes derived from MSCs will be another important issue. Therefore, a robust, scalable, and highly renewable cell source will be central to the development of a commercially viable process for the production of "Good Manufacturing Practices"-grade MSC-derived exosomes. Interestingly, immortalization of the ESC-MSCs by Myc has been shown not to compromise the quality or yield of exosomes [109].

In summary, exosomes derived from MSCs have great potentials for treating various cardiovascular diseases. Carefully designed, engineering-modified MSCs would allow production of special exosomes, which target specifically molecular signaling and exert therapeutic effects. A lot of work needs to be done to elucidate the biology of exosomes, which will facilitate its clinical applications in the future.

ABBREVIATIONS

Akt	Protein kinase B
Bcl-2	B-cell lymphoma 2
HPC	Hypoxia-preconditioning
HSP	Heat shock proteins
I/R	Ischemia/reperfusion
MAPK	mitogen-activated protein kinase
miR	MicroRNA-
MSCs	Mesenchymal stem (stromal) cells
MVs	Microvesicles
Shh	Sonic hedgehog
TGF	Transforming growth factor
TMZ	Trimetazidine (1-[2,3,4-trimethoxybenzyl]piperazine)
VEGF	Vascular endothelial growth factor

REFERENCES

[1] Williams AR, Hare JM. Mesenchymal stem cells: biology, pathophysiology, translational findings, and therapeutic implications for cardiac disease. Circ Res 2011;109(8):923–40.

[2] Lee RH, Kim B, Choi I, Kim H, Choi HS, Suh K, et al. Characterization and expression analysis of mesenchymal stem cells from human bone marrow and adipose tissue. Cell Physiol Biochem 2004;14(4–6):311–24.

[3] Giordano A, Galderisi U, Marino IR. From the laboratory bench to the patient's bedside: an update on clinical trials with mesenchymal stem cells. J Cell Physiol 2007;211(1): 27–35.

[4] Le Blanc K, Pittenger M. Mesenchymal stem cells: progress toward promise. Cytotherapy 2005;7(1):36–45.

[5] Uccelli A, Moretta L, Pistoia V. Mesenchymal stem cells in health and disease. Nat Rev Immunol 2008;8(9):726–36.

[6] Pittenger MF, Mackay AM, Beck SC, Jaiswal RK, Douglas R, Mosca JD, et al. Multilineage potential of adult human mesenchymal stem cells. Science 1999;284(5411):143–7.

[7] Toma C, Pittenger MF, Cahill KS, Byrne BJ, Kessler PD. Human mesenchymal stem cells differentiate to a cardiomyocyte phenotype in the adult murine heart. Circulation 2002;105(1):93–8.

[8] Tomita S, Li RK, Weisel RD, Mickle DA, Kim EJ, Sakai T, et al. Autologous transplantation of bone marrow cells improves damaged heart function. Circulation 1999;100(19 Suppl):II247–56.

[9] Kopen GC, Prockop DJ, Phinney DG. Marrow stromal cells migrate throughout forebrain and cerebellum, and they differentiate into astrocytes after injection into neonatal mouse brains. Proc Natl Acad Sci USA 1999;96(19):10711–16.

[10] Amado LC, Saliaris AP, Schuleri KH, St John M, Xie JS, Cattaneo S, et al. Cardiac repair with intramyocardial injection of allogeneic mesenchymal stem cells after myocardial infarction. Proc Natl Acad Sci USA 2005;102(32):11474–9.

[11] Valina C, Pinkernell K, Song YH, Bai X, Sadat S, Campeau RJ, et al. Intracoronary administration of autologous adipose tissue-derived stem cells improves left ventricular function, perfusion, and remodelling after acute myocardial infarction. Eur Heart J 2007;28(21):2667–77.

[12] Karp JM, Leng Teo GS. Mesenchymal stem cell homing: the devil is in the details. Cell Stem Cell 2009;4(3):206–16.

[13] Freyman T, Polin G, Osman H, Crary J, Lu M, Cheng L, et al. A quantitative, randomized study evaluating three methods of mesenchymal stem cell delivery following myocardial infarction. Eur Heart J 2006;27(9):1114–22.

[14] Gnecchi M, He H, Noiseux N, Liang OD, Zhang L, Morello F, et al. Evidence supporting paracrine hypothesis for Akt-modified mesenchymal stem cell-mediated cardiac protection and functional improvement. FASEB J 2006;20(6):661–9.

[15] Haynesworth SE, Baber MA, Caplan AI. Cytokine expression by human marrow-derived mesenchymal progenitor cells in vitro: effects of dexamethasone and IL-1 alpha. J Cell Physiol 1996;166(3):585–92.

[16] Togel F, Hu Z, Weiss K, Isaac J, Lange C, Westenfelder C. Administered mesenchymal stem cells protect against ischemic acute renal failure through differentiation-independent mechanisms. Am J Physiol Renal Physiol 2005;289(1):F31–42.

[17] Kinnaird T, Stabile E, Burnett MS, Epstein SE. Bone-marrow-derived cells for enhancing collateral development: mechanisms, animal data, and initial clinical experiences. Circ Res 2004;95(4):354–63.

[18] Nakagami H, Maeda K, Morishita R, Iguchi S, Nishikawa T, Takami Y, et al. Novel autologous cell therapy in ischemic limb disease through growth factor secretion by cultured adipose tissue-derived stromal cells. Arterioscler Thromb Vasc Biol 2005;25(12): 2542–7.

[19] Van Overstraeten-Schlogel N, Beguin Y, Gothot A. Role of stromal-derived factor-1 in the hematopoietic-supporting activity of human mesenchymal stem cells. Eur J Haematol 2006;76(6):488–93.

[20] Ortiz LA, Dutreil M, Fattman C, Pandey AC, Torres G, Go K, et al. Interleukin 1 receptor antagonist mediates the antiinflammatory and antifibrotic effect of mesenchymal stem cells during lung injury. Proc Natl Acad Sci USA 2007;104(26):11002–7.

[21] Miyahara Y, Nagaya N, Kataoka M, Yanagawa B, Tanaka K, Hao H, et al. Monolayered mesenchymal stem cells repair scarred myocardium after myocardial infarction. Nat Med 2006;12(4):459–65.

[22] Shake JG, Gruber PJ, Baumgartner WA, Senechal G, Meyers J, Redmond JM, et al. Mesenchymal stem cell implantation in a swine myocardial infarct model: engraftment and functional effects. Ann Thorac Surg 2002;73(6):1919–25. discussion 26.

[23] Kinnaird T, Stabile E, Burnett MS, Lee CW, Barr S, Fuchs S, et al. Marrow-derived stromal cells express genes encoding a broad spectrum of arteriogenic cytokines and promote in vitro and in vivo arteriogenesis through paracrine mechanisms. Circ Res 2004;94(5):678–85.

[24] Thery C, Ostrowski M, Segura E. Membrane vesicles as conveyors of immune responses. Nat Rev Immunol 2009;9(8):581–93.

[25] Chargaff E, West R. The biological significance of the thromboplastic protein of blood. J Biol Chem 1946;166(1):189–97.

[26] Wolf P. The nature and significance of platelet products in human plasma. Br J Haematol 1967;13(3):269–88.

[27] Pan BT, Teng K, Wu C, Adam M, Johnstone RM. Electron microscopic evidence for externalization of the transferrin receptor in vesicular form in sheep reticulocytes. J Cell Biol 1985;101(3):942–8.

[28] Harding C, Heuser J, Stahl P. Endocytosis and intracellular processing of transferrin and colloidal gold-transferrin in rat reticulocytes: demonstration of a pathway for receptor shedding. Eur J Cell Biol 1984;35(2):256–63.

[29] Johnstone RM, Adam M, Hammond JR, Orr L, Turbide C. Vesicle formation during reticulocyte maturation. Association of plasma membrane activities with released vesicles (exosomes). J Biol Chem 1987;262(19):9412–20.

[30] Clayton A, Turkes A, Navabi H, Mason MD, Tabi Z. Induction of heat shock proteins in B-cell exosomes. J Cell Sci 2005;118(Pt 16):3631–8.

[31] Taylor DD, Gercel-Taylor C. Tumour-derived exosomes and their role in cancer-associated T-cell signalling defects. Br J Cancer 2005;92(2):305–11.

[32] Skokos D, Goubran-Botros H, Roa M, Mecheri S. Immunoregulatory properties of mast cell-derived exosomes. Mol Immunol 2002;38(16–18):1359–62.

[33] van Niel G, Raposo G, Candalh C, Boussac M, Hershberg R, Cerf-Bensussan N, et al. Intestinal epithelial cells secrete exosome-like vesicles. Gastroenterology 2001;121(2):337–49.

[34] Faure J, Lachenal G, Court M, Hirrlinger J, Chatellard-Causse C, Blot B, et al. Exosomes are released by cultured cortical neurones. Mol Cell Neurosci 2006;31(4):642–8.

[35] Lai RC, Arslan F, Lee MM, Sze NS, Choo A, Chen TS, et al. Exosome secreted by MSC reduces myocardial ischemia/reperfusion injury. Stem Cell Res 2010;4(3):214–22.

[36] Lee C, Mitsialis SA, Aslam M, Vitali SH, Vergadi E, Konstantinou G, et al. Exosomes mediate the cytoprotective action of mesenchymal stromal cells on hypoxia-induced pulmonary hypertension. Circulation 2012;126(22):2601–11.

[37] Lakkaraju A, Rodriguez-Boulan E. Itinerant exosomes: emerging roles in cell and tissue polarity. Trends Cell Biol 2008;18(5):199–209.

[38] Pisitkun T, Shen RF, Knepper MA, Identification. Identification and proteomic profiling of exosomes in human urine. Proc Natl Acad Sci USA 2004;101(36):13368–73.

[39] Mathivanan S, Ji H, Simpson RJ. Exosomes: extracellular organelles important in intercellular communication. J Proteomics 2010;73(10):1907–20.

[40] Pant S, Hilton H, Burczynski ME. The multifaceted exosome: biogenesis, role in normal and aberrant cellular function, and frontiers for pharmacological and biomarker opportunities. Biochem Pharmacol 2012;83(11):1484–94.

[41] Conde-Vancells J, Rodriguez-Suarez E, Embade N, Gil D, Matthiesen R, Valle M, et al. Characterization and comprehensive proteome profiling of exosomes secreted by hepatocytes. J Proteome Res 2008;7(12):5157–66.

[42] Thery C, Amigorena S, Raposo G, Clayton A. Isolation and characterization of exosomes from cell culture supernatants and biological fluids. Curr Protoc Cell Biol 2006;. Chapter 3: Unit 3.22.

[43] Pfeffer SR. Two Rabs for exosome release. Nat Cell Biol 2010;12(1):3–4.

[44] Tian T, Wang Y, Wang H, Zhu Z, Xiao Z. Visualizing of the cellular uptake and intracellular trafficking of exosomes by live-cell microscopy. J Cell Biochem 2010;111(2):488–96.

[45] Dragovic RA, Gardiner C, Brooks AS, Tannetta DS, Ferguson DJ, Hole P, et al. Sizing and phenotyping of cellular vesicles using nanoparticle tracking analysis. Nanomedicine 2011;7(6):780–8.

[46] Vlassov AV, Magdaleno S, Setterquist R, Conrad R. Exosomes: current knowledge of their composition, biological functions, and diagnostic and therapeutic potentials. Biochim Biophys Acta 2012;1820(7):940–8.

[47] Simpson RJ, Jensen SS, Lim JW. Proteomic profiling of exosomes: current perspectives. Proteomics 2008;8(19):4083–99.

[48] Kim HS, Choi DY, Yun SJ, Choi SM, Kang JW, Jung JW, et al. Proteomic analysis of microvesicles derived from human mesenchymal stem cells. J Proteome Res 2012;11(2):839–49.

[49] Choi DS, Kim DK, Kim YK, Gho YS. Proteomics, transcriptomics and lipidomics of exosomes and ectosomes. Proteomics 2013;13(10–11):1554–71.

[50] Subra C, Grand D, Laulagnier K, Stella A, Lambeau G, Paillasse M, et al. Exosomes account for vesicle-mediated transcellular transport of activatable phospholipases and prostaglandins. J Lipid Res 2010;51(8):2105–20.

[51] Valadi H, Ekstrom K, Bossios A, Sjostrand M, Lee JJ, Lotvall JO. Exosome-mediated transfer of mRNAs and microRNAs is a novel mechanism of genetic exchange between cells. Nat Cell Biol 2007;9(6):654–9.

[52] Gibbings DJ, Ciaudo C, Erhardt M, Voinnet O. Multivesicular bodies associate with components of miRNA effector complexes and modulate miRNA activity. Nat Cell Biol 2009;11(9):1143–9.

[53] Blanchard N, Lankar D, Faure F, Regnault A, Dumont C, Raposo G, et al. TCR activation of human T cells induces the production of exosomes bearing the TCR/CD3/zeta complex. J Immunol 2002;168(7):3235–41.

[54] Wolfers J, Lozier A, Raposo G, Regnault A, Thery C, Masurier C, et al. Tumor-derived exosomes are a source of shared tumor rejection antigens for CTL cross-priming. Nat Med 2001;7(3):297–303.

[55] Rana S, Zoller M. Exosome target cell selection and the importance of exosomal tetraspanins: a hypothesis. Biochem Soc Trans 2011;39(2):559–62.

[56] Lancaster GI, Febbraio MA. Exosome-dependent trafficking of HSP70: a novel secretory pathway for cellular stress proteins. J Biol Chem 2005;280(24):23349–55.

[57] Gastpar R, Gehrmann M, Bausero MA, Asea A, Gross C, Schroeder JA, et al. Heat shock protein 70 surface-positive tumor exosomes stimulate migratory and cytolytic activity of natural killer cells. Cancer Res 2005;65(12):5238–47.

[58] Sreekumar PG, Kannan R, Kitamura M, Spee C, Barron E, Ryan SJ, et al. alphaB crystallin is apically secreted within exosomes by polarized human retinal pigment epithelium and provides neuroprotection to adjacent cells. PLoS One 2010;5(10):e12578.

[59] Lai RC, Chen TS, Lim SK. Mesenchymal stem cell exosome: a novel stem cell-based therapy for cardiovascular disease. Regen Med 2011;6(4):481–92.

[60] Lai RC, Arslan F, Tan SS, Tan B, Choo A, Lee MM, et al. Derivation and characterization of human fetal MSCs: an alternative cell source for large-scale production of cardioprotective microparticles. J Mol Cell Cardiol 2010;48(6):1215–24.

[61] Shiojima I, Sato K, Izumiya Y, Schiekofer S, Ito M, Liao R, et al. Disruption of coordinated cardiac hypertrophy and angiogenesis contributes to the transition to heart failure. J Clin Invest 2005;115(8):2108–18.

[62] Mineo M, Garfield SH, Taverna S, Flugy A, De Leo G, Alessandro R, et al. Exosomes released by K562 chronic myeloid leukemia cells promote angiogenesis in a Src-dependent fashion. Angiogenesis 2012;15(1):33–45.

[63] Martinez MC, Andriantsitohaina R. Microparticles in angiogenesis: therapeutic potential. Circ Res 2011;109(1):110–19.

[64] Sahoo S, Klychko E, Thorne T, Misener S, Schultz KM, Millay M, et al. Exosomes from human CD34(+) stem cells mediate their proangiogenic paracrine activity. Circ Res 2011;109(7):724–8.

[65] Salomon C, Ryan J, Sobrevia L, Kobayashi M, Ashman K, Mitchell M, et al. Exosomal signaling during hypoxia mediates microvascular endothelial cell migration and vasculogenesis. PLoS One 2013;8(7):e68451.

[66] Wang JA, Chen TL, Jiang J, Shi H, Gui C, Luo RH, et al. Hypoxic preconditioning attenuates hypoxia/reoxygenation-induced apoptosis in mesenchymal stem cells. Acta Pharmacol Sin 2008;29(1):74–82.

[67] Wang JA, He A, Hu X, Jiang Y, Sun Y, Jiang J, et al. Anoxic preconditioning: a way to enhance the cardioprotection of mesenchymal stem cells. Int J Cardiol 2009;133(3):410–12.

[68] Hu X, Yu SP, Fraser JL, Lu Z, Ogle ME, Wang JA, et al. Transplantation of hypoxia-preconditioned mesenchymal stem cells improves infarcted heart function via enhanced survival of implanted cells and angiogenesis. J Thorac Cardiovasc Surg 2008;135(4):799–808.

[69] Haider KH, Kim HW, Ashraf M. Hypoxia-inducible factor-1alpha in stem cell preconditioning: mechanistic role of hypoxia-related micro-RNAs. J Thorac Cardiovasc Surg 2009;138(1):257.

[70] Rosova I, Dao M, Capoccia B, Link D, Nolta JA. Hypoxic preconditioning results in increased motility and improved therapeutic potential of human mesenchymal stem cells. Stem Cells 2008;26(8):2173–82.

[71] Wisel S, Khan M, Kuppusamy ML, Mohan IK, Chacko SM, Rivera BK, et al. Pharmacological preconditioning of mesenchymal stem cells with trimetazidine (1-[2,3,4-trimethoxybenzyl]piperazine) protects hypoxic cells against oxidative stress and enhances recovery of myocardial function in infarcted heart through Bcl-2 expression. J Pharmacol Exp Ther 2009;329(2):543–50.

[72] Herrmann JL, Wang Y, Abarbanell AM, Weil BR, Tan J, Meldrum DR. Preconditioning mesenchymal stem cells with transforming growth factor-alpha improves mesenchymal stem cell-mediated cardioprotection. Shock 2010;33(1):24–30.

[73] Gnecchi M, He H, Liang OD, Melo LG, Morello F, Mu H, et al. Paracrine action accounts for marked protection of ischemic heart by Akt-modified mesenchymal stem cells. Nat Med 2005;11(4):367–8.

[74] Mirotsou M, Zhang Z, Deb A, Zhang L, Gnecchi M, Noiseux N, et al. Secreted frizzled related protein 2 (Sfrp2) is the key Akt-mesenchymal stem cell-released paracrine factor mediating myocardial survival and repair. Proc Natl Acad Sci USA 2007;104(5): 1643–8.

[75] He W, Zhang L, Ni A, Zhang Z, Mirotsou M, Mao L, et al. Exogenously administered secreted frizzled related protein 2 (Sfrp2) reduces fibrosis and improves cardiac function in a rat model of myocardial infarction. Proc Natl Acad Sci USA 2010;107(49): 21110–15.

[76] Sheldon H, Heikamp E, Turley H, Dragovic R, Thomas P, Oon CE, et al. New mechanism for Notch signaling to endothelium at a distance by Delta-like 4 incorporation into exosomes. Blood 2010;116(13):2385–94.

[77] Taraboletti G, D'Ascenzo S, Borsotti P, Giavazzi R, Pavan A, Dolo V. Shedding of the matrix metalloproteinases MMP-2, MMP-9, and MT1-MMP as membrane vesicle-associated components by endothelial cells. Am J Pathol 2002;160(2):673–80.

[78] Soleti R, Benameur T, Porro C, Panaro MA, Andriantsitohaina R, Martinez MC. Microparticles harboring sonic hedgehog promote angiogenesis through the upregulation of adhesion proteins and proangiogenic factors. Carcinogenesis 2009;30(4):580–8.

[79] Benameur T, Soleti R, Porro C, Andriantsitohaina R, Martinez MC. Microparticles carrying sonic hedgehog favor neovascularization through the activation of nitric oxide pathway in mice. PLoS One 2010;5(9):e12688.

[80] Agouni A, Mostefai HA, Porro C, Carusio N, Favre J, Richard V, et al. Sonic hedgehog carried by microparticles corrects endothelial injury through nitric oxide release. FASEB J 2007;21(11):2735–41.

[81] Yang C, Mwaikambo BR, Zhu T, Gagnon C, Lafleur J, Seshadri S, et al. Lymphocytic microparticles inhibit angiogenesis by stimulating oxidative stress and negatively regulating VEGF-induced pathways. Am J Physiol Regul Integr Comp Physiol 2008;294(2):R467–76.

[82] Mackie AR, Klyachko E, Thorne T, Schultz KM, Millay M, Ito A, et al. Sonic hedgehog-modified human CD34+ cells preserve cardiac function after acute myocardial infarction. Circ Res 2012;111(3):312–21.

[83] Gallo A, Tandon M, Alevizos I, Illei GG. The majority of microRNAs detectable in serum and saliva is concentrated in exosomes. PLoS One 2012;7(3):e30679.

[84] Hergenreider E, Heydt S, Treguer K, Boettger T, Horrevoets AJ, Zeiher AM, et al. Atheroprotective communication between endothelial cells and smooth muscle cells through miRNAs. Nat Cell Biol 2012;14(3):249–56.

[85] Montecalvo A, Larregina AT, Shufesky WJ, Stolz DB, Sullivan ML, Karlsson JM, et al. Mechanism of transfer of functional microRNAs between mouse dendritic cells via exosomes. Blood 2012;119(3):756–66.

 [86] Eldh M, Ekstrom K, Valadi H, Sjostrand M, Olsson B, Jernas M, et al. Exosomes communicate protective messages during oxidative stress; possible role of exosomal shuttle RNA. PLoS One 2010;5(12):e15353.

 [87] Kim HW, Haider HK, Jiang S, Ashraf M. Ischemic preconditioning augments survival of stem cells via miR-210 expression by targeting caspase-8-associated protein 2. J Biol Chem 2009;284(48):33161–8.

 [88] Suzuki Y, Kim HW, Ashraf M, Haider H. Diazoxide potentiates mesenchymal stem cell survival via NF-kappaB-dependent miR-146a expression by targeting Fas. Am J Physiol Heart Circ Physiol 2010;299(4):H1077–82.

 [89] Liu J, van Mil A, Vrijsen K, Zhao J, Gao L, Metz CH, et al. MicroRNA-155 prevents necrotic cell death in human cardiomyocyte progenitor cells via targeting RIP1. J Cell Mol Med 2011;15(7):1474–82.

 [90] Kuwabara Y, Ono K, Horie T, Nishi H, Nagao K, Kinoshita M, et al. Increased microRNA-1 and microRNA-133a levels in serum of patients with cardiovascular disease indicate myocardial damage. Circ Cardiovasc Genet 2011;4(4):446–54.

 [91] Drawnel FM, Wachten D, Molkentin JD, Maillet M, Aronsen JM, Swift F, et al. Mutual antagonism between IP(3)RII and miRNA-133a regulates calcium signals and cardiac hypertrophy. J Cell Biol 2012;199(5):783–98.

 [92] Vrijsen KR, Sluijter JP, Schuchardt MW, van Balkom BW, Noort WA, Chamuleau SA, et al. Cardiomyocyte progenitor cell-derived exosomes stimulate migration of endothelial cells. J Cell Mol Med 2010;14(5):1064–70.

 [93] Cantaluppi V, Gatti S, Medica D, Figliolini F, Bruno S, Deregibus MC, et al. Microvesicles derived from endothelial progenitor cells protect the kidney from ischemia-reperfusion injury by microRNA-dependent reprogramming of resident renal cells. Kidney Int 2012;82(4):412–27.

 [94] Ranghino A, Cantaluppi V, Grange C, Vitillo L, Fop F, Biancone L, et al. Endothelial progenitor cell-derived microvesicles improve neovascularization in a murine model of hindlimb ischemia. Int J Immunopathol Pharmacol 2012;25(1):75–85.

 [95] Deregibus MC, Cantaluppi V, Calogero R, Lo Iacono M, Tetta C, Biancone L, et al. Endothelial progenitor cell derived microvesicles activate an angiogenic program in endothelial cells by a horizontal transfer of mRNA. Blood 2007;110(7):2440–8.

 [96] Majno G, Joris I. Apoptosis, oncosis, and necrosis: an overview of cell death. Am J Pathol 1995;146(1):3–15.

 [97] Aupeix K, Hugel B, Martin T, Bischoff P, Lill H, Pasquali JL, et al. The significance of shed membrane particles during programmed cell death in vitro, and in vivo, in HIV-1 infection. J Clin Invest 1997;99(7):1546–54.

 [98] Fevrier B, Raposo G. Exosomes: endosomal-derived vesicles shipping extracellular messages. Curr Opin Cell Biol 2004;16(4):415–21.

 [99] Heijnen HF, Schiel AE, Fijnheer R, Geuze HJ, Sixma JJ. Activated platelets release two types of membrane vesicles: microvesicles by surface shedding and exosomes derived from exocytosis of multivesicular bodies and alpha-granules. Blood 1999;94(11): 3791–9.

[100] Yellon DM, Davidson SM. Exosomes: nanoparticles involved in cardioprotection? Circ Res 2014;114(2):325–32.

[101] Dignat-George F, Boulanger CM. The many faces of endothelial microparticles. Arterioscler Thromb Vasc Biol 2011;31(1):27–33.

[102] Viera AJ, Mooberry M, Key NS. Microparticles in cardiovascular disease pathophysiology and outcomes. J Am Soc Hypertens 2012;6(4):243–52.

[103] Mallat Z, Benamer H, Hugel B, Benessiano J, Steg PG, Freyssinet JM, et al. Elevated levels of shed membrane microparticles with procoagulant potential in the peripheral circulating blood of patients with acute coronary syndromes. Circulation 2000;101(8):841–3.

[104] Porto I, Biasucci LM, De Maria GL, Leone AM, Niccoli G, Burzotta F, et al. Intra-coronary microparticles and microvascular obstruction in patients with ST elevation myocardial infarction undergoing primary percutaneous intervention. Eur Heart J 2012;33(23):2928–38.

[105] Nozaki T, Sugiyama S, Koga H, Sugamura K, Ohba K, Matsuzawa Y, et al. Significance of a multiple biomarkers strategy including endothelial dysfunction to improve risk stratification for cardiovascular events in patients at high risk for coronary heart disease. J Am Coll Cardiol 2009;54(7):601–8.

[106] Loyer X, Vion AC, Tedgui A, Boulanger CM. Microvesicles as cell-cell messengers in cardiovascular diseases. Circ Res 2014;114(2):345–53.

[107] Alvarez-Erviti L, Seow Y, Yin H, Betts C, Lakhal S, Wood MJ. Delivery of siRNA to the mouse brain by systemic injection of targeted exosomes. Nat Biotechnol 2011;29(4):341–55.

[108] Abid Hussein MN, Nieuwland R, Hau CM, Evers LM, Meesters EW, Sturk A. Cell-derived microparticles contain caspase 3 *in vitro* and *in vivo*. J Thromb Haemost 2005;3(5):888–96.

[109] Chen TS, Arslan F, Yin Y, Tan SS, Lai RC, Choo AB, et al. Enabling a robust scalable manufacturing process for therapeutic exosomes through oncogenic immortalization of human ESC-derived MSCs. J Transl Med 2011;9:47.

CHAPTER 8

Exosome-Based Translational Nanomedicine: The Therapeutic Potential for Drug Delivery

Lei Lv, Qingtan Zeng, Shenjun Wu, Hui Xie, Jiaquan Chen,
Xiang Jiang Guo, Changning Hao, Xue Zhang, Meng Ye, Lan Zhang
Department of Vascular Surgery, Shanghai Jiao Tong University, School of Medicine, Renji Hospital,
Shanghai, China

Contents

1 OVERVIEW

The advent of nanotechnology has reignited interest in the field of pharmaceutical science for the development of the application to medicine. A major purpose of the utilization of nanoscale technologies for drug delivery is to improve clinical therapeutic effect by increasing its potency at specific sites while simultaneously reducing overall toxicity [1]. This kind of nanoscale drug delivery system by the US Food and Drug Administration in cancer treatments is currently available, for example, Doxil® (doxorubicin encapsulated in liposomes) [2] and Abraxane® (paclitaxel attached to nanoparticles) [3]. There is increasing evidence that nanometric-scale systems, for example, virus-like particles, inert nanobeads, and immunostimulating complexes, are being used in cancer vaccine development due to their effectiveness at eliciting powerful and long-lasting cellular and humoral immune responses [4].

Synthetic nanoparticles could be linked to biological molecules and could potentially be utilized as an effective vaccine delivery system [5]. Moreover, there is a growing interest in the utilization of biological molecule delivery systems that serve as conduits to deliver drugs into the body, and one of the successful candidates is exosomes [6]. Exosomes are naturally occurring biological nanoparticles 30–200 nm in size with unique structure, surface biochemistry, and mechanical characteristics [7–10]. Nanovesicles, known as exosomes, have received much attention because of their apparent ability to incorporate functional drugs and to transfer drugs to recipient cells [11]. Exosomes have the potential to serve as the "ideal drug delivery vehicle" owing to their desirable properties such as low immunogenicity, ability to effectively transport a range of biomolecules, and interact with a host of target cells, but most importantly, their amenability to manipulation for personalized medicine.

2 SAFETY AND TOXICITY OF EXOSOMES

Exosomes are endogenous nanoparticles that retain critical nanoparticle characteristics, such as the enhanced permeability and retention (EPR) effect and passive targeting. They contain various molecular constituents of their cell of origin, including proteins, RNA/microRNAs (miRNAs), and lipids [12,13]. However, unlike nanoparticles synthesized *in vitro*, exosomes are membrane vesicles of endocytic origin produced by numerous cells, so they are believed to be not cytotoxic and can transfer information from one cell to another via membrane vesicle trafficking based on their composition and the substance in/on the exosome. Therefore, the exosomal drug delivery system emerged as a new important method to transport small molecule pharmaceuticals to specifically target tissues or cells, typically in a noncytotoxic manner. There is accumulating evidence that exosome-mediated drug delivery is feasible, safe, and may be an effective procedure for prevention or treatment of different diseases in the future.

3 PASSIVE TARGETING OF EXOSOMES

Nanoparticles can escape from the vasculature through the leaky endothelial tissue by the so-called EPR effect [14,15], which constitutes an important mechanism for size-dependent "passive targeting." To get into the tissue, nanoparticles must circulate in the blood stream. Macrophages and neutrophils, the major phagocytic innate immune cell types, engulf large-sized particles from 250 nm to 1000 nm [16]. Nanoparticles must be small enough to avoid being eliminated by phagocytosis. At the same time, nanoparticles must reach a sufficient size to prevent flow through endothelial gaps. Depending on the experimental system used, particles between 20 nm and 100 nm in diameter are very efficaciously delivered [17–19] to target tissues. Exosomes are stable in the circulation and resistant to storage handling. Circulating exosomal miRNAs as sensitive and specific biomarkers for diagnosis and monitoring of human diseases due to this feature [20–22]. Retention of exosomes may also be affected by the local microenvironment regardless of exosome size. For example, abnormal tumor angiogenesis results in defective endothelial cell lining of vessels, which leads to larger endothelial gaps and allows larger nanoparticles or exosomes to move into tissues. In that case, transmission of much larger amounts of drugs and longer periods of time in the affected area can be reached by the larger exosomes.

4 THE BIOAVAILABILITY OF EXOSOMES

Nanoparticle-based drug delivery systems have considerable potential for increasing systemic bioavailability of drugs with low solubility. For instance, insulin–loaded nanoparticles modified with a CSKSSDYQC targeting peptide have a higher bioavailability than unmodified insulin [23]. The oral bioavailability of the low molecular weight heparin is largely increased using chitosan-shelled nanoparticles and has great therapeutic potential for the clinical treatment of cardiovascular disease [24]. Moreover, when exosomes are used to deliver therapeutic concentrations of curcumin *in vivo*, the systemic bioavailability was dramatically increased as evidenced by higher serum concentrations. Moreover, this study provides evidence that, unlike liposome as a carrier, exosomes, such as dendritic and tumor cell-derived exosomes, exhibit strong tendencies to regulate immune responses and tumor progression. Selecting different types of exosomes in combination with therapeutic drugs achieves targetable and additive/syngeneic therapeutic effects. The specificity of using exosomes as a drug delivery system creates opportunities for treatments of many inflammation-related diseases without significant side effects. However, these results have proposed exosomes as an ideal carrier for drugs and may not be limited to target only inflammation-related diseases [25]. All in all, exosomes increase the bioavailability of therapeutic agents and can be used as a delivery vehicle that is far superior to synthetic nanoparticles by enhancing its therapeutic effect through

(i) increasing the solubility, stability, and bioavailability of drugs and (ii) increasing delivery of drugs to recipient cells.

5 ACTIVE TARGETING OF EXOSOMES

Active targeting involves the use of peripherally conjugated targeting moieties for enhanced delivery of nanoparticles. The ideal scheme for drug delivery is selective targeting of tissues to be treated and accumulation of the therapeutic agent at a bioactive concentration with limited or no collateral tissue or systemic side effects. Recent studies indicate that such a strategy may be feasible. To specifically deliver small interfering RNA (siRNA) to the brain, neuron acetylcholine receptor binding rabies viral glycoprotein (RVG) peptides have been fused to the exosomal protein LAMP-2b. Engineered exosomes carrying β-site APP-cleaving enzyme siRNA lead to the knockdown of mRNA levels in cortical tissue and decreased beta-amyloid levels in RVG-exosome-treated mice [26]. Another experiment confirms that targeting antigen localization to exosomes is a viable approach for improving the therapeutic potential of MVA-BN-PRO (BN ImmunoTherapeutics) in humans [27]. Past work has shown that targeted delivery of therapeutic agents can be achieved by chemically coupling specific ligands to nanoparticles so that agent delivery is based on ligand–receptor interaction with tissue. But the targeted receptors are not solely expressed on the abnormal cells. Therefore, adverse effects to normal tissues or cells are almost unavoidable. However, exosomes are cell-derived vesicles that have a highly selective homing specificity, which eliminates the need to manipulate exosomes to achieve tissue targeting specificity. For example, exosomes released from T cells are preferentially taken up by CD11bGr1 cells from inflammatory tissues [28]. Dysregulation of CD11bGr1 myeloid cells is associated with development of many inflammatory-related diseases, and therefore we can use T cell-derived exosomes as drug nanocarriers to target CD11bGr1 myeloid cells.

6 EXOSOMES CAN CROSS THE BLOOD–BRAIN BARRIER

The blood–brain barrier (BBB) protects the brain from harmful chemicals but also makes it difficult for drugs to reach the target cells. The development of vehicles for brain drug delivery is largely obscured by hurdles of the BBB. The BBB is a cellular and metabolic barrier located at the capillaries in the brain that alters permeability, restricting the passage of harmful particles, such as bacteria and large hydrophilic molecules, into the brain cerebrospinal fluids. There are two possible pathways for entry of agents into the brain from the nasal cavity: (i) systemic pathway by which drugs are absorbed through the nasal cavity into the systemic circulation via the rich vasculature of the respiratory system circulation and subsequently reach the brain by crossing the BBB and (ii) the olfactory region, which is situated in the top of the nasal cavity and provides the olfactory pathway by which nanoparticles are directly delivered to brain tissue, bypassing the BBB [29]. Delivering drugs across the BBB is one of the most promising applications of nanotechnology in clinical neuroscience. A recent study has shown that intranasal administration of exosomes leads to highly efficient delivery of curcumin and antisignal transducer and activator of transcription 3 inhibitor into the brain. [30] The fact that exosomes are typically not cytotoxic in the brain and are more efficient in the delivery of agents to the brain makes them the delivery vehicle of choice when compared to other nanoparticles. A recent study shows that exosomes, injected into the blood, act as "drugs vehicles" and can cross the normally impermeable BBB to the brain where they are needed [26]. The ability of exosomes to be transported into the brain, not to be cytotoxic, and to efficiently deliver agents has raised hopes of more effective treatments for diseases like Alzheimer's and Parkinson's.

7 EXOSOMES CAN STIMULATE THE IMMUNE SYSTEM

Depending on the cell type from which they are released, exosomes have different intrinsic biological effects [31,32]. Therefore, different exosomes might be chosen depending on the therapeutic effect being sought. There are a great number of studies suggesting that exosomes play an important role in modulating immune responses. A number of reports demonstrate that exosomes frequently target immune cells, which include B cells, macrophages, and dendritic cells (DCs) [33–41], all of which are effective antigen-presenting cells (APCs). By virtue of how exosomes are generated, composed of unique sets of proteins, they are vastly different from polymeric or liposomal nanoparticles. Exosomes display both major histocompatibility complex (MHC)-I and MHC-II molecules on their surface, assembled with antigenic peptides, suggesting that exosomes pulsed with tumor antigen could stimulate antitumor immune responses. Consequently, immune stimulation is a principal function of exosomes depending on the cellular source and target of the vesicles. Research [41] shows that exosomes released from DCs represent a novel source of tumor-rejection antigens for

T-cell cross-priming, relevant for immunointerventions. DC-derived exosomes (Dex) pulsed with a Human leukocyte antigen -A2/Mart1 peptide complex or synthetic adjuvant CpG oligonucleotides can activate cytotoxic T-lymphocytes (CTLs) and promote the immune system to recognize and kill cancer cells [42,43]. Envelope glycoprotein GP120 (or gp120) is a glycoprotein expressed at the surface of the human immunodeficiency virus (HIV) envelope. Dex engineered with ovalbumin and pcDNAgp120 can induce ovalbumin and pcDNAgp120specific CTL responses, and can be used to treat HIV patients [44]. Coupling the extracellular domain of the tumor-associated antigen carcinoembryonic antigen and human epidermal growth factor receptor 2 to the C1C2 domain of lactadherin conjugated to exosomes enhances the vaccine potency of exosomes. These results suggest that exosomal targeting could be adapted for the development of more potent vaccines in some viral and parasitic diseases where the classical vaccine approach has demonstrated limitations [45].

DC exosomes carrying specific tumor antigens have also been demonstrated to cross present to different tumor hosts and induce immunogenicity [41]. This strategy raises the possibility that pulsed or natural exosomes can be used to treat many different cancers. The optimized loading conditions and the ability to transfer both MHC-I and MHC-II antigens to APC have led to the development of exosomes as an "acellular" immunotherapy approach being tested in clinical trials [46]. Dex bearing MHCs promotes T-cell-dependent antitumor effects in mice. The clinical use of Dex has demonstrated the feasibility and safety of peptide-pulsed Dex in vaccinating cancer patients [47]. The first clinical trials using immature Dex as a cell-free vaccine against advanced melanoma and lung cancer-bearing patients failed to detect vaccine-specific T-cell responses but observed potent Dex-related natural killer (NK) cell activation [48], suggesting that Dex can also stimulate the innate immune response. After injecting patients with Dex loaded with antigenic peptides from human melanoma during Phase I clinical trials, an increased number of NK cells was observed and natural-killer group 2, member D (NKG2D) expression was restored in the NK cells and CD T cells in some patients [49].

Exosomes can inhibit an immune response. They can not only induce active immune response but can also cause immune inhibition. Exosomes released from tumor cells provoke a marked reduction in NK cytotoxicity by displaying NKG2D ligands leading to downregulation of the NKG2D receptor [50] or downregulation of the Janus kinases 3 protein [37]. Exosomes released from the thymus cause an expansion of regulatory T cells

(Tregs) [51]. Exosomes from tumor-bearing patients have been demonstrated to lead to T-cell apoptosis *in vitro* via FasL [52] and galectin-9 [53], promote differentiation of T cells into Tregs, and inhibit CD8[+] T-cell growth [33]. Immature DCs secrete tolerogenic exosomes whereas maturated DCs release immunogenic exosomes. By manipulating the maturation of DCs, exosomes can be used as a potential therapeutic agent to treat autoimmune diseases. TGF-β1 gene-modified DCs produce exosomes capable of inducing immune tolerance [54], and exosomes from DCs overexpressing indoleamine 2,3-dioxygenase inhibit arthritis in a murine arthritis model [55]. Interest in exosomes was rekindled when B-cell and dendritic cell-derived exosomes were shown to mediate MHC-dependent immune responses. Recent data suggest that the immunogenicity of DC exosomes can be enhanced by prestimulation of the DCs with interferon-γ, as a consequence, and promotes direct activation of CD8[+] T cells and NK cells *in vitro* and the priming of CD8[+] T cells *in vivo* [48].

8 THE HYDROPHOBICITY FEATURE OF EXOSOMES

Exosomes are membrane-bound nanovesicles so they are lipid enriched and hydrophobic. Exploiting this feature with specific agents may be a potentially economical strategy for ligating hydrophobicity-sensing therapeutic agents to exosomes. Using the chemical properties of exosomes to our advantage may prevent costly and timely conjugation schemes for recombinant protein ligands or synthetic ligands.

9 STORAGE OF EXOSOMES

Exosome-mediated immunotherapy could be referred to as a type of cellular therapy because exosomes are biological products. However, they are more convenient to handle than a cell because they are stable vesicles that keep their biological activities for more than 2 years at −80°C. After storage, there is no need to expand them, and they can be used directly, either alone or in combination with other pharmacological agents. In addition, they maintain the antigen presentation within a lymph node twice as long as an antigen-presenting cell, which indicates that they potentiate the immune response [56]. The only limitations are that they must be autologous and that the yield of the tumor antigen-loaded exosomes prepared from DCs feature large variations between individuals.

10 THERAPEUTIC APPLICATION OF EXOSOMES

Increasing knowledge about the advantages of exosomes as a drug delivery system provides exciting prospects for exosome development in therapy. More recently, it has been demonstrated that resting and activated NK cells, freshly isolated from the blood of healthy donors, release exosomes expressing both typical protein markers of NK cells and killer proteins. These killer nanovesicles display cytotoxic activity against several tumor cell lines and activated, not resting, immune cells. Consistently, exosomes purified from the plasma of healthy donors express NK cell markers and exert cytotoxic activity against different human tumor target cells and activated immune cells as well [57]. These data demonstrate that NK cell-derived exosomes play a critical role in immune surveillance and homeostasis. Moreover, this study also supports the use of exosomes as an almost perfect example of biomimetic nanovesicles possibly useful in future therapeutic approaches against various diseases. More recently, adenoviral vectors associated with exosomes have shown higher transduction efficiency as compared with conventionally purified adeno-associated virus vectors [58], indicating that exosomes represent a unique entity, which may improve gene delivery. The study also shows that vector-exosomes bound to magnetic beads can be attracted to a magnetized area in cultured cells. There are many other studies reporting the potential role of exosomes as an ideal vector for therapeutic use focused on miRNAs. Exosomes from human and mouse mast cells were found to be natural carriers of nucleic acids, including over 1300 mRNAs and 121 noncoding miRNAs. Furthermore, it was established that the RNA content of exosomes did not strictly reflect that of the parent cell, suggesting the existence of actively selective pathways involved in the loading of exosomes with RNA within the parent cell [59]. Exosome-mediated transfer of proteins, mRNAs, miRNAs, and signaling molecules offers the promise that they may be used for therapeutic purposes [60]. In fact, mesenchymal stem cell-derived exosomes display cardioprotective effects by reducing myocardial ischemia/reperfusion injury [61]. Comparable results have been provided with cardiomyocyte progenitor cell-derived exosomes [62] and in experimental stem cell therapy of acute tubular injury [63].

However, a groundbreaking experiment showing that exosomes may be used for delivery was performed by Alvarez-Erviti et al. [64] who explored the ability of these nanovesicles to transfer nucleic acids to cells in a very specific manner. In this study, the result was achieved using elf exosomes loaded with chemically modified siRNAs and displaying specific targeting molecules. Thus, the target cells (DCs) were engineered to express

LAMP-2b fused to the central nervous system–specific RVG peptide that specifically binds to the acetylcholine receptor. Exosomes were loaded with exogenous siRNA by electroporation and, after *in vitro* experiments, were injected into mice together with a series of control exosome preparations. The injection of RVG exosomes resulted in a significant knockdown of GAPDH mRNA in the brain regions expressing the target of the RVG ligand-nicotinic acetylcholine receptors. These results not only showed the therapeutic potential of RVG exosome technology for new therapeutic approaches against neurodegenerative diseases but also strongly supported the use of implemented and innovative exosome technologies for targeting therapies for a number of diseases.

Collectively, exosomes contain protein and lipid determinants that allow them to interact with target cells, thus avoiding their dilution in the intercellular space. This vectorized signaling appears more efficient than soluble agonists that can be diluted in the extracellular medium. Moreover, it appears conceivable that molecules of various origins may be more stable and functional when expressed on a membrane than in a free soluble state. Overall, exosomes appear as a multisignaling device that can signal target cells at the cell periphery or bring information to the cytosol and translation machinery and possibly to the nucleus as well.

11 CHALLENGES OF EXOSOMES AS DRUG DELIVERY CARRIERS

Many studies suggest that exosomes have a great potential for use as drug carriers and for treatment of cancer and autoimmune diseases by selecting immunogenic or toleragenic exosomes. Moreover, we can optimize the therapeutic effect through personalizing the immunogenicity or toleragenicity of exosomes with chemotherapeutic drugs, specific miRNAs, or lipids.

Currently, some exosome technologies as a therapeutic approach are not extremely encouraging. In fact, in the clinical trials undertaken so far, the amount of immunocompetent exosomes produced *ex vivo* by DCs originating from patients with melanoma was highly variable and constituted a limiting step in immunotherapy. Decreasing the amount of tumor exosomes released by modifying the pH of multivesicular bodies (MVBs) with amiloride enhances the efficacy of chemotherapeutic agents [65]. This is also supported by experiments performed with another class of anti-acidic molecules directed against proton pumps [66]. Thus, inhibiting the release of tumor exosomes would be a therapeutic strategy to prevent tumor growth and metastasis. In this respect, proton exchanger inhibitors may well represent a pharmacological class of agents to block tumor exosome release. It is interesting to identify other drugs that are potentially useful in exosome release inhibition using a large-scale systematic screening of a wide spectrum of drugs that inhibit tumor exosome release. Nevertheless, a general treatment avoiding exosome release in cancer patients might have side effects because the release of immunocompetent exosomes, which can enhance tumor recognition, by the immune system [67] would be inhibited as well. Therefore, an optimal treatment would require differential therapeutic targets between tumor and immunocompetent cells.

An efficient cancer treatment would be considered to trigger an enhanced immunocompetent exosome biogenesis or release while inhibiting tumor exosome production. For that reason, drugs used to cure cancers should be screened in terms of their effect on the ratio between immunocompetent MHC-bearing exosomes and tumor immunosuppressive-containing molecule exosomes to bring new perspectives on the efficacy of the various treatments. Of note, the activity and production of exosomes might be improved by manipulating the biosynthesis of a fusogenic lipid [68–70]. However, the modification of the numerous exosome molecules (proteins, lipids, mRNA/miRNA) seems to be the most promising therapeutic strategy for both increasing the efficacy of existing molecules and reducing the side effects. Another way is to interfere with the exosome formation by modifying the molecular content of typical exosome compartments such as MVBs.

Despite many advantages, there are still many challenges ahead before exosomes can be used on a routine basis clinically. One issue is that the response that exosomes elicit depends on many factors, including the state of maturation of the exosome-producing cell and how the exosomes are taken up and processed by a cell. Antigens presented by exosomes in the context of MHC class I molecules are recognized by $CD8^+$ T cells, whereas those bound to MHC class II molecules are recognized by $CD4^+$ T cells. Exosomes released from DCs, classically $CD8^+$ DCs, can present antigen in the context of both classes of MHC molecules. Therefore, if the purpose is to stimulate both CD8 as well as CD8 T-cell responses, $CD8^+$ DCs would appear to be the cell of choice for production of exosomes. The cytokine environment is critical and can affect communication between exosomes and T cells and therefore must be carefully considered, especially in the context of the therapeutic agent carried by the exosomes. Approaches that manipulate or make use of the cytokine environment are important design considerations. For these reasons, an important task in exosome-mediated delivery system development is the design of specific delivery strategies by which to present antigen along with pathogen-associated molecular patterns to ligate particular pattern-recognition receptors so as to induce a desirable T-cell response to elucidate an immune response against tumor. In addition, design considerations to enhance exosome uptake by APCs are important to ensure recognition by and recruitment of the appropriate cells. With regard to recognition, some subclasses of DCs possess an endocytic receptor, DEC205 [71]; therefore, it is possible to coat exosomes with anti-DEC205 antibody to enhance DC uptake of exosomes.

Once the exosomes enter the cells, several barriers must be overcome to ensure activity: (i) entry into the endolysosomal compartment of the cell after endocytosis and (ii) entry into the cytosol and presented by the endosomal membrane. Utilizing the mechanism's viruses use to escape these components of the cell may be a way to achieve cellular compartmental delivery of agents carried by exosomes. Furthermore, the role of exosome size and shape may also have an effect on the therapeutic effect. As discussed previously, exosome size can affect biological transport and the bioavailability of the agent being carried, as is the case for the tissue interstitium and mucosal barriers. However, exosome size characteristics may also be an important determinant of exosomal immunological activity. Conjugation of exosomes with the most common adjuvants in clinical use may result in increasing the size of exosomes. Altering the size of the exosomes may be a way to affect tissue targeting and enhance immune responses. Therefore, carefully choosing the agent or ligand to be carried is another critical factor. For example, exosomes ∼100-nm size are transported across interstitial boundaries and gain access to lymph nodes [72], where large numbers of DCs reside. Careful selection of the agent carried will ensure that exosome size is not altered in such a way that would prevent delivery of the agent to lymph nodes. Exosome shape may also be another determining factor in exosome applications. Macrophages phagocytose particles as a function of particle size and shape. This phenomenon is greatly influenced by actin mechanics at the points of particle contact with a macrophage [73]. Along with particle size, morphology also influences the uptake of particles from the extracellular environment by macrophages. Modifying the shape of exosomal cargoes opens up the possibility of potentially altering exosome uptake. However, even with the wealth of information already known regarding exosome biology much work remains to be done to ensure the safe and effective use of exosomes in therapeutic situations. For example, the identity of exosomal components that are essential to ensure that it is an effective vehicle must be determined so that mass production of synthesized exosomes can occur. As our understanding of exosomal biology grows, so will the range of principles for the design of exosomes and exosomal conjugates used in immunotherapeutics.

Moreover, the design principles will be different for various contexts regarding formulations, administration routes, doses, and cost. As proteomic and lipidomic technologies are advanced, these factors and components will be identified. Furthermore, despite the growing amount of data on the changes induced by exosomes on target cells, most of these studies were conducted *in vitro* with purified exosomes. More *in vivo* studies on the potency and toxicology of exosomes would be imperative to bring us a step closer to the development of the clinical practice of exosomal nanomedicine.

ABBREVIATIONS

APCs	Antigen-presenting cells
BBB	Blood–brain barrier
CTL	Cytotoxic T lymphocytes
DCs	Dendritic cells
Dex	DC-derived exosome
EPR	Enhanced permeability and retention
HIV	Human immunodeficiency virus
MHC	Major histocompatibility complex
miRNA	MicroRNA
MVB	Multivesicular body
NK	Natural killer
NKG2D	Natural-killer group 2, member D
RVG	Rabies viral glycoprotein
siRNA	Small interfering RNA
Tregs	Regulatory T cells

REFERENCES

[1] Mukherjee B. Nanosize drug delivery system. Curr Pharm Biotechnol 2013;14(15):1221.
[2] Leonard RC1, Williams S, Tulpule A, Levine AM, Oliveros S. Improving the therapeutic index of anthracycline chemotherapy: focus on liposomal doxorubicin (Myocet). Breast 2009;18(4):218–24.
[3] Petrelli F, Borgonovo K, Barni S. Targeted delivery for breast cancer therapy: the history of nanoparticle-albumin-bound paclitaxel. Expert Opin Pharmacother 2010;11(8):1413–32.
[4] Zolnik B.S., Gonzalez-Fernandez A., Sadrieh N., Dobrovolskaia M.A. Nanoparticles and the immune system. Endocrinology. 2010;151(2):458–65; Scheerlinck J.P., Greenwood D.L. Virus-sized vaccine delivery systems. Drug Discov Today 2008; 13(19–20):882–7.
[5] Rieger J, Freichels H, Imberty A, Putaux JL, Delair T, Jérôme C, et al. Polyester nanoparticles presenting mannose residues: toward the development of new vaccine delivery systems combining biodegradability and targeting properties. Biomacromolecules 2009;10(3):651–7.

[6] Seow Y, Wood MJ. Biological gene delivery vehicles: beyond viral vectors. Mol Ther 2009;17(5):767–77.

[7] Lakkaraju A, Rodriguez-Boulan E. Itinerant exosomes: emerging roles in cell and tissue polarity. Trends Cell Biol 2008;18:199–209.

[8] Iero M, Valenti R, Huber V, Filipazzi P, Parmiani G, Fais S, et al. Tumour-released exosomes and their implications in cancer immunity. Cell Death Differ 2008;15:80–8.

[9] Taylor DD, Gercel-Taylor C. Tumour-derived exosomes and their role in cancer-associated T-cell signalling defects. Br J Cancer 2005;92:305–11.

[10] Oshima K, Aoki N, Kato T, Kitajima K, Matsuda T. Secretion of a peripheral membrane protein, MFG-E8, as a complex with membrane vesicles. Eur J Biochem 2002;269:1209–18.

[11] Pegtel DM, Cosmopoulos K, Thorley-Lawson DA, van Eijndhoven MA, Hopmans ES, Lindenberg JL, et al. Functional delivery of viral miRNAs via exosomes. Proc Natl Acad Sci USA 2010;107:6328–33.

[12] Skog J, Würdinger T, van Rijn S, Meijer DH, Gainche L, Sena-Esteves M, et al. Glioblastoma microvesicles transport RNA and proteins that promote tumour growth and provide diagnostic biomarkers. Nat Cell Biol 2008;10(12):1470–6.

[13] Azmi AS, Bao B, Sarkar FH. Exosomes in cancer development, metastasis, and drug resistance: a comprehensive review. Cancer Metastasis Rev 2013;32(3–4):623–42.

[14] Matsumura Y, Oda T, Maeda H. General mechanism of intratumor accumulation of macromolecules: advantage of macromolecular therapeutics. Gan To Kagaku Ryoho 1987;14:821–9.

[15] Matsumura Y, Maeda H. A new concept for macromolecular therapeutics in cancer chemotherapy: mechanism of tumoritropic accumulation of proteins and the antitumor agent smancs. Cancer Res 1986;46:6387–92.

[16] Geiser M, Casaulta M, Kupferschmid B, Schulz H, Semmler-Behnke M, Kreyling W. The role of macrophages in the clearance of inhaled ultrafine titanium dioxide particles. Am J Respir Cell Mol Biol 2008;38:371–6.

[17] Kuhlbusch TA, Asbach C, Fissan H, Gohler D, Stintz M. Nanoparticle exposure at nanotechnology workplaces: a review. Part Fibre Toxicol 2011;8:22.

[18] Almeida JP, Chen AL, Foster A, Drezek R. In vivo biodistribution of nanoparticles. Nanomedicine (Lond) 2011;6:815–35.

[19] Saha RN, Vasanthakumar S, Bende G, Snehalatha M. Nanoparticulate drug delivery systems for cancer chemotherapy. Mol Membr Biol 2010;27:215–31.

[20] Bryant RJ, Pawlowski T, Catto JW, Marsden G, Vessella RL, Rhees B, et al. Changes in circulating microRNA levels associated with prostate cancer. Br J Cancer 2012;106:768–74.

[21] Wittmann J, Jack HM. Serum microRNAs as powerful cancer biomarkers. Biochim Biophys Acta 2010;1806:200–7.

[22] Kosaka N, Iguchi H, Ochiya T. Circulating microRNA in body fluid: a new potential biomarker for cancer diagnosis and prognosis. Cancer Sci 2010;101:2087–92.

[23] Jin Y, Song Y, Zhu X, Zhou D, Chen C, Zhang Z, et al. Goblet cell-targeting nanoparticles for oral insulin delivery and the influence of mucus on insulin transport. Biomaterials 2012;33:1573–82.

[24] Paliwal R, Paliwal SR, Agrawal GP, Vyas SP. Chitosan nanoconstructs for improved oral delivery of low molecular weight heparin: in vitro and in vivo evaluation. Int J Pharm 2012;422:179–84.

[25] Sun D, Zhuang X, Xiang X, Liu Y, Zhang S, Liu C, et al. A novel nanoparticle drug delivery system: the anti-inflammatory activity of curcumin is enhanced when encapsulated in exosomes. Mol Ther 2010;18:1606–14.

[26] Alvarez-Erviti L, Seow Y, Yin H, Betts C, Lakhal S, Wood MJ. Delivery of siRNA to the mouse brain by systemic injection of targeted exosomes. Nat Biotechnol 2011;29:341–5.

[27] Malam Y, Loizidou M, Seifalian AM. Exosome targeting of tumor antigens expressed by cancer vaccines can improve antigen immunogenicity and therapeutic efficacy. Cancer Res 2011;71(15):5235–44.

[28] Zhuang X, Xiang X, Grizzle W, Sun D, Zhang S, Axtell RC, et al. Treatment of brain inflammatory diseases by delivering exosome encapsulated anti-inflammatory drugs from the nasal region to the brain. Mol Ther 2011;19:1769–79.

[29] Illum L. Transport of drugs from the nasal cavity to the central nervous system. Eur J Pharm Sci 2000;11:1–18.

[30] Zhang Y, Luo CL, He BC, Zhang JM, Cheng G, Wu XH. Exosomes derived from IL-12-anchored renal cancer cells increase induction of specific antitumor response *in vitro*: a novel vaccine for renal cell carcinoma. Int J Oncol 2010;36:133–40.

[31] Xiang X, Liu Y, Zhuang X, Zhang S, Michalek S, Taylor DD, et al. TLR2-mediated expansion of MDSCs is dependent on the source of tumor exosomes. Am J Pathol 2010;177:1606–10.

[32] Chalmin F, Ladoire S, Mignot G, Vincent J, Bruchard M, Remy-Martin JP, et al. Membrane-associated Hsp72 from tumor derived exosomes mediates STAT3-dependent immunosuppressive function of mouse and human myeloid-derived suppressor cells. J Clin Invest 2010;120:457–71.

[33] Szajnik M, Czystowska M, Szczepanski MJ, Mandapathil M, Whiteside TL. Tumor-derived microvesicles induce, expand and up-regulate biological activities of human regulatory T cells (Treg). PLoS One 2010;5:e11469.

[34] Xiang X, Poliakov A, Liu C, Liu Y, Deng ZB, Wang J, et al. Induction of myeloid-derived suppressor cells by tumor exosomes. Int J Cancer 2009;124:2621–33.

[35] Valenti R, Huber V, Iero M, Filipazzi P, Parmiani G, Rivoltini L. Tumor-released micro vesicles as vehicles of immuno suppression. Cancer Res 2007;67:2912–15.

[36] Zhang HG, Liu C, Su K, Yu S, Zhang L, Zhang S, et al. A membrane form of TNF-alpha presented by exosomes delays T cell activation-induced cell death. J Immunol 2006;176:7385–93.

[37] Liu C, Yu S, Zinn K, Wang J, Zhang L, Jia Y, et al. Murine mammary carcinoma exosomes promote tumor growth by suppression of NK cell function. J Immunol 2006;176:1375–85.

[38] Whiteside TL. Tumour-derived exosomes or microvesicles: another mechanism of tumour escape from the host immune system? Br J Cancer 2005;92:209–11.

[39] Altieri SL, Khan AN, Tomasi TB. Exosomes from plasmacytoma cells as a tumorvaccine. J Immunother 2004;27:282–8.

[40] Andre F, Schartz NE, Chaput N, Flament C, Raposo G, Amigorena S, et al. Tumor-derived exosomes: a new source of tumor rejection antigens. Vaccine 2002;20(Suppl 4):A28–31.

[41] Wolfers J, Lozier A, Raposo G, Regnault A, Thery C, Masurier C, et al. Tumor-derived exosomes are a source of shared tumor rejection antigens for CTL cross-priming. Nat Med 2001;7:297–303.

[42] Andre F, Chaput N, Schartz NE, Flament C, Aubert N, Bernard J, et al. Exosomes as potent cell-free peptide-based vaccine. I. Dendritic cell-derived exosomes transfer functional MHC class I/peptide complexes to dendritic cells. J Immunol 2004;172:2126–36.

[43] Hsu DH, Paz P, Villaflor G, Rivas A, Mehta-Damani A, Angevin E, et al. Exosomes as a tumor vaccine: enhancing potency through direct loading of antigenic peptides. J Immunother 2003;26:440–50.

[44] Nanjundappa RH, Wang R, Xie Y, Umeshappa CS, Chibbar R, Wei Y, et al. GP120-specific exosome-targeted T cell-based vaccine capable of stimulating DC- and CD4(+) T-independent CTL responses. Vaccine 2011;29:3538–47.

[45] Hartman ZC, Wei J, Glass OK, Guo H, Lei G, Yang XY, et al. Increasing vaccine potency through exosome antigen targeting. Vaccine 2011;29:9361–7.

[46] Leonard RC, Williams S, Tulpule A, Levine AM, Oliveros S. Exosomes as a tumor vaccine: enhancing potency through direct loading of antigenic peptides. J Immunother 2003;26(5):440–50.

[47] Viaud S, Ploix S, Lapierre V, Thery C, Commere PH, Tramalloni D, et al. Updated technology to produce highly immunogenic dendritic cell-derived exosomes of clinical grade: a critical role of interferon-gamma. J Immunother 2011;34:65–75.

[48] Viaud S, Thery C, Ploix S, Tursz T, Lapierre V, Lantz O, et al. Dendritic cell derived exosomes for cancer immunotherapy: what next? Cancer Res 2010;70:1281–5.

[49] Escudier B, Dorval T, Chaput N, Andr F, Caby MP, Novault S, et al. Vaccination of metastatic melanoma patients with autologous dendritic cell (DC) derived exosomes: results of the first phase I clinical trial. J Transl Med 2005;3:1–3.

[50] Ashiru O, Boutet P, Fernandez-Messina L, Aguera-Gonzalez S, Skepper JN, Vales-Gomez M, et al. Natural killer cell cytotoxicity is suppressed by exposure to the human NKG2D ligand MICA*008 that is shed by tumor cells in exosomes. Cancer Res 2010;70:481–9.

[51] Wang GJ, Liu Y, Qin A, Shah SV, Deng ZB, Xiang X, et al. Thymus exosomes-like particles induce regulatory T cells. J Immunol 2008;181:5242–8.

[52] Abusamra AJ, Zhong Z, Zheng X, Li M, Ichim TE, Chin JL, et al. Tumor exosomes expressing Fas ligand mediate CD8+ T-cell apoptosis. Blood Cells Mol Dis 2005;35:169–73.

[53] Klibi J, Niki T, Riedel A, Pioche-Durieu C, Souquere S, Rubinstein E, et al. Blood diffusion and Th1-suppressive effects of galectin-9-containing exosomes released by Epstein–Barr virus-infected nasopharyngeal carcinoma cells. Blood 2009;113:1957–66.

[54] Cai Z, Zhang W, Yang F, Yu L, Yu Z, Pan J, et al. Immunosuppressive exosomes from TGF-beta1 gene-modified dendritic cells attenuate Th17-mediated inflammatory autoimmune disease by inducing regulatory T cells. Cell Res 2011;22:607–10.

[55] Bianco NR, Kim SH, Ruffner MA, Robbins PD. Therapeutic effect of exosomes from indoleamine 2,3-dioxygenase-positive dendritic cells in collagen-induced arthritis and delayed-type hypersensitivity disease models. Arthritis Rheum 2009;60:380–9.

[56] Luketic L, Delanghe J, Sobol PT, Yang P, Frotten E, Mossman KL, et al. Antigen presentation by exosomes released from peptide-pulsed dendritic cells is not suppressed by the presence of active CTL. J Immunol 2007;179:5024–32.

[57] Lugini L, Cecchetti S, Huber V, Luciani F, Macchia G, Spadaro F, et al. Immune surveillance properties of human NK cell-derived exosomes. J Immunol 2012;89:2833–42.

[58] Maguire CA, Balaj L, Sivaraman S, Crommentuijn MH, Ericsson M, Mincheva-Nilsson L, et al. Microvesicle-associated AAV vector as a novel gene delivery system. Mol Ther 2012;20:960–71.

[59] Valadi H, Ekström K, Bossios A, Sjöstrand M, Lee JJ, Lötvall JO. Exosome mediated transfer of mRNAs and microRNAs is a novel mechanism of genetic exchange between cells. Nat Cell Biol 2007;9:654–9.

[60] van Balkom BW, Pisitkun T, Verhaar MC, Knepper MA. Exosomes and the kidney: prospects for diagnosis and therapy of renal diseases. Kidney Int 2011;80:1138–45.

[61] Lai RC, Arslan F, Lee MM, Sze NS, Choo A, Chen TS, et al. Exosome secreted by MSC reduces myocardial ischemia/reperfusion injury. Stem Cell Res 2010;4:214–22.

[62] Vrijsen KR, Sluijter JP, Schuchardt MW, van Balkom BW, Noort WA, Chamuleau SA, et al. Cardiomyocyte progenitor cell-derived exosomes stimulate migration of endothelial cells. J Cell Mol Med 2010;14:41064–70.

[63] Bruno S, Grange C, Deregibus MC, Calogero RA, Saviozzi S, Collino F, et al. Mesenchymal stem cell-derived microvesicles protect against acute tubular injury. J Am Soc Nephrol 2009;20:1053–67.

[64] Alvarez-Erviti L, Seow Y, Yin H, Betts C, Lakhal S, Wood MJ. Delivery of siRNA to the mouse brain by systemic injection of targeted exosomes. Nat Biotechnol 2011;9:341–5.

[65] Chalmin F, Ladoire S, Mignot G, Vincent J, Bruchard M, Remy-Martin JP, et al. Membrane-associated Hsp72 from tumor-derived exosomes mediates STAT3-dependent immuno-suppressive function of mouse and human myeloid-derived suppressor cells. J Clin Invest 2010;20:457–71.

[66] Parolini I, Federici C, Raggi C, Lugini L, Palleschi S, De Milito A, et al. Microenvironmental pH is a key factor for exosome traffic in tumor cells. J Biol Chem 2009;84:34211–22.

[67] Zitvogel L, Regnault A, Lozier A, Wolfers J, Flament C, Tenza D, et al. Eradication of established murine tumors using a novel cell-free vaccine: dendritic cell-derived exosomes. Nat Med 1998;4:594–600.

[68] Laulagnier K, Grand D, Dujardin A, Hamdi S, Vincent-Schneider H, Lankar D, et al. PLD2 is enriched on exosomes and its activity is correlated to the release of exosomes. FEBS Lett 2004;572:11–14.

[69] Laulagnier K, Motta C, Hamdi S, Roy S, Fauvelle F, Pageaux JF, et al. Mast cell and dendritic cell-derived exosomes display a specific lipid composition and an unusual membrane organization. Biochem J 2004,80.161–71.

[70] Scott SA, Selvy PE, Buck JR, Cho HP, Criswell TL, Thomas AL, et al. Design of isoform-selective phospholipase D inhibitors that modulate cancer cell invasiveness. Nat Chem Biol 2009;5:108–17.

[71] Kamphorst AO, Guermonprez P, Dudziak D, Nussenzweig MC. Route of antigen uptake differentially impacts presentation by dendritic cells and activated monocytes. J Immunol 2010;185:3426–35.

[72] Oussoren C, Zuidema J, Crommelin DJ, Storm G. Lymphatic uptake and biodistribution of liposomes after subcutaneous injection. II. Influence of liposomal size, lipid composition and lipid dose. Biochim Biophys Acta 1997;1328:261–72.

[73] Champion JA, Mitragotri S. Role of target geometry in phagocytosis. Proc Natl Acad Sci USA 2006;103:4930–4.

CHAPTER 9

Effect of Exosomes from Mesenchymal Stem Cells on Angiogenesis

Susmita Sahoo*, Feng Dong**, Lola DiVincenzo**, William Chilian**, Liya Yin**
*Department of Cardiovascular Research Center, Icahn School of Medicine at Mount Sinai, New York, NY, USA
**Department of Integrative Medical Science, Northeast Ohio Medical University, Rootstown, OH, USA

Contents

1 INTRODUCTION

In the realm of regenerative medicine, the emergence of stem cell-based therapies has engendered a new area of research focusing on promising treatment strategies for patients with ischemic cardiovascular diseases, a leading cause of death. Mesenchymal stem cells (MSCs), also referred to as multipotent stem cells, have been extensively studied since they were first identified in the 1970s [1]. MSCs have been isolated from a variety of tissues and are capable of *ex vivo* expansion and differentiation into the mesodermal cell lineage. More recently, MSCs have been shown to transdifferentiate into endodermal or ectodermal lineages [2]. MSCs have also exhibited a profound anti-inflammatory and immunoregulatory function *in vitro* by inhibiting monocyte maturation, decreasing expression of major histocompatibility complex (MHC) and costimulatory molecules, inhibiting dendritic cell (DC) production of tumor necrosis factor (TNF), and up-regulating plasmacytoid dendritic cell production of the anti-inflammatory cytokine interleukin (IL)-10, thus impairing DC antigen presentation and proinflammatory function [3]. Because of their plasticity, poor immunogenicity, and allogeneic application, MSCs have emerged as a promising cell source for regenerative medicine.

MSCs are currently used in more than 400 clinical trials – 23 of which focus MSC use on coronary artery disease, 52 trials focus MSC use on ischemia, and 25 focus on infarction (http://www.clinicaltrials.gov). Such trials aiming at proangiogenic therapy used MSCs to repair myocardial ischemic injury and/or restore heart function in patients with chronic myocardial ischemia or patients who have experienced myocardial infarction (MI). It was first believed that MSCs may ameliorate these ischemic diseases via a direct angiogenic and arteriogenic effect; however, the function of MSCs in tissue regeneration has not shown the capacity to differentiate into multiple cell types and then directly replace and engraft in the damaged tissue [4,5]. It is becoming apparent that MSCs' regenerative capacity is more indirect and may function through secretion and modulation rather than direct engraftment. Indeed, evidence supports that immune/inflammatory suppression and the paracrine mechanism of MSCs play a critical role in MSC-mediated therapeutic effects in tissue regeneration [3].

An increasing number of clinical trials have focused on the paracrine and immunomodulatory effects of MSCs that secrete a variety of bioactive molecules such as stromal cell-derived factor 1 (SDF-1), vascular endothelial growth factor (VEGF), fibroblast growth factor (FGF), hepatocyte growth factor (HGF), insulin-like growth factor 1, IL-6, IL-1, and TNF-α [6–9].

These secreted bioactive factors can induce endogenous stem/progenitor cell recruitment, and increase their retention, survival, proliferation, and differentiation, therefore contributing to tissue repair [10].

Using adipose tissue-derived MSCs, Katsuda et al. demonstrated that transplantation of undifferentiated MSCs into the injured liver resulted in higher levels of serological recovery than that of the transplantations of MSC-differentiated hepatocytes, suggesting that MSCs may not only repair damaged cells through replacement but may also exhibit a paracrine like effect of the undifferentiated MSCs on their surroundings [2]. Furthermore, in 2007, using a murine model of wound healing, Wu et al. showed beneficial effects of MSCs with 27% of MSCs engrafted into the wound 7 days post injection, and only 2.5% of MSCs engrafted 28 days post injection, implying that MSCs exerted a therapeutic effect even with poor engraftment [11]. Prior to this, Horwitz et al. demonstrated that MSC implantation significantly improved the bone mineral density and growth velocity of children with osteogenesis imperfecta while the engrafted MSC population remained less than 1%, further implying MSCs' function in the absence of significant engraftment [12]. Collectively, it appears that MSCs exert their regenerative capacity not by differentiation or engraftment, but rather through their ability to alter the host microenvironment via paracrine factors or secretome [13].

Extracellular vesicles (EV) and microvesicles (MVs) – termed exosomes – comprise a component of the MSCs' paracrine effect and secretome; in fact, MSCs are known as a suitable source for mass production of exosomes. These EV are small (30–100 nm) membrane-bound vesicles, actively secreted by many cell types. They are derived from the luminal membranes of multivesicular bodies and constitutively released by fusion of multivesicular bodies with the cell membrane. Exosomes are considered to be the most potent of intercellular communicators. Being released in all body fluids, they can communicate with neighboring as well as distant cells. It is clear that exosomes carry and transfer complex biological information such as proteins and RNAs from the cell of their origin to elicit a pleiotropic response in recipient cells with potential relevance in physiology, cancer, and other pathological conditions. A recent surge in exosomal research coupled with the discovery that the signaling status of the donor endosomal compartment may largely determine the biological output in recipient cells via exosomal uptake has sparked new questions regarding their potential in modulating angiogenic signals, not only between the cells in the vicinity but also between distant tissues. Whereas more conventional modes of cell–cell communication mostly depend upon extracellular ligand concentration and cell surface receptor availability, the magnitude of exosome signaling response relies on the capture and uptake by the target cells, allowing release of the exosome's content. Increasing evidence suggests a key role for the exosomal-mediated stimulation of angiogenic communication in numerous cellular processes involved in growth, development, cancer metastasis, and onset of other pathological manifestations.

Angiogenesis, the process by which new blood vessels are formed from pre-existing vessels, is a critical phenomenon that is activated during various stages of mammalian growth, development, and tissue repair. In this chapter, we will discuss the role of MSCs in angiogenesis and its underlying mechanism, provide an overview of the current knowledge on the transfer, spread, and modulation of angiogenic signals via exosomes, and offer a new perspective on the emerging role of exosomally circulated proteins and microRNAs (miRNAs) on regulation of angiogenesis at the molecular level. Finally, we will discuss the perspective of biomarker and therapeutic application of exosomes.

2 ANGIOGENIC EFFECT FROM MESENCHYMAL STEM CELLS

2.1 General Effects of Mesenchymal Stem Cells in Tissue Repair

MSCs are pluripotent adult stem cells and are defined by their ability to differentiate into mesenchymal lineages, such as bone, cartilage, or fat [7,14]. MSCs can be easily obtained through standard procedures and expanded in culture for transplantation. Besides bone marrow-derived MSCs, endogenous MSCs may also be derived from pericytes, which are activated and released after tissue injury [15]. MSCs have been very well studied and their effects in tissue repair confirmed. MSCs delivery can increase angiogenesis, improve cardiac function, decrease scar size by increasing cardiac myocyte preservation in the infarct zone, and induce endogenous stem cell homing [16–19]. Both endogenous and transplanted MSCs can function as "drugstores" and provide a microenvironment for tissue repair by secreting a variety of bioactive molecules [6].

MSCs may enhance angiogenesis and contribute to endogenous repair by the releasing of proangiogenic and antiapoptotic cytokines [20,21]. Besides their multilineage potential and paracrine effects, MSCs have immunomodulatory effects, not only by secreting bioactive molecules, but also by cell–cell contact involving a variety of lymphocytes [6]. MSCs have the ability to affect inflammatory cytokine/chemokine expression [22]. Their intravenous injection can increase the secretion of TNF-α-induced protein 6, which can inhibit neutrophil and monocyte infiltration at sites of injury [23,24]. The safety of MSCs in cell therapy of acute myocardial infarction (AMI) and ischemic limbs has been confirmed in multiple clinical trials and studies [25,26]. Previous studies have found that allogeneic human MSC transplantation is safe and well tolerated even after multiple infusions [6,27,28]. Moreover, allogeneic adult human MSC therapy demonstrated significant benefit in decreases of heart failure and rehospitalizations following AMI [29]. That is, more and more solid evidence has demonstrated that due to lacking MHC-II and low immunogenicity, MSCs could be safely used in allogeneic therapy [30]. The unique characters of MSC make it a promising source and a good candidate in stem cell-based tissue repair [15,31,32].

2.2 Angiogenic Properties of MSCs

Ideal vascular regeneration should include the restoration of original vasculature, the correction of vascular dysfunction, and the development of new blood vessels. This requires increased endothelial cell (EC) migration, engraftment, survival, and proliferation [20,33]. MSCs have angiogenic potential and are more advantageous for angiogenesis stimulation when compared with VEGF and FGF-1 gene therapy. Both endogenous and transplanted MSCs can release multiple growth factors and cytokines including SDF-1 and VEGF, and also provide a microenvironment for angiogenesis and tissue regeneration [15]. MSC transplantation demonstrated better efficiency compared to gene therapy of angiogenic growth factor for improving myocardial performance in AMI [32,34]. Transplanted MSC can secret VEGF that induces vessel sprouting and SDF-1 that can mobilize bone marrow-derived stem/progenitor cells to the site of injury [16,35]. Recent studies demonstrated that SDF-1 release by MSCs leads to the recruitment of chemokine (C-X-C motif) receptor (CXCR4)-positive cardiac progenitor cells that may further release SDF-1 [16]. Moreover, as a releasing source of growth factors and cytokines, MSCs are renewable [26]. In addition, MSCs can be easily harvested from a variety of tissue sources, including bone marrow, peripheral blood, fat tissue, and may also be derived from pericytes [6,36].

2.3 Mechanisms of MSC-Induced Angiogenesis

Multiple mechanisms may be involved in MSC-induced angiogenesis. MSCs may promote angiogenesis by inducing EC chemoattraction and engraftment, decreasing EC apoptosis, as well as increasing mitotic activity in

ECs [20]. Recent studies showed that in normoxic or hypoxic conditions, the factors secreted by MSCs supported EC adhesion, but not proliferation, while the factors secreted by endothelial progenitor cells (EPCs) supported EC proliferation, but did not sustain EC adhesion [20].

2.3.1 VEGF and FGF-1 and Angiogenesis

MSCs can release multiple growth factors and cytokines including VEGF and FGF-1, which stimulate angiogenesis and tissue regeneration [15,35]. Several angiogenic growth factors, including VEGF and FGF-1, showed angiogenesis capabilities in some Phase I studies [37,38]. However, VEGF gene or recombinant protein therapy failed to stimulate collateral growth and angiogenesis in patients with ischemic heart disease [39–41] and showed no significant improvement in a Phase II clinical trial of peripheral arterial occlusive disease [42,43]. Similarly, FGF-1 gene therapy did not show significant differences in amputation and death rate between the FGF-1 group and the placebo group in a Phase III Therapeutic Angiogenesis for the Management of Arteriosclerosis in a Randomized International Study [44].

2.3.2 Angiogenic Properties of SDF-1

Although VEGF may stimulate ECs proliferation [45] and is regarded as an important factor in the development of vasculature, VEGF-induced vessels are often leaky and unstable, which indicates that VEGF alone cannot achieve functional and mature blood vessels [46,47]. Instead, SDF-1 is important for the development of mature blood vessels and plays an important role in MSC-induced angiogenesis [32]. Studies demonstrated that SDF-1 can promote angiogenesis and collateral blood flow in limb ischemia [48]; SDF-1 may prevent uncontrolled EC proliferation and less generation of hyperpermeable vessels, which is seen in VEGF-induced angiogenesis [48,49]. MSCs overexpressing SDF-1a promote angiogenesis and improve heart function in AMI [50].

SDF-1-induced angiogenesis includes several important steps through multiple mechanisms: (i) SDF-1a can induce mobilization of CXCR4-positive bone marrow-derived stem cells out of bone marrow to circulation; (ii) SDF-1a has the ability to recruit CXCR4-positive progenitor cells including bone marrow-derived stem cells, EPCs, and pericytes that contribute to the development of mature blood vessels at the sites of injury [16,51]; (iii) SDF-1 can retain and correctly position recruited cells around the growing vessels, therefore promoting further angiogenesis [52]; (iv) several studies demonstrated the antiapoptosis effect of SDF-1 on various cells including stem cells, cardiomyocytes, ECs, and EPCs [16,19,52]; and (v) SDF-1 has direct angiogenic effects on ECs. SDF-1 may increase the production of nitric oxide (NO) and promote both vasculogenesis and angiogenesis through the endothelial nitric oxide synthase/Phospho-Akt pathway [53].

2.3.3 Extracellular Vesicles Derived from MSCs (MSC-EVs) Promote Angiogenesis

MSC-EVs might be a novel mechanism in MSC-based therapy in MI. MSC-EVs are a mixture of MVs and exosomes; MVs are large extracellular membrane vesicles (100–1000 nm in diameter) and exosomes are small membrane vesicles of 30–100 nm in diameter [54]. EVs can be released by various stem cells including MSCs. Exosomes contain genetic material, such as mRNAs and miRNAs, originate from multivesicular bodies, and are released by a variety of cell types. MSC-derived exosomes may serve as significant mediators of cell-to-cell communication [55]. Exosomes secreted by MSCs can decrease myocardial ischemia/reperfusion injury [56], and MSC-EVs can promote neoangiogenesis and improve cardiac function following AMI, which may play a role in the maintenance of stem cell homeostasis [57,58]. MSCs express hypoxia-inducible factors (HIFs) and HIF-1 alpha may contribute to the release of exosomes in cardiomyocytes [59,60]. Studies suggest that miR150 may be a key molecule for the proangiogenic activity of MSC-EVs [61,62]. MSC-EVs promote neoangiogenesis and might play a critical role in cardiac repair following ischemic injury [63].

2.4 Therapeutic Strategy for Neovascularization with MSCs and Angiogenic Genes

MSCs' capacity to repair declines with age and other risk factors. There is a difference in the MSC pool between young and old individuals [15,64], as children respond better than adults in MSC therapy [65]. MSC quantity, mobilization, angiogenic potential, and proliferation all decrease with age [15,64]. Endogenous MSCs are also limited by insufficient bioactive molecule signals that are needed to activate and induce endogenous MSC homing. For example, the expression of SDF-1 increases immediately within the injured myocardium after MI, but return to the baseline in 7 days [19,66]. MSC-induced angiogenesis can also be restricted by a hostile microenvironment [67,68]. MSCs isolated from diabetic animals are dysfunctional [69], and diabetes impairs MSC function in limb ischemia [70].

In order to overcome the limitations of MSC-induced angiogenesis and improve vasculogenic function, MSCs may be genetically modified with an angiogenic factor. Although MSCs already express SDF-1 and VEGF genes [34], SDF-1 and VEGF gene overexpression can induce more SDF-1 and VEGF protein synthesis. Paracrine factors such as VEGF and SDF-1 can activate Akt in MSCs and improve MSC survival at the site of injury and promote differentiation of MSCs into vascular lineages [32,71,72]. VEGF, SDF-1a, angiopoietin-1, and Akt gene overexpression in MSCs have demonstrated improved angiogenic capability of MSCs and improved cardiac function after AMI in rats [32,73,74]. Increased vascular density along with smaller infarct size is found in SDF–VEGF-overexpressed MSC treatment compared to MSC infusion alone [19,32]. Endogenous CSCs are recruited to the infarct border zone in response to MSC engraftment, which can be enhanced by the local overexpression of SDF-1 [75]. That is, the combination of chemokine and angiogenic factor genes with stem cells such as MSCs is more potent than stem cells alone in promoting vascular regeneration, increased capillary density, and may also be a potent strategy for therapeutic neovascularization in severe ischemic diseases [64,76]. In addition, SDF-1 overexpression in MSCs releases extra stem cell homing factor–SDF-1, which not only sends angiogenic signals, but also induces amplification of this local response by the recruitment of endogenous stem cells, which also secrete angiogenic factors such as VEGF and SDF-1a [52].

3 ROLE OF EXOSOMES IN REGULATION OF ANGIOGENESIS

3.1 Exosomes in Developmental and Physiological Angiogenesis

Surprisingly, much of what we know about the angiogenic function of exosomes in cell determination is gathered from pathologically transformed cancer cells and wound healing, while data about their biogenesis and biology in normal growth and development in adult tissue are lacking. In the past few years, exosomes have been shown to play a significant role in the induction and modulation of cell fate-inducing signaling pathways, such as the Hedgehog (Hh) [77,78], Wnts [79], Notch [80,81], transforming growth factor (TGF-β) [82], epidermal growth factor (EGF) [83], and FGF [84] pathways, which places them in a wider context of development [85]. These protein families induce signaling cascades responsible for tissue specification, homeostasis, and maintenance; besides, they orchestrate angiogenic activity through direct and indirect regulation of quiescence, migration, and proliferation of ECs. Exosomes contribute to cell fate signal secretion, and conversely, these proteins can induce exosome secretion. The possibility of involvement of argosomes or exosomes or some form of vesicular communication in angiogenesis was first reported in vesicle-bound Wingless (Wg, a *Drosophila* Wnt) protein [86], and later a similar mechanism was discovered in *Xenopous laevis* [79]. Many groups found that both *Drosophila* and human cell lines can release functional Wg (Wnt)-carrying exosomes [87,88]. Interestingly, a decision on whether exosomes or any of their cargo-containing miRNAs and proteins directly contribute to developmental angiogenesis is still pending.

3.2 Exosomes in Tumor Angiogenesis

Angiogenesis is a hallmark of cancer development that has long been considered an attractive therapeutic target for the disease. In order to grow beyond microscopic size, tumors depend on angiogenesis. The cancer cells react with the microenvironment, which provides the interface between parenchymal tumor cells and surrounding stroma. For tumor suppression or progression, cross-talk between different components of the microenvironment, such as blood vessel-forming cells, immune and inflammatory cells and fibroblasts, are important. Such communication is executed by growth factors, including chemokines, cytokines, and cell fate signaling molecules. Many reports suggest that membrane vesicles including tumor-derived exosomes and extracellularly secreted vesicles (EVs) are new candidates with important roles in the exchange of information that induces tumor growth and metastasis and can promote endothelial angiogenic responses. Exosomes can be readily isolated from body fluids of cancer patients, including blood, lymph, urine, saliva, cerebrospinal fluid, and ascites. In fact, the number of circulating exosomes in cancer patients seems to be higher than in healthy individuals [89,90] and has been found to correlate with poor prognosis. Further, the melanoma tumor exosomes were found to contain tumor-specific proteins, such as TYRP2, VLA-4, HSP70, and the MET-oncoprotein or tumor-specific mutations as demonstrated by the pioneering work of Lyden and his team. These oncosomes (exosomes from tumor cells) transfer oncogenic tumor-specific molecules not only to the cancer microenvironment, but also to healthy cells in distant organs like lungs and bone marrow to create new metastatic and angiogenic niches for the growth of the tumor. The authors proposed a novel mechanism that circulating exosomes from a tumor could reprogram, cross-talk, and permanently educate the bone marrow progenitor cells to mobilize out of the bone marrow. This process, they propose, contributes to a switch from a localized disease to a disseminated, metastatic disease. While the specifics of this work related to cancer biology are intriguing, we believe these findings have much broader implications, providing evidence of the complexity of the cargo carried by exosomes, and their enormous potential to directly influence the biology of distant microenvironments.

In addition, there are several reports of exosomes/EVs released from different cancer cells that can stimulate angiogenesis and form a metastatic niche in distant organs. EVs derived from squamous carcinoma cells can transfer oncogenic EGFR to ECs. This leads to the activation of mitogen-activated protein kinases (MAPK) and Akt pathways and increases expression of endogenous VEGF and autocrine induction of VEGFR2 to induce angiogenesis [91]. Glioblastoma EVs are enriched in angiogenic proteins, such as FGF, interleukin (IL)-6, and VEGF, and stimulate angiogenesis *in vitro* in a brain microvascular endothelial tubule formation assay [92]. Similarly, B16-F10 melanoma-derived EVs induce production of proangiogenic cytokines including IL-1α, FGF, and TNF-α in 2F-2B ECs, which results in increased formation of endothelial spheroids and sprouts [93]. In this study, however, the stimulatory components of the vesicles were not identified. In another study, MVs derived from CD105-positive human renal cancer stem cells stimulate blood vessel formation of ECs and induce metastatic diffusion of the tumor and lung premetastatic niche formation [94]. A recent study suggests that hypoxia, which has been associated with tumor aggressiveness and acidification of tumor microenvironment, induces tumor exosome release and tumor growth, angiogenesis and pericyte coverage of the vessels [95]. Collectively, these findings open up new avenues in tumor biology, where exosomes are proposed as a new way of proangiogenic communication within and between other organs. The tumor-specific proteins or nucleic acids carried by the exosomes can potentially be used as a biomarker for monitoring the disease, or can be targeted for therapeutic purposes.

Recently, it has been shown for the first time that the presence of double-stranded DNA (dsDNA) in chronic myeloid leukemia (CML) cell-derived exosomes represents the whole genomic DNA [90]. It was reported that exoDNA could be used to identify mutations present in parental tumor cells. Although the dsDNA was present in about 10% of the total exosome-isolated population, this illustrates the significant translational potential of dsDNA as a circulating biomarker for cancer in the clinic. ExoDNA is an attractive, potential biomarker candidate in the early detection of cancers and the monitoring of treatment response for several reasons: its protection and thus inherent stability within exosomes; the possibility to isolate or enrich tumor-derived exosomes in complex plasma samples via exosomal surface markers; and its easy and fast preparation.

3.3 Exosomes in Pathological Angiogenesis and Tissue Homeostasis

It has been postulated that the molecular players and mechanism of vascular branching in pathological angiogenesis are parallel to developmental angiogenesis, but have dysregulated expression. However, some molecules have different functions during physiological and pathological angiogenesis. For example, in developmental angiogenesis, VEGFR1 has a negative role by trapping VEGF [96], but this model does not explain its disease-restricted proangiogenic activity [97]. Although it is superfluous for vascular development, VEGF-B promotes contextual enlargement of myocardial capillaries [98] or growth of coronary vessels [99] in a hypertension-induced/pressure-overload model of cardiac hypertrophy. These examples (and others) suggest that part of the molecular basis of pathological angiogenesis is different from that of vascular development. Moreover, insights obtained from developmental angiogenesis models may not completely recapitulate the mechanisms that drive human pathological angiogenesis.

3.3.1 Platelet-Derived Microparticles

It has been shown that platelet-derived EVs are able to modify steps involved in angiogenesis such as proliferation, migration, and adhesion of ECs. Kim et al. [100] were pioneers in demonstrating that platelet-derived EVs increase proliferation, migration, and formation of capillary-like tubes of human ECs. It has also been reported that human platelet-derived MVs/exosomes play an important role in tumor progression, angiogenesis, metastasis, and invasive potential in lung cancer [101] and breast cancer [102]. The platelet-derived MVs modulate the proliferation, adhesion, and metastatic potential of tumor cells via increasing expression of proangiogenic factors such as VEGF, HGF, matrix metalloproteinases (MMP)-2, MMP-9, IL-8, and tissue inhibitor of metalloproteinase-2. The increased level of platelet-derived MVs/exosomes circulating in peripheral blood of individuals with tumors is a strong predictor of metastasis in patients with gastric cancer [103].

3.3.2 Exosomes/EVs Released from Apoptotic Endothelial Cells are Angiogenic

ECs undergoing apoptosis are able to release microparticles (MPs) that display vascular protective effects, promote proliferation of ECs, and inhibit endothelial apoptosis [104]. This elegant study shows that MPs from apoptotic

ECs abundantly express miRNA-126, which triggers chemokine (C-X-C motif) ligand (CXCL) 12 production through the CXCR4 pathway; thus, MPs released from apoptotic ECs may act as paracrine elements necessary to protect ECs during atherosclerosis.

3.3.3 Cardiomyocyte-Derived Exosomes

It has been shown that exosomes actively released from cultured rat cardiomyocytes carry Hsp20, an angiogenic heat shock protein that can induce tube formation of human umbilical vein endothelial cells (HUVECs) and may regulate myocardial angiogenesis via its interaction with VEGFR2, both under steady state and under ischemic stress conditions [105].

3.3.4 Stem Cell-Derived Exosomes in Angiogenesis

Finally, it has been recently described that stem cells can also release exosomes or EVs that may contain both autocrine and paracrine angiogenic factors [77,106,107]. Lim and coworkers have detected RNAs and miRNAs in the supernatant of human mesenchymal stem cells [108] that were possibly associated with exosomes. They had identified a cardioprotective effect of these exosomes when injected into the ischemic myocardium [109].

Collectively, these facts indicate that exosomes/EVs from several different sources can mediate pathological angiogenesis in different tissues and exosomes/EVs can be a suitable target for the development of novel therapeutics. In general, the clinical application of antiangiogenic therapeutics for cancer has been more successful than that of proangiogenic therapeutics for cardiovascular diseases, since the interruption of angiogenic signaling is technically feasible compared with reconstruction of vascular structure. Nevertheless, a more profound understanding of the spatiotemporal interactions of regulatory signaling cascades and advances in genetic profiling are required for optimization of any targeted therapy.

3.4 Molecular Regulation of Exosome-Mediated Angiogenesis

Exosomes can affect angiogenesis by inducing changes in the secretome of ECs, either increasing the production of proangiogenic factors or decreasing the production of antiangiogenic factors. Furthermore, MPs can modify all steps leading to angiogenesis. Here, we primarily discuss the proangiogenic factors that the exosomes carry that have been shown to affect EC gene expression, cell cycle change, proliferation, and migration properties which are associated with induction of angiogenesis.

3.4.1 Uptake of Exosomes

Function of exosomes in physiological and pathological angiogenesis depends on their ability to interact with recipient cells to deliver their content of proteins, lipids, and RNAs. There is a certain degree of specificity of target cell binding that is illustrated by the finding that EVs released by a human intestinal epithelial cell line interacted preferentially with dendritic cells rather than with B or T lymphocytes [110]. A recent study has demonstrated that cancer cell-derived exosomes use their surface heparin sulfate proteoglycans (HSPGs) for their internalization and functional activity [111], indicating that HSPGs are one of the key receptors of this macromolecular cargo. The cellular and molecular basis for exosome/EV targeting is still undetermined, but several target cell-dependent and -conditional aspects are beginning to emerge. Target cell selectivity for binding of exosomes is likely to be determined by adhesion molecules, such as integrins, that are present on exosomes.

After binding to the recipient cells, exosomes may directly fuse with the plasma membrane or be internalized through distinct endocytic pathways [112]. Detection of fusion of small exosomes with target cell membrane by fluorescence microscopy in live cells is limited by resolution and the fast dynamics of the fusion events. Nevertheless, direct evidence of fusion has been obtained by labeling exosomes with the lipofilic dye R18, in which self quenching is relieved as a consequence of fusion, resulting in increase in fluorescence of target cells [113].

3.4.2 Exosomal Proteins/mRNAs in Angiogenesis

With angiogenesis induction being one of the hallmarks of cancer, intense efforts have been taken to elaborate the contribution of tumor-derived exosomes. Tumor exosomes containing TNF-α, IL1-β, TGF-β, and TNF receptor1 recruit endothelial progenitors, promote angiogenesis [114], and stimulate ECs by paracrine signaling [93]. Delta-like 4 bearing tumor exosomes confer a tip cell phenotype to ECs with filipodia formation, enhancing vessel density and branching [81]. However, the pathways and angiogenic mechanisms differ significantly depending on the exosomes source [115,116]. One of the leading pathways proceeds via peroxisome proliferator-activated receptor alpha, extracellular-signal-regulated kinases, and nuclear factor kappa-light-chain-enhancer of activated B cells activation [115,117]. Platelets interact via lipid components of endothelial proliferation, the effect being abolished by a stripper of bioactive lipids [100]. Sonic hedgehog (Shh) expressing T cell exosomes display multiple effects on ECs attributed to Shh binding its receptor, thereby inducing activation of NO synthase and the endogenous Shh pathway with increased FGF and VEGF expression, suggesting that Shh$^+$ exosomes induce reparative vascularization in ischemic tissues [115]. MPs from human atherosclerotic plaques promote endothelial intercellular adhesion molecules-dependent monocyte adhesion and transendothelial migration. These MPs stimulate endothelial proliferation and angiogenesis, which was suggested to proceed via CD40 ligand expression on exosomes with CD40 on ECs [118]. In a feedback, there is a cross-talk between fibroblasts and tumor cells in which prostate cancer cell line-derived MVs lead to activation of fibroblasts via MMPs, which then shed exosomes that increase tumor cell migration via CX3C-CX3CR1 [119]. Furthermore, EC-derived exosomes harbor MMPs that favor invasion and capillary tube formation [101].

Tumor exosome uptake is known to induce modulation of endothelial gene expression and function and is discussed in detail here [114]. Colorectal cancer cell-derived exosomes are enriched in cell cycle-related mRNA, which promote EC proliferation [120]. Glioblastoma-derived exosomes induce angiogenesis by transferring exosomal proteins and mRNA to ECs [92]. Further, EGFR-positive tumor-derived exosomes are taken up by ECs and stimulate EGFR-dependent responses including activation of the MAPK and Akt pathway and VEGFR2 expression [114]. Angiogenesis was also reported to be induced by the transfer of Notch-ligand-delta-like 4, which inhibits Notch signaling and increases angiogenesis [81]. Exosome uptake initiates endothelial progenitor maturation as well as endothelial activation including VEGFR transcription [121]. CML exosomes induce angiogenic activity in ECs, where an Src inhibitor affects exosome production as well as vascular differentiation [122].

3.4.3 Exosomal miRNAs in Angiogenesis

A major breakthrough in the study of exosome function was the demonstration that their cargo includes both mRNA and miRNA and that exosome-associated mRNAs and miRNAs are functional and can be either translated into proteins or regulate gene expression in the target cells [123]. Many RNAs that were isolated with exosomes were found to be enriched relative to the RNA profiles of the originating cells indicating that RNA molecules are selectively incorporated into the exosomes. The exosomes released from leukemia cells carry miR-92a – taken up by ECs (HUVECs) when exogenously applied – which suppresses the target gene integrin a5, and affects EC migration and tube formation abilities [124]. The miR-17-92 cluster was initially found amplified in diffuse cell lymphomas; however, an ectopically overexpressed, truncated version that lacked miR-92 showed the cluster's role as an oncogene [125]. Mir-17-92 cluster is also known as an oncomiR. MiR-210 secretion was found be increased significantly under hypoxia from leukemic cells. The exosomes containing increased miR-210 induced angiogenic activity in ECs [126]. Both miR-17-92 cluster and miR-210 target hypoxia-inducible factor (HIF1a), which is considered an adaptor gene because it not only promotes angiogenic signaling during hypoxia, but also contributes to the regulation of human metabolism [127]. Further, miR-1, which is known to suppress seryl-transfer RNA synthetase and regulate VEGF-A expression, was found to be loaded in the glioblastoma-derived vesicles that targeted multiple pro-oncogenic signals in cells within the glioblastoma microenvironment [128].

Zampetaki et al. [129] provided further insights as to the role of exosomal/vesicular miRNAs in angiogenesis. They found miR-126 to be significantly downregulated in the blood of patients with type-2 diabetes (T-2D), a disease characterized by systematic angiogenic abnormalities. Patients with T-2D showed a significant reduction in the level of vesicular miR-126, with little change in the level of nonvesicle-associated miR-126. High glucose reduced the miR-126 enrichment in EC-derived apoptotic bodies, which has been shown to reduce tubule formation as well as expression of key angiogenic factors including VEGF and FGF in human islets [130]. Another study by Landmesser's group revealed that loss of miR-126 release via exosomes impaired the proangiogenic function of human CD34$^+$ stem cells and that high glucose or diabetes reduced miR-126 levels in the CD34$^+$ cells as well as exosomes. Taken together, these findings suggest a link between glucose levels and systemic angiogenic signaling that is mediated by exosome/vesicle-encapsulated miR-126. Further investigation is needed to determine whether the cellular export of proangiogenic miRNAs is inhibited by disease or their cellular uptake is enhanced (as proposed here [131]) – either of which could result in decreased levels of circulating exosomal/vesicular proangiogenic miRNA.

While reports of exosomes-encapsulated miRNAs in the systemic transduction of angiogenic signals are abundant, the possibility of targeted, vesicle-mediated miRNA transfer cannot be ruled out. Exosomes, for example, which are released by cells under various stimulatory conditions, express membrane proteins that vary depending on their cellular origin [132]. EC-derived exosomes express specific exosomal markers, including MHC class I/II. Only a few specialized cells have receptors for these molecules, including

T cells, and it is tempting to speculate that the incorporation of miRNAs into exosomes produced by ECs is reflective of a process of specifically targeting angiogenic or other signaling pathways in specific cell types [127].

Finally, exosomes represent a delivery system that confers stability to mRNA and miRNA packaged by them and protects from external RNases by the surrounding membrane. Thus, miRNA-150 is actively secreted from monocytes into exosomes and evokes *in vitro* EC migration, whereas injecting mice with those exosomes increases miRNA-150 in mouse blood vessels [133]. Taken together, generation of exosomes may represent an alternative approach to favor angiogenesis in pathological conditions.

The exosomal angiogenic effectors and activities are summarized in Table 9.1.

Table 9.1 Angiogenic *effectors* and *activities* associated with exosomes (EVs)

Angiogenic factors (proteins)	Angiogenic activities induced	Exosomes (EV) source	Reference
Shh	Induce tube formation, differentiation	T lymphocytes	[77,78]
Wnt	Induce Wnt signaling	*Xenopus laevis*, *Drosophila* Kc167 cells	[79,88]
DlI4-Notch	Induction of tip cell phenotype, philopodia formation, and increased branching	Endothelial/tumor cells	[80,81]
EGF	–	Oligodendrological cells	[83]
EGFR	Activation of MAPK and Akt pathways	Tumor cells	[91]
FGF	Exosomal secretion of FGFR	Hepatoma/breast cancer cells	[84]
VEGF	Increased MMP expression, EC sprouting, vessel density	Tumor cells	[134]
IL-8	Increased endothelial migration, tube, and vessel formation	CML cells, human glioblastoma cells	[92,135]

Table 9.1 Angiogenic *effectors* and *activities* associated with exosomes (EVs) *(cont.)*

Angiogenic factors (proteins)	Angiogenic activities induced	Exosomes (EV) source	Reference
Angiogenin, IL-6	Induce tube formation	Human glioblastoma cells	[92]
Ephrin B2	–	Colorectal cancer cells	[136]
MMP-9	Induce EC sprouting and vessel density	Mesoangioblasts/ tumor cells	[134]
Hsp20	Induce VEGFR2, tube formation, and myocardial angiogenesis	Rat cardiomyocytes	[105]

Angiogenic factors (RNA/miRNAs)	Angiogenic activities induced	Exosomes (EV) source	Reference
mir-17-92 cluster	Migration and tube formation of ECs	Leukemia cells	[124]
miR-210	Endothelial tube formation	Leukemic K562 cells	[126]
miR-126	Endothelial tube formation, *in vivo* vessel density	Human CD34+ stem cells	[106,107]
miR-130	Tube formation, vessel density	Human CD34+ stem cells	[106]
miR-146a	Endothelial tube formation	ECs (HUVECs)	[137]
mir-1	Induce VEGF-A expression	Glioblastoma cells	[128]

4 BIOMARKER OPPORTUNITIES AND THERAPEUTIC POTENTIALS OF EXOSOMES

Angiogenesis, the process of forming new blood vessels, is a critical phenomenon that is activated during various stages of development and maintains adult tissue homeostasis and carcinogenesis. The abnormal angiogenesis leads to a variety of diseases [138]. During physiological state and pathological conditions, cells can affect other cells and pass on their messengers to reflect themselves or environmental change by exosomes that have been shown to contain genetic materials and proteins. The role of exosomes as a cellular messenger has been elucidated by genetic and protein profiling of these components. The dual role of exosomes as biomarkers and cellular messenger creates opportunities to measure cellular temporal–spatial status and to gain more insight into the possibility of interfering with the biological targets of these released MVs [139]. Isolating exosomes and detecting expression of their contents can develop biomarkers for a pathological condition or even the severity or stage of the condition. Exosomes have been extensively studied in cancer and inflammation as a potential biomarker. However, this process is not reserved only for cancer research. Similar studies based on exosome-enriched proteomics also extended to other cell types and various body fluids. It will be interesting to understand the role of cardiovascular exosomes in cardiovascular physiology, which might lead to a biomarker discovery in cardiovascular diseases [140].

Either overexpression or downregulation of exosome contents could serve as a pathological biomarker. miRNA-18a, a known component of the oncogenic cluster miR-17-92, which resists ribonuclease activity in the plasma by the protective packaging of the exosome, has a much higher concentration in cancer patients than in healthy patients. Concentrations of miR-18a may serve as a noninvasive diagnostic tool of various types of cancer through the mere plasma isolation of the exosomes carrying it [141]. miR-145, a known tumor suppressor packaged in exosomes, was found to be significantly lower in malignant cancer than benign and is dependent on the protective carrier function of exosomes [142]. The contents of exosomes are not exclusively miRNAs, but can be various proteins indicative of various conditions. Transcription factors in urinary EVs (uEVs) are readily detectable in a patient with acute kidney injury (AKI) but not at all in whole urine or uEVs of patients without AKI implying that transcription factors in uEVs are indicative of an AKI [143]. *In vitro*, the exosomes released from cardiomyocytes almost doubled after 2 h hypoxia and some of these inducers were shown to have cardioprotective effects. Changes in MP release and subsequent miRNA content changes led to less miRNA transfer to other cells [139].

Stem cells have been transplanted with the main goal of differentiation of the progenitor cells for cardiovascular regeneration; however, the results are mixed and limited due to low efficient engraftment of stem cells and low number of cells regenerated from the stem cells. The beneficial paracrine effect of stem cells has opened a new door for therapeutic application without cell injection, possibly providing a solution to the results seen with poor engraftment and regeneration. The secretome of MSCs is of great interest and synonymous with the EVs produced by stem cells used for regenerative therapy. EV such as exosomes may deliver critical information to target cells for tissue regeneration after injuries. Thus, exosomes may provide a new tool for overcoming poor bioavailability of drugs and lead to specific targeting of new therapeutics.

The fact that exosomes mirror the phenotype and function of their parent cell implies that MSC-derived exosomes may be able to carry out the beneficial therapeutic functions of MSCs in their absence [2]. The intravenous infusion of MSCs results in over 80% of the MSCs being trapped in the lungs with less than 1% detected in the ischemic heart or brain [24]. Furthermore, success of intra-arterial infusion is strictly dependent on MSC size due to their entrapment at the precapillary level [144]. If not cultured in a 3D model or a 3D model following successive monolayer culture, infusion causes occlusion of the coronary arteries resulting in ischemia of myocardial tissue or occlusion of the carotid artery resulting in stroke [145]. The use of MSC-derived exosomes for therapeutic treatment as opposed to the MSCs themselves proves an attractive method for avoiding vascular occlusion and subsequent tissue ischemia and stroke specifically in the heart and brain, respectively. On the other hand, compared to viral delivery for gene therapy, exosomes have more potential for gene therapy in terms of safety issue, immune response, and gene carrying capacity because it is a membrane-bound vesicle and naturally released from cells [146].

Following an MI, conditioned media (CM) collected from engineered mesenchymal stem cells overexpressing Akt (MSC-Akt) cells was found to protect ventricular cardiomyocytes from hypoxia-induced apoptosis *in vitro* and limit the infarct size while protecting ventricular function *in vivo*. The proposed mechanism involves the paracrine function of MSC-Akt cell secretion of secreted frizzled related protein (Sfrp2), which was found to protect against hypoxia-induced apoptosis via suppression of Wnt signaling. Apoptotic precursor caspase showed a 33% increase after knockdown of the paracrine signal Sfrp2 [140]. The results support the hypothesis that paracrine exosome signaling is having a profound effect on the microenvironment maintaining functional phenotype [147]. Tim and colleagues demonstrated that intravenous and intracoronary injection of MSC-CM significantly restored ventricular performance in a porcine model of ischemia/reperfusion injury [148].

Human CD34+ peripheral blood mononuclear cells were found to secrete exosomes, which can promote angiogenesis both *in vitro* in HUVECs and *in vivo* in mice cornea. These exosomes, which were found to carry the $CD34^+$ protein on their surface, were enriched with proangiogenic miR-NAs such as miR-126 and miR-130a [106], which were possibly related to the proangiogenic function of the exosomes [107]. Further, in a recent work, Ibrahim et al. identified cardiosphere-derived exosomes as critical agents of cardiac regeneration triggered by cardiosphere-mediated cell therapies [149]. The cardiosphere-derived exosomes were found to be enriched with miR-146a, which mediated the therapeutic function of the exosomes.

5 FUTURE PERSPECTIVES

Human diseases, such as cancer, cardiovascular diseases, diabetes, and many others, are associated with the exosomal dissemination of angiogenesis. Exosomes, with their selective molecular content, are increasingly recognized as important regulators of angiogenesis, not only in the vicinity of a cell, but also in distant tissues. Proteins and nucleic acids circulated via exosomes are now considered as novel mediators or even as biomarkers of angiogenesis-associated intercellular signaling linked to a disease. As more details regarding the regulation of exosomes release and uptake are uncovered, the roles that different exosomal miRNA/protein transport modalities play in essential physiological processes will be identified. Although exosomes have potentials as novel biomarkers or therapeutic targets due to the decreased risk and noninvasive nature of the procedures, limitations must be realized, analyzed, and overcome in one fashion or another. Future endeavors still need to further understand how they are produced and how this "paracrine effect" affects target cells. The mere presence of an exosome is not always entirely useful as a biomarker or pharmacological tool due to its ubiquitous role in cell–cell communication; however, the contents provide insight for these future designs. Our understanding of signal transduction between cells will deepen as more is uncovered about the role of circulating exosomes and exosomal molecules as this will likely impact aspects of disease diagnostics, prognostics, and treatment. A more profound understanding of the spatiotemporal interactions of regulatory signaling cascades and advances in profiling of molecules transferred by exosomes is required to exploit and optimize the exosomes for therapeutic angiogenesis or targeted therapy.

The therapeutic potential will also expand if methods for manipulating exosomes are developed. Currently, perhaps the most significant hurdle to overcome is the isolation and purification of exosomes, which is predominantly ultracentrifugation. While it seems working, to isolate exosomes, it may exclude some 40% of vesicles [143]. Additional methods may be used to retrieve the latter 40%, but this process is not used across the board. If sufficient progress is to be made, standardization in exosome retrieval and purification must advance so as to be assured that the exosomes being administered in therapeutics are carrying the desired contents to elicit the desired results from target tissue; and the exosomes retrieved are of sufficient purity to serve as a biomarker for diagnosis and prognosis. Finally, the therapeutic potential of (progenitor) cell-derived EV is promising, as they are naturally occurring, efficient, therapeutic delivery vehicles that might be used to deliver therapeutics and drugs to specific cell types.

ABBREVIATIONS

AKI	Acute kidney injury
AMI	Acute myocardial infarction
CXCL	Chemokine (C-X-C motif) ligand
CXCR	Chemokine (C-X-C motif) receptor
CM	Conditioned media
CML	Chronic myeloid leukemia
DC	Dendritic cell
EC	Endothelial cell
EGF	Epidermal growth factor
EPC	Endothelial progenitor cell
EV	Extracellular vesicle
FGF	Fibroblast growth factor
HGF	Hepatocyte growth factor
Hh	Hedgehog
HIF	Hypoxia-inducible factor
HSPGs	Heparin sulfate proteoglycans
HUVECs	Human umbilical vein endothelial cells
IL	Interleukin
MAPK	Mitogen-activated protein kinases
MHC	Major histocompatibility complex
MI	Myocardial infarction
miRNAs	MicroRNAs
MMP	Matrix metalloproteinase
MSC	Mesenchymal stem cell
MSC-EVs	Mesenchymal stem cell-derived extracellular vesicle
NO	Nitric oxide
SDF	Stromal cell-derived factor
Shh	Sonic hedgehog
Sfrp2	Secreted frizzled related protein
T-2D	Type-2 diabetes
TGF	Transforming growth factor
TIMP	Tissue inhibitor of metalloproteinase

TNF   Tumor necrosis factor
uEV   Urinary extracellular vesicle
VEGF   Vascular endothelial growth factor
VEGFR   Vascular endothelial growth factor receptor
Wg   Wingless

REFERENCES

[1] Lavoie JR, Rosu-Myles M. Uncovering the secretes of mesenchymal stem cells. Biochimie 2013;95(12):2212–21.

[2] Katsuda T, Kosaka N, Takeshita F, Ochiya T. The therapeutic potential of mesenchymal stem cell-derived extracellular vesicles. Proteomics 2013;13(10–11):1637–53.

[3] Uccelli A, Moretta L, Pistoia V. Mesenchymal stem cells in health and disease. Nat Rev Immunol 2008;8(9):726–36.

[4] Waldenstrom A, Ronquist G. Role of exosomes in myocardial remodeling. Circ Res 2014;114(2):315–24.

[5] Bruno S, Camussi G. Role of mesenchymal stem cell-derived microvesicles in tissue repair. Pediatr Nephrol 2013;28(12):2249–54.

[6] Caplan AI, Correa D. The M.S.C.: an injury drugstore. Cell Stem Cell 2011;9(1):11–15.

[7] Murphy MB, Moncivais K, Caplan AI. Mesenchymal stem cells: environmentally responsive therapeutics for regenerative medicine. Exp Mol Med 2013;45:e54.

[8] Singer NG, Caplan AI. Mesenchymal stem cells: mechanisms of inflammation. Annu Rev Pathol 2011;6:457–78.

[9] Wen Z, Zheng S, Zhou C, Wang J, Wang T. Repair mechanisms of bone marrow mesenchymal stem cells in myocardial infarction. J Cell Mol Med 2011;15(5):1032–43.

[10] Joyce N, Annett G, Wirthlin L, Olson S, Bauer G, Nolta JA. Mesenchymal stem cells for the treatment of neurodegenerative disease. Regen Med 2010;5(6):933–46.

[11] Wu Y, Chen L, Scott PG, Tredget EE. Mesenchymal stem cells enhance wound healing through differentiation and angiogenesis. Stem Cells 2007;25(10):2648–59.

[12] Horwitz EM, Gordon PL, Koo WK, Marx JC, Neel MD, McNall RY, et al. Isolated allogeneic bone marrow-derived mesenchymal cells engraft and stimulate growth in children with osteogenesis imperfecta: implications for cell therapy of bone. Proc Natl Acad Sci USA 2002;99(13):8932–7.

[13] Bronckaers A, Hilkens P, Martens W, Gervois P, Ratajczak J, Struys T, et al. Mesenchymal stem/stromal cells as a pharmacological and therapeutic approach to accelerate angiogenesis. Pharmacol Ther 2014;143:181–96.

[14] Finan A, Dong F, Penn MS. Regenerative strategies for preserving and restoring cardiac function. Front Biosci (Elite Ed) 2013;5:232–48.

[15] Dong F, Caplan AI. Cell transplantation as an initiator of endogenous stem cell-based tissue repair. Curr Opin Organ Transplant 2012;17(6):670–4.

[16] Dong F, Harvey J, Finan A, Weber K, Agarwal U, Penn M. Myocardial CXCR4 expression is required for mesenchymal stem cell mediated repair following acute myocardial infarction. Circulation 2012;126:314–24.

[17] Hatzistergos KE, Quevedo H, Oskouei BN, Hu Q, Feigenbaum GS, Margitich IS, et al. Bone marrow mesenchymal stem cells stimulate cardiac stem cell proliferation and differentiation. Circ Res 2010;107(7):913–22.

[18] Schenk S, Mal N, Finan A, Zhang M, Kiedrowski M, Popovic Z, et al. Monocyte chemotactic protein-3 is a myocardial mesenchymal stem cell homing factor. Stem Cells 2007;25(1):245–51.

[19] Zhang M, Mal N, Kiedrowski M, Chacko M, Askari AT, Popovic ZB, et al. SDF-1 expression by mesenchymal stem cells results in trophic support of cardiac myocytes after myocardial infarction. Faseb J 2007;21(12):3197–207.

[20] Burlacu A, Grigorescu G, Rosca AM, Preda MB, Simionescu M. Factors secreted by mesenchymal stem cells and endothelial progenitor cells have complementary effects on angiogenesis *in vitro*. Stem Cells Dev 2013;22(4):643–53.

[21] Ruvinov E, Harel-Adar T, Cohen S. Bioengineering the infarcted heart by applying bio-inspired materials. J Cardiovasc Transl Res 2011;4(5):559–74.

[22] Colnot C. Cell sources for bone tissue engineering: insights from basic science. Tissue Eng Part B Rev 2011;17(6):449–57.

[23] Forteza R, Casalino-Matsuda SM, Monzon ME, Fries E, Rugg MS, Milner CM, et al. TSG-6 potentiates the antitissue kallikrein activity of inter-alpha-inhibitor through bikunin release. Am J Respir Cell Mol Biol 2007;36(1):20–31.

[24] Lee RH, Pulin AA, Seo MJ, Kota DJ, Ylostalo J, Larson BL, et al. Intravenous hMSCs improve myocardial infarction in mice because cells embolized in lung are activated to secrete the anti-inflammatory protein TSG-6. Cell Stem Cell 2009;5(1):54–63.

[25] Rodrigo SF, van Ramshorst J, Hoogslag GE, Boden H, Velders MA, Cannegieter SC, et al. Intramyocardial injection of autologous bone marrow-derived *ex vivo* expanded mesenchymal stem cells in acute myocardial infarction patients is feasible and safe up to 5 years of follow-up. J Cardiovasc Transl Res 2013;6(5):816–25.

[26] Yang SS, Kim NR, Park KB, Do YS, Roh K, Kang KS, et al. A phase I study of human cord blood-derived mesenchymal stem cell therapy in patients with peripheral arterial occlusive disease. Int J Stem Cells 2013;6(1):37–44.

[27] Hare JM, Fishman JE, Gerstenblith G, DiFede Velazquez DL, Zambrano JP, Suncion VY, et al. Comparison of allogeneic vs autologous bone marrow-derived mesenchymal stem cells delivered by transendocardial injection in patients with ischemic cardiomyopathy: the POSEIDON randomized trial. Jama 2012;308(22):2369–79.

[28] Hare JM, Traverse JH, Henry TD, Dib N, Strumpf RK, Schulman SP, et al. A random-ized, double-blind, placebo-controlled, dose-escalation study of intravenous adult hu-man mesenchymal stem cells (prochymal) after acute myocardial infarction. J Am Coll Cardiol 2009;54(24):2277–86.

[29] Puliafico SB, Penn MS, Silver KH. Stem cell therapy for heart disease. J Gen Intern Med 2013;28:1353–63.

[30] Barry FP, Murphy JM, English K, Mahon BP. Immunogenicity of adult mesenchymal stem cells: lessons from the fetal allograft. Stem Cells Dev 2005;14(3):252–65.

[31] Schuleri KH, Boyle AJ, Hare JM. Mesenchymal stem cells for cardiac regenerative ther-apy. Handb Exp Pharmacol 2007;180:195–218.

[32] Tang J, Wang J, Zheng F, Kong X, Guo L, Yang J, et al. Combination of chemokine and angiogenic factor genes and mesenchymal stem cells could enhance angiogenesis and improve cardiac function after acute myocardial infarction in rats. Mol Cell Biochem 2010;339(1–2):107–18.

[33] Leeper NJ, Hunter AL, Cooke JP. Stem cell therapy for vascular regeneration: adult, embryonic, and induced pluripotent stem cells. Circulation 2010;122(5):517–26.

[34] Shyu KG, Wang BW, Hung HF, Chang CC, Shih DT. Mesenchymal stem cells are su-perior to angiogenic growth factor genes for improving myocardial performance in the mouse model of acute myocardial infarction. J Biomed Sci 2006;13(1):47–58.

[35] Asahara T, Takahashi T, Masuda H, Kalka C, Chen D, Iwaguro H, et al. VEGF contrib-utes to postnatal neovascularization by mobilizing bone marrow-derived endothelial progenitor cells. Embo J 1999;18(14):3964–72.

[36] Caplan AI, Correa D. PDGF in bone formation and regeneration: new insights into a novel mechanism involving MSCs. J Orthop Res 2011;29(12):1795–803.

[37] Comerota AJ, Throm RC, Miller KA, Henry T, Chronos N, Laird J, et al. Naked plasmid DNA encoding fibroblast growth factor type 1 for the treatment of end-stage unreconstructible lower extremity ischemia: preliminary results of a phase I trial. J Vasc Surg 2002;35(5):930–6.

[38] Mohler ER 3rd, Rajagopalan S, Olin JW, Trachtenberg JD, Rasmussen H, Pak R, et al. Adenoviral-mediated gene transfer of vascular endothelial growth factor in critical limb ischemia: safety results from a phase I trial. Vasc Med 2003;8(1):9–13.

[39] Emanueli C, Madeddu P. Angiogenesis gene therapy to rescue ischaemic tissues: achievements and future directions. Br J Pharmacol 2001;133(7):951–8.

[40] Losordo DW, Vale PR, Isner JM. Gene therapy for myocardial angiogenesis. Am Heart J 1999;138(2 Pt 2):S132–41.

[41] Simons M, Bonow RO, Chronos NA, Cohen DJ, Giordano FJ, Hammond HK, et al. Clinical trials in coronary angiogenesis: issues, problems, consensus: an expert panel summary. Circulation 2000;102(11):E73–86.

[42] Kusumanto YH, van Weel V, Mulder NH, Smit AJ, van den Dungen JJ, Hooymans JM, et al. Treatment with intramuscular vascular endothelial growth factor gene compared with placebo for patients with diabetes mellitus and critical limb ischemia: a double-blind randomized trial. Hum Gene Ther 2006;17(6):683–91.

[43] Makinen K, Manninen H, Hedman M, Matsi P, Mussalo H, Alhava E, et al. Increased vascularity detected by digital subtraction angiography after VEGF gene transfer to human lower limb artery: a randomized, placebo-controlled, double-blinded phase II study. Mol Ther 2002;6(1):127–33.

[44] Belch J, Hiatt WR, Baumgartner I, Driver IV, Nikol S, Norgren L, et al. Effect of fibroblast growth factor NV1FGF on amputation and death: a randomised placebo-controlled trial of gene therapy in critical limb ischaemia. Lancet 2011;377(9781):1929–37.

[45] Ungvari Z, Pacher P, Csiszar A. Can simvastatin promote tumor growth by inducing angiogenesis similar to VEGF? Med Hypotheses 2002;58(1):85–6.

[46] Carmeliet P. VEGF gene therapy: stimulating angiogenesis or angioma-genesis? Nat Med 2000;6(10):1102–3.

[47] Zentilin L, Tafuro S, Zacchigna S, Arsic N, Pattarini L, Sinigaglia M, et al. Bone marrow mononuclear cells are recruited to the sites of VEGF-induced neovascularization but are not incorporated into the newly formed vessels. Blood 2006;107(9):3546–54.

[48] Frangogiannis NG. Stromal cell-derived factor-1-mediated angiogenesis for peripheral arterial disease: ready for prime time? Circulation 2011;123(12):1267–9.

[49] Ho TK, Tsui J, Xu S, Leoni P, Abraham DJ, Baker DM. Angiogenic effects of stromal cell-derived factor-1 (SDF-1/CXCL12) variants in vitro and the in vivo expressions of CXCL12 variants and CXCR4 in human critical leg ischemia. J Vasc Surg 2010;51(3):689–99.

[50] Tang J, Wang J, Yang J, Kong X, Zheng F, Guo L, et al. Mesenchymal stem cells overexpressing SDF-1 promote angiogenesis and improve heart function in experimental myocardial infarction in rats. Eur J Cardiothorac Surg 2009;36(4):644–50.

[51] Grunewald M, Avraham I, Dor Y, Bachar-Lustig E, Itin A, Jung S, et al. VEGF-induced adult neovascularization: recruitment, retention, and role of accessory cells. Cell 2006;124(1):175–89.

[52] Zhou B, Han ZC, Poon MC, Pu W. Mesenchymal stem/stromal cells (MSC) transfected with stromal derived factor 1 (SDF-1) for therapeutic neovascularization: enhancement of cell recruitment and entrapment. Med Hypotheses 2007;68(6):1268–71.

[53] Hiasa K, Ishibashi M, Ohtani K, Inoue S, Zhao Q, Kitamoto S, et al. Gene transfer of stromal cell-derived factor-1alpha enhances ischemic vasculogenesis and angiogenesis via vascular endothelial growth factor/endothelial nitric oxide synthase-related pathway: next-generation chemokine therapy for therapeutic neovascularization. Circulation 2004;109(20):2454–61.

[54] Hugel B, Martinez MC, Kunzelmann C, Freyssinet JM. Membrane microparticles: two sides of the coin. Physiology (Bethesda) 2005;20:22–7.

[55] Lee JK, Park SR, Jung BK, Jeon YK, Lee YS, Kim MK, et al. Exosomes derived from mesenchymal stem cells suppress angiogenesis by down-regulating VEGF expression in breast cancer cells. PLoS One 2013;8(12):e84256.

[56] Lai RC, Arslan F, Lee MM, Sze NS, Choo A, Chen TS, et al. Exosome secreted by MSC reduces myocardial ischemia/reperfusion injury. Stem Cell Res 2010;4(3):214–22.

[57] Camussi G, Deregibus MC, Tetta C. Paracrine/endocrine mechanism of stem cells on kidney repair: role of microvesicle-mediated transfer of genetic information. Curr Opin Nephrol Hypertens 2010;19(1):7–12.

[58] Quesenberry PJ, Dooner MS, Aliotta JM. Stem cell plasticity revisited: the continuum marrow model and phenotypic changes mediated by microvesicles. Exp Hematol 2010;38(7):581–92.

[59] Nekanti U, Dastidar S, Venugopal P, Totey S, Ta M. Increased proliferation and analysis of differential gene expression in human Wharton's jelly-derived mesenchymal stromal cells under hypoxia. Int J Biol Sci 2010;6(5):499–512.

[60] Yu X, Deng L, Wang D, Li N, Chen X, Cheng X, et al. Mechanism of TNF-alpha autocrine effects in hypoxic cardiomyocytes: initiated by hypoxia inducible factor 1alpha, presented by exosomes. J Mol Cell Cardiol 2012;53(6):848–57.

[61] Aoki N, Yokoyama R, Asai N, Ohki M, Ohki Y, Kusubata K, et al. Adipocyte-derived microvesicles are associated with multiple angiogenic factors and induce angiogenesis *in vivo* and *in vitro*. Endocrinology 2010;151(6):2567–76.

[62] Svensson KJ, Kucharzewska P, Christianson HC, Skold S, Lofstedt T, Johansson MC, et al. Hypoxia triggers a proangiogenic pathway involving cancer cell microvesicles and PAR-2-mediated heparin-binding EGF signaling in endothelial cells. Proc Natl Acad Sci USA 2011;108(32):13147–52.

[63] Bian S, Zhang L, Duan L, Wang X, Min Y, Yu H. Extracellular vesicles derived from human bone marrow mesenchymal stem cells promote angiogenesis in a rat myocardial infarction model. J Mol Med (Berl) 2014;92(4):387–97.

[64] Roobrouck VD, Ulloa-Montoya F, Verfaillie CM. Self-renewal and differentiation capacity of young and aged stem cells. Exp Cell Res 2008;314(9):1937–44.

[65] Wernicke CM, Grunewald TG, Juenger H, Kuci S, Kuci Z, Koehl U, et al. Mesenchymal stromal cells for treatment of steroid-refractory GvHD: a review of the literature and two pediatric cases. Int Arch Med 2011;4(1):27.

[66] Askari AT, Unzek S, Popovic ZB, Goldman CK, Forudi F, Kiedrowski M, et al. Effect of stromal-cell-derived factor 1 on stem-cell homing and tissue regeneration in ischaemic cardiomyopathy. Lancet 2003;362(9385):697–703.

[67] DiPersio JF. Diabetic stem-cell "mobilopathy". N Engl J Med 2011;365(26):2536–8.

[68] Ferraro F, Lymperi S, Mendez-Ferrer S, Saez B, Spencer JA, Yeap BY, et al. Diabetes impairs hematopoietic stem cell mobilization by altering niche function. Sci Transl Med 2011;3(104). 104ra1.

[69] Jin P, Zhang X, Wu Y, Li L, Yin Q, Zheng L, et al. Streptozotocin-induced diabetic rat-derived bone marrow mesenchymal stem cells have impaired abilities in proliferation, paracrine, antiapoptosis, and myogenic differentiation. Transplant Proc 2010;42(7):2745–52.

[70] Yan J, Tie G, Wang S, Messina KE, DiDato S, Guo S, et al. Type 2 diabetes restricts multipotency of mesenchymal stem cells and impairs their capacity to augment postischemic neovascularization in db/db mice. J Am Heart Assoc 2012;1(6):e002238.

[71] Chen M, Xie HQ, Deng L, Li XQ, Wang Y, Zhi W, et al. Stromal cell-derived factor-1 promotes bone marrow-derived cells differentiation to cardiomyocyte phenotypes *in vitro*. Cell Prolif 2008;41(2):336–47.

[72] Pasha Z, Wang Y, Sheikh R, Zhang D, Zhao T, Ashraf M. Preconditioning enhances cell survival and differentiation of stem cells during transplantation in infarcted myocardium. Cardiovasc Res 2008;77(1):134–42.

[73] Mangi AA, Noiseux N, Kong D, He H, Rezvani M, Ingwall JS, et al. Mesenchymal stem cells modified with Akt prevent remodeling and restore performance of infarcted hearts. Nat Med 2003;9(9):1195–201.

[74] Yang J, Zhou W, Zheng W, Ma Y, Lin L, Tang T, et al. Effects of myocardial transplantation of marrow mesenchymal stem cells transfected with vascular endothelial growth

factor for the improvement of heart function and angiogenesis after myocardial infarction. Cardiology 2007;107(1):17–29.

[75] Unzek S, Zhang M, Mal N, Mills WR, Laurita KR, Penn MS. SDF-1 recruits cardiac stem cell-like cells that depolarize *in vivo*. Cell Transplant 2007;16(9):879–86.

[76] Chen HK, Hung HF, Shyu KG, Wang BW, Sheu JR, Liang YJ, et al. Combined cord blood stem cells and gene therapy enhances angiogenesis and improves cardiac performance in mouse after acute myocardial infarction. Eur J Clin Invest 2005;35(11):677–86.

[77] Mackie AR, Klyachko E, Thorne T, Schultz KM, Millay M, Ito A, et al. Sonic hedgehog-modified human CD34+ cells preserve cardiac function after acute myocardial infarction. Circ Res 2012;111(3):312–21.

[78] Martinez MC, Larbret F, Zobairi F, Coulombe J, Debili N, Vainchenker W, et al. Transfer of differentiation signal by membrane microvesicles harboring hedgehog morphogens. Blood 2006;108(9):3012–20.

[79] Danilchik M, Williams M, Brown E. Blastocoel-spanning filopodia in cleavage-stage *Xenopus laevis*: potential roles in morphogen distribution and detection. Dev Biol 2013;382(1):70–81.

[80] Sharghi-Namini S, Tan E, Ong LL, Ge R, Asada HH. Dll4-containing exosomes induce capillary sprout retraction in a 3D microenvironment. Sci Rep 2014;4:4031.

[81] Sheldon H, Heikamp E, Turley H, Dragovic R, Thomas P, Oon CE, et al. New mechanism for Notch signaling to endothelium at a distance by delta-like 4 incorporation into exosomes. Blood 2010;116(13):2385–94.

[82] Borges FT, Melo SA, Ozdemir BC, Kato N, Revuelta I, Miller CA, et al. TGF-beta1-containing exosomes from injured epithelial cells activate fibroblasts to initiate tissue regenerative responses and fibrosis. J Am Soc Nephrol 2013;24(3):385–92.

[83] Trajkovic K, Hsu C, Chiantia S, Rajendran L, Wenzel D, Wieland F, et al. Ceramide triggers budding of exosome vesicles into multivesicular endosomes. Science 2008;319(5867):1244–7.

[84] Baietti MF, Zhang Z, Mortier E, Melchior A, Degeest G, Geeraerts A, et al. Syndecan-syntenin-ALIX regulates the biogenesis of exosomes. Nat Cell Biol 2012;14(7):677–85.

[85] Wendler F, Bota-Rabassedas N, Franch-Marro X. Cancer becomes wasteful: emerging roles of exosomes in cell-fate determination. J Extracell Vesicles 2013;2.

[86] Greco V, Hannus M, Eaton S. Argosomes: a potential vehicle for the spread of morphogens through epithelia. Cell 2001;106(5):633–45.

[87] Beckett K, Monier S, Palmer L, Alexandre C, Green H, Bonneil E, et al. *Drosophila* S2 cells secrete wingless on exosome-like vesicles but the wingless gradient forms independently of exosomes. Traffic 2013;14(1):82–96.

[88] Gross JC, Chaudhary V, Bartscherer K, Boutros M. Active Wnt proteins are secreted on exosomes. Nat Cell Biol 2012;14(10):1036–45.

[89] Peinado H, Aleckovic M, Lavotshkin S, Matei I, Costa-Silva B, Moreno-Bueno G, et al. Melanoma exosomes educate bone marrow progenitor cells toward a pro-metastatic phenotype through MET. Nat Med 2012;18(6):883–91.

[90] Thakur BK, Zhang H, Becker A, Matei I, Huang Y, Costa-Silva B, et al. Double-stranded DNA in exosomes: a novel biomarker in cancer detection. Cell Res 2014;24:766–9.

[91] Al-Nedawi K, Meehan B, Kerbel RS, Allison AC, Rak J. Endothelial expression of autocrine VEGF upon the uptake of tumor-derived microvesicles containing oncogenic EGFR. Proc Natl Acad Sci USA 2009;106(10):3794–9.

[92] Skog J, Wurdinger T, van Rijn S, Meijer DH, Gainche L, Sena-Esteves M, et al. Glioblastoma microvesicles transport RNA and proteins that promote tumour growth and provide diagnostic biomarkers. Nat Cell Biol 2008;10(12):1470–6.

[93] Hood JL, Pan H, Lanza GM, Wickline SA. Consortium for Translational Research in Advanced Imaging and Nanomedicine. Paracrine induction of endothelium by tumor exosomes. Laboratory investigation. J Tech Meth Pathol 2009;89(11):1317–28.

[94] Grange C, Tapparo M, Collino F, Vitillo L, Damasco C, Deregibus MC, et al. Microvesicles released from human renal cancer stem cells stimulate angiogenesis and formation of lung premetastatic niche. Cancer Res 2011;71(15):5346–56.

[95] Kucharzewska P, Christianson HC, Welch JE, Svensson KJ, Fredlund E, Ringner M, et al. Exosomes reflect the hypoxic status of glioma cells and mediate hypoxia-dependent activation of vascular cells during tumor development. Proc Natl Acad Sci USA 2013;110(18):7312–17.

[96] Chappell JC, Taylor SM, Ferrara N, Bautch VL. Local guidance of emerging vessel sprouts requires soluble Flt-1. Dev Cell 2009;17(3):377–86.

[97] Boscolo E, Mulliken JB, Bischoff J. VEGFR-1 mediates endothelial differentiation and formation of blood vessels in a murine model of infantile hemangioma. Am J Pathol 2011;179(5):2266–77.

[98] Serpi R, Tolonen AM, Huusko J, Rysa J, Tenhunen O, Yla-Herttuala S, et al. Vascular endothelial growth factor-B gene transfer prevents angiotensin II-induced diastolic dysfunction via proliferation and capillary dilatation in rats. Cardiovasc Res 2011;89(1):204–13.

[99] Bry M, Kivela R, Holopainen T, Anisimov A, Tammela T, Soronen J, et al. Vascular endothelial growth factor-B acts as a coronary growth factor in transgenic rats without inducing angiogenesis, vascular leak, or inflammation. Circulation 2010;122(17):1725–33.

[100] Kim HK, Song KS, Chung JH, Lee KR, Lee SN. Platelet microparticles induce angiogenesis in vitro. Brit J Haematol 2004;124(3):376–84.

[101] Janowska-Wieczorek A, Wysoczynski M, Kijowski J, Marquez-Curtis L, Machalinski B, Ratajczak J, et al. Microvesicles derived from activated platelets induce metastasis and angiogenesis in lung cancer. Int J Cancer 2005;113(5):752–60.

[102] Janowska-Wieczorek A, Marquez-Curtis LA, Wysoczynski M, Ratajczak MZ. Enhancing effect of platelet-derived microvesicles on the invasive potential of breast cancer cells. Transfusion 2006;46(7):1199–209.

[103] Kim HK, Song KS, Park YS, Kang YH, Lee YJ, Lee KR, et al. Elevated levels of circulating platelet microparticles, VEGF, IL-6 and RANTES in patients with gastric cancer: possible role of a metastasis predictor. Eur J Cancer 2003;39(2):184–91.

[104] Zernecke A, Bidzhekov K, Noels H, Shagdarsuren E, Gan L, Denecke B, et al. Delivery of microRNA-126 by apoptotic bodies induces CXCL12-dependent vascular protection. Sci Signal 2009;2(100). ra81.

[105] Zhang X, Wang X, Zhu H, Kranias EG, Tang Y, Peng T, et al. Hsp20 functions as a novel cardiokine in promoting angiogenesis via activation of VEGFR2. PLoS One 2012;7(3):e32765.

[106] Sahoo S, Klychko E, Thorne T, Misener S, Schultz KM, Millay M, et al. Exosomes from human CD34(+) stem cells mediate their proangiogenic paracrine activity. Circ Res 2011;109(7):724–8.

[107] Mocharla P, Briand S, Giannotti G, Dorries C, Jakob P, Paneni F, et al. AngiomiR-126 expression and secretion from circulating CD34(+) and CD14(+) PBMCs: role for proangiogenic effects and alterations in type 2 diabetics. Blood 2013;121(1):226–36.

[108] Chen TS, Lai RC, Lee MM, Choo AB, Lee CN, Lim SK. Mesenchymal stem cell secretes microparticles enriched in pre-microRNAs. Nucleic Acids Res 2010;38(1):215–24.

[109] Lai RC, Arslan F, Lee MM, Sze NS, Choo A, Chen TS, et al. Exosome secreted by MSC reduces myocardial ischemia/reperfusion injury. Stem Cell Res 2010;4(3):214–22.

[110] Mallegol J, Van Niel G, Lebreton C, Lepelletier Y, Candalh C, Dugave C, et al. T84-intestinal epithelial exosomes bear MHC class II/peptide complexes potentiating antigen presentation by dendritic cells. Gastroenterology 2007;132(5):1866–76.

[111] Christianson HC, Svensson KJ, van Kuppevelt TH, Li JP, Belting M. Cancer cell exosomes depend on cell-surface heparan sulfate proteoglycans for their internalization and functional activity. Proc Natl Acad Sci USA 2013;110(43):17380–5.

[112] Raposo G, Stoorvogel W. Extracellular vesicles: exosomes, microvesicles, and friends. J Cell Biol 2013;200(4):373–83.

[113] Montecalvo A, Larregina AT, Shufesky WJ, Stolz DB, Sullivan ML, Karlsson JM, et al. Mechanism of transfer of functional microRNAs between mouse dendritic cells via exosomes. Blood 2012;119(3):756–66.

[114] Thuma F, Zoller M. Outsmart tumor exosomes to steal the cancer initiating cell its niche. Semin Cancer Biol 2014;28:39–50.

[115] Benameur T, Soleti R, Porro C, Andriantsitohaina R, Martinez MC. Microparticles carrying Sonic hedgehog favor neovascularization through the activation of nitric oxide pathway in mice. PLoS One 2010;5(9):e12688.

[116] Tual-Chalot S, Gagnadoux F, Trzepizur W, Priou P, Andriantsitohaina R, Martinez MC. Circulating microparticles from obstructive sleep apnea syndrome patients induce endothelin-mediated angiogenesis. Biochim Biophys Acta 2014;1842(2):202–27.

[117] Cai Z, Yang F, Yu L, Yu Z, Jiang L, Wang Q, et al. Activated T cell exosomes promote tumor invasion via Fas signaling pathway. J Immunol 2012;188(12):5954–61.

[118] Leroyer AS, Rautou PE, Silvestre JS, Castier Y, Leseche G, Devue C, et al. CD40 ligand+ microparticles from human atherosclerotic plaques stimulate endothelial proliferation and angiogenesis, a potential mechanism for intraplaque neovascularization. J Am Coll Cardiol 2008;52(16):1302–11.

[119] Castellana D, Zobairi F, Martinez MC, Panaro MA, Mitolo V, Freyssinet JM, et al. Membrane microvesicles as actors in the establishment of a favorable prostatic tumoral niche: a role for activated fibroblasts and CX3CL1-CX3CR1 axis. Cancer Res 2009;69(3):785–93.

[120] Hong BS, Cho JH, Kim H, Choi EJ, Rho S, Kim J, et al. Colorectal cancer cell-derived microvesicles are enriched in cell cycle-related mRNAs that promote proliferation of endothelial cells. BMC Genom 2009;10:556.

[121] Nazarenko I, Rana S, Baumann A, McAlear J, Hellwig A, Trendelenburg M, et al. Cell surface tetraspanin Tspan8 contributes to molecular pathways of exosome-induced endothelial cell activation. Cancer Res 2010;70(4):1668–78.

[122] Mineo M, Garfield SH, Taverna S, Flugy A, De Leo G, Alessandro R, et al. Exosomes released by K562 chronic myeloid leukemia cells promote angiogenesis in a Src-dependent fashion. Angiogenesis 2012;15(1):33–45.

[123] Valadi H, Ekstrom K, Bossios A, Sjostrand M, Lee JJ, Lotvall JO. Exosome-mediated transfer of mRNAs and microRNAs is a novel mechanism of genetic exchange between cells. Nat Cell Biol 2007;9(6):654–9.

[124] Umezu T, Ohyashiki K, Kuroda M, Ohyashiki JH. Leukemia cell to endothelial cell communication via exosomal miRNAs. Oncogene 2013;32(22):2747–55.

[125] He L, Thomson JM, Hemann MT, Hernando-Monge E, Mu D, Goodson S, et al. A microRNA polycistron as a potential human oncogene. Nature 2005;435(7043):828–33.

[126] Tadokoro H, Umezu T, Ohyashiki K, Hirano T, Ohyashiki JH. Exosomes derived from hypoxic leukemia cells enhance tube formation in endothelial cells. J Biol Chem 2013;288(48):34343–51.

[127] Finn NA, Searles CD. Intracellular and extracellular miRNAs in regulation of angiogenesis signaling. Curr Angiogen 2012;4(102):299–307.

[128] Bronisz A, Wang Y, Nowicki MO, Peruzzi P, Ansari KI, Ogawa D, et al. Extracellular vesicles modulate the glioblastoma microenvironment via a tumor suppression signaling network directed by miR-1. Cancer Res 2014;74(3):738–50.

[129] Zampetaki A, Kiechl S, Drozdov I, Willeit P, Mayr U, Prokopi M, et al. Plasma microRNA profiling reveals loss of endothelial miR-126 and other microRNAs in type 2 diabetes. Circ Res 2010;107(6):810–17.

[130] Dubois S, Madec AM, Mesnier A, Armanet M, Chikh K, Berney T, et al. Glucose inhibits angiogenesis of isolated human pancreatic islets. J Mol Endocrinol 2010;45(2):99–105.

[131] De Rosa S, Fichtlscherer S, Lehmann R, Assmus B, Dimmeler S, Zeiher AM. Transcoronary concentration gradients of circulating microRNAs. Circulation 2011;124(18):1936–44.

[132] Simpson RJ, Jensen SS, Lim JW. Proteomic profiling of exosomes: current perspectives. Proteomics 2008;8(19):4083–99.

[133] Zhang Y, Liu D, Chen X, Li J, Li L, Bian Z, et al. Secreted monocytic miR-150 enhances targeted endothelial cell migration. Mol Cell 2010;39(1):133–44.

[134] Gesierich S, Berezovskiy I, Ryschich E, Zoller M. Systemic induction of the angiogenesis switch by the tetraspanin d6.1a/co-029. Cancer Res 2006;66:7083–94.

[135] Taverna S, Flugy A, Saieva L, Kohn EC, Santoro A, Meraviglia S, et al. Role of exosomes released by chronic myelogenous leukemia cells in angiogenesis. Int J Cancer 2012;130:2033–43.

[136] Mathivanan S, Lim JW, Tauro BJ, Ji H, Moritz RL, Simpson RJ. Proteomics analysis of a33 immunoaffinity-purified exosomes released from the human colon tumor cell line lim1215 reveals a tissue-specific protein signature. Mol Cell Proteomics 2010;9:197–208.

[137] Halkein J, Tabruyn SP, Ricke-Hoch M, Haghikia A, Nguyen NQ, Scherr M, et al. MicroRNA-146a is a therapeutic target and biomarker for peripartum cardiomyopathy. J Clin Invest 2013;123:2143–54.

[138] Katoh M. Therapeutics targeting angiogenesis: genetics and epigenetics, extracellular miRNAs and signaling networks (Review). Int J Mol Med 2013;32(4):763–7.

[139] Sluijter JP, Verhage V, Deddens JC, van den Akker F, Doevendans PA. Microvesicles and exosomes for intracardiac communication. Cardiovasc Res 2014;102(2):302–11.

[140] Cosme J, Liu PP, Gramolini AO. The cardiovascular exosome: current perspectives and potential. Proteomics 2013;13(10–11):1654–9.

[141] Komatsu S, Ichikawa D, Takeshita H, Morimura R, Hirajima S, Tsujiura M, et al. Circulating miR-18a: a sensitive cancer screening biomarker in human cancer. In Vivo (Athens Greece) 2014;28(3):293–7.

[142] Boufraqech M, Zhang L, Jain M, Patel D, Ellis R, Xiong Y, et al. miR-145 suppresses thyroid cancer growth and metastasis and targets AKT3. Endocr Relat Cancer 2014;21:517–31.

[143] Salih M, Zietse R, Hoorn EJ. Urinary extracellular vesicles and the kidney: biomarkers and beyond. Am J Physiol Renal Physiol 2014;306:F1251–9.

[144] Toma C, Wagner WR, Bowry S, Schwartz A, Villanueva F. Fate of culture-expanded mesenchymal stem cells in the microvasculature: in vivo observations of cell kinetics. Circ Res 2009;104(3):398–402.

[145] Ge J, Guo L, Wang S, Zhang Y, Cai T, Zhao RC, et al. The size of mesenchymal stem cells is a significant cause of vascular obstructions and stroke. Stem Cell Rev 2014;10(2):295–303.

[146] O'Loughlin AJ, Woffindale CA, Wood MJ. Exosomes and the emerging field of exosome-based gene therapy. Curr Gene Ther 2012;12(4):262–74.

[147] Mirotsou M, Zhang Z, Deb A, Zhang L, Gnecchi M, Noiseux N, et al. Secreted frizzled related protein 2 (Sfrp2) is the key Akt-mesenchymal stem cell-released paracrine factor mediating myocardial survival and repair. Proc Natl Acad Sci USA 2007;104(5):1643–8.

[148] Timmers L, Lim SK, Arslan F, Armstrong JS, Hoefer IE, Doevendans PA, et al. Reduction of myocardial infarct size by human mesenchymal stem cell conditioned medium. Stem Cell Res 2007;1(2):129–37.

[149] Ibrahim AG, Cheng K, Marban E. Exosomes as critical agents of cardiac regeneration triggered by cell therapy. Stem Cell Rep 2004;2(5):606–19.

CHAPTER 10

Exosomes for Bone Diseases

Paulomi Sanghavi, Porter Young, Sunil Upadhyay, Mark W. Hamrick
Department of Cellular Biology and Anatomy, Medical College of Georgia, Georgia Regents University, Augusta, GA, USA

Contents

1 INTRODUCTION: INTEGRATIVE PHYSIOLOGY OF THE SKELETON

The maintenance of bone strength throughout life has important clinical implications, as bone fractures are a significant cause of morbidity and mortality among the elderly. Hip fractures are particularly debilitating, and it is estimated that approximately 40% of those experiencing a hip fracture will require assisted living and 20% will never walk again [1]. Strategies to increase bone formation both early and late in life, and reduce bone loss with aging, can therefore have a major impact on public health. Bone is a dynamic tissue that has historically been viewed as playing a primary role in weight bearing and in serving as a system of levers about which muscles can act to produce movement. In this way the overall morphology and metabolism of bone were seen primarily as being related to this core structural, mechanical function. Several key assumptions of this paradigm are that if mechanical loads on bone are increased then the normal stresses and strains on bone tissue will also increase, and bone formation will be activated to bring these strains back into a normal physiological window. On the other hand, if loads are significantly reduced, for example in the case of weight loss or bedrest, then bone resorption will occur in order to remove unnecessary bone tissue. These general principles form the cornerstone of models such as the "mechanostat" concept that relate bone loading and bone strain to bone formation and bone resorption (e.g., [2]). The characterization above is overly simplified, as there are obviously other confounding variables that impact the response of bone tissue (e.g., sex steroid status) to loading, but the basic premise is that the mechanical environment plays an important role in the regulation of bone micro- and macromorphology.

There are, however, a number of recent studies showing that bone formation and resorption are mediated by signals other than just mechanically derived stimuli, and that the skeleton may itself secrete factors that can impact other organs. For example, there is evidence to suggest that the brain plays a key role in bone metabolism via the activation of beta-adrenergic pathways in bone through sympathetic nervous innervation [3,4], and that restoring leptin signaling in the brain of leptin-deficient mice can increase bone mass [5,6]. In fact, data from rats with vestibular lesions of the inner ear show that these lesions can induce significant bone loss, and that this loss is reversible with the beta-adrenergic antagonist propranolol [7]. In addition, skeletal muscle secretes a number of peptides collectively termed myokines [8], many of which are anabolic for bone [9]. Neural and muscular systems therefore appear to influence bone tissue through pathways that may be independent of mechanical loading. Finally, bone itself releases peptides that can affect physiological processes such as insulin sensitivity, the most well known of these factors being osteocalcin [10,11] Oury et al., 2013. Thus, bone not only serves a mechanical function within the body but can also function in a paracrine or even endocrine manner.

2 CELL–CELL INTERACTIONS IN THE BONE MARROW MICROENVIRONMENT WITH AGING AND DISEASE

As discussed earlier, bone plays an important role in regulating a number of metabolic functions in the body. The most obvious of these functions is that bone serves as a reservoir of calcium, so that when calcium levels are low parathyroid hormone (PTH) levels are increased, activating osteoclasts to degrade bone tissue and release calcium stores into the bloodstream. Bone is also a primary site for hematopoiesis, and the bone marrow microenvironment supports not only hematopoietic stem cells (HSCs) but also derivatives of these cells such as erythrocytes, lymphocytes, and neutrophils. The hematopoietic lineage cells, at least in humans, are generally positive for cell surface markers such as CD14, 20, 34, and 45 and negative for CD 271

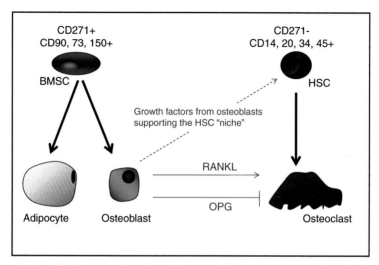

Figure 10.1 *Cell Types in the Bone Marrow Microenvironment, Showing Surface Markers for BMSCs and HSCs.* BMSCs give rise to both adipocytes and osteoblasts, and the former mediate the differentiation of osteoclasts and also secrete growth factors that maintain the HSC "niche."

([12]; Figure 10.1). Maintenance of the hematopoietic "stem cell niche" is mediated in part by bone marrow stromal cells (BMSCs), which in humans are CD271 positive, negative for the hematopoietic markers, and positive for CD90, 105, and 73 ([12,13]; Figure 10.1). BMSCs not only support the HSC niche but also give rise to bone forming cells, the osteoblasts, and can also differentiate to form bone marrow adipocytes, which are known to increase with aging, estrogen deficiency, and disuse [14–16]. The converse also appears to be true, namely that HSCs can stimulate osteoblastic differentiation of BMSCs [17].

Circulating HSCs appear to play an important role in normal tissue repair [18–20]. Like BMSCs, osteoblasts are also known to produce hematopoietic growth factors [21,22] and it has been shown by Calvi et al. [23] and Zhang et al. [24] that osteoblast cells regulate HSC number in bone marrow (see also [25–27]; Figure 10.1). For example, transgenic mice expressing an osteoblast-specific PTH/PTHrP receptor show a significant increase in trabecular bone mass and HSC number. This is further indicated by the fact that mice given donor marrow after myoablation showed a rapid increase in HSC number with PTH treatment, whereas mice that did not receive PTH showed a marked increase in marrow adipocyte number and a significant decrease in HSCs [23]. Maintenance of the HSC niche by factors secreted by osteoblasts and/or BMSCs can therefore affect regenerative processes in other organs, revealing that changes in bone have much broader impacts on multiple organs and tissues.

BMSCs and osteoblasts not only support the HSC stem cell niche, but also interact to regulate bone resorption (Figure 10.1). Osteoclasts, which resorb bone, are derived from the monocyte-macrophage progenitor cell lineage and receptor activator of nuclear factor kappa-B (RANK) (NF-κB) ligand is required for osteoclast differentiation. RANK-ligand (RANK-L) is produced by both osteoblasts and BMSCs, providing a mechanistic coupling of bone formation and bone resorption. Osteoprotegerin, a decoy receptor for RANK-L, is also produced by osteoblasts and can inhibit osteoclastogenesis. Thus, bone resorption is regulated in part by signals derived from bone forming cells (Figure 10.1). Adipocytes within bone marrow also play a role in regulating bone metabolism, as they can secrete osteoclastogenic cytokines such as interleukin-6 [28,29] and can also inhibit osteoblast activity in culture [29]. Bone marrow adipogenesis is therefore linked directly with decreased osteogenesis and increased osteoclastogenesis. Finally, BMSCs express high levels of stromal-derived factor 1 (SDF-1) and also express the SDF-1 receptor CXCR4, which play important roles in cell migration and engraftment [30].

These studies highlight the important interactions that exist among different types of cells in the bone marrow microenvironment; moreover, they also demonstrate that, as these interactions are altered with aging, there are consequences that may not only impact bone metabolism and bone mass but also tissue repair in other organs that require circulating, bone-derived progenitors. More importantly, these studies demonstrate that a better understanding of the mechanisms by which cells communicate within the bone marrow microenvironment can open up new therapeutic opportunities to prevent bone loss, stimulate bone formation, and enhance the normal process of tissue repair.

3 ROLE OF EXTRACELLULAR VESICLES IN BONE BIOLOGY

Cell-derived microvesicles (MVs) and exosomes have recently been described as a new mechanism of cell–cell communication [31,32]. Exosomes are small (40–100 nm) and MVs are larger (>100 nm) membrane-derived structures that are released into the extracellular space by a variety of cell types [33,34]. These lipid-based carriers are now known to shuttle miRNAs between cells, delivering their miRNAs to target cells via endocytosis and membrane fusion [35]. Microvesicle- and exosome-derived transport of miRNAs therefore represents one cellular and molecular pathway for epigenetic reprogramming of target cells [36].

As mentioned earlier, cell-derived MVs and exosomes are a novel source of cell–cell communication [37]. These vesicles mediate the transfer of key proteins, and genetic material such as DNA, mRNAs, mi-RNAs, and other noncoding RNAs from the donor to the recipient cell [38]. MVs derived from adult human bone marrow mesenchymal stem cells (hBMSCs), also called BMSCs, contain a variety of ribonucleoproteins. Proteins such as Staufen-1, Staufen-2, and human antigen-R that are required for mRNA trafficking and stabilization, and processing protein Argonaute, which is involved in processing of miRNAs, are well-known cargoes of these vesicles. These findings suggest that RNAs from different cellular compartments are shuttled into these MVs [35]. Because miRNAs function as regulators of gene expression, a lot of work has been done studying the miRNA content of these MVs, and this work reveals that the miRNA repertoire in MVs is quite different from that of the parent cells. Although some of the miRNAs are present with the same abundance in parent BMSCs as their secreted MVs, many of the signature miRNAs such as miR-223, miR-451, and miR-564 were specifically seen to accumulate in MVs. This hints toward an active and organized sorting process that loads these miRNAs into MVs.

The exact molecular pathways by which these miRNAs are selected for transport into these vesicles mostly remain unclear. Recently, short sequence motifs called EXOmotifs have been identified within these miRNA sequences that are essential for their enrichment into these extracellular vesicles (EVs). These EXOmotifs basically mediate the association of cargo miRNAs with various RNA-binding proteins; one of the reported ones is heterogeneous nuclear ribonucleoproteins A2/B1 [39]. Additionally, release of these MVs requires actin polymerization as well as GTPase Ras-related protein Rab-27A [35,38]. These MVs released by BMSCs act on different target cells to bring about different physiological responses [40,41] (Rocarro et al., 2013). The other progenitor cell type in the bone marrow that secretes these MVs is the HSC. Similar to BMSC-derived MVs, MVs released by HSCs also show enrichment of miRNAs such as miR-10b, miR-378, and miR-95. These miRNAs, enriched in the MVs of adult BMSCs as well as HSCs, influence a number of cellular processes such as development, cell differentiation, transcriptional regulation, and immune system processes. Also, both the stem cell populations seem to share

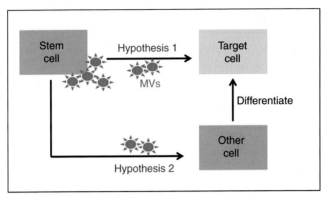

Figure 10.2 *Schematic Diagram Showing the Mechanism by Which Stem Cell Functions to Bring About the Desired Response in the Target Cells.* After the target cells signal the stem cells (via release of MVs or other factors), stem cells secrete regenerative MVs and bring about the appropriate response either by acting directly on the target cells (Hypothesis 1) or these MVs act on other local cell types, which in turn differentiate and re-enter the cell cycle to bring about the required action (Hypothesis 2).

more than 90% of their miRNAs consistent with their common ancestral origin [35].

One intriguing area that requires some study is regulation of secretion and mode of action of these vesicles. How exactly do these stem cells in the bone microenvironment function to carry out these *in vivo* processes to affect their respective target cells? According to one hypothesis, MVs released from the target cells signal the stem cells to release MVs containing various factors required for their growth, repair, or differentiation. These MVs either directly act on the target cell to bring about their epigenetic reprogramming (Figure 10.2) or, alternatively, these MVs might act on the neighboring cells and cause them to dedifferentiate and re-enter the cell cycle in order to regenerate or differentiate into the required target cell type (Figure 10.2 [37]). Consistent with this model, BMSC-derived vesicles have been successfully shown to reverse the radiation toxicity of bone marrow stem cells [41].

Apart from the marrow cells, bone cells such as osteocytes and osteoblasts are also known to secrete MVs containing various bone regulatory proteins in the extracellular medium. Fu et al. [42] observed that MVs from osteoblast cell line UAMS32 facilitate the formation of osteoclasts via the RANK/RANKL OPG pathway and modulate bone remodeling. According to their results, RANKL was found to be enriched on

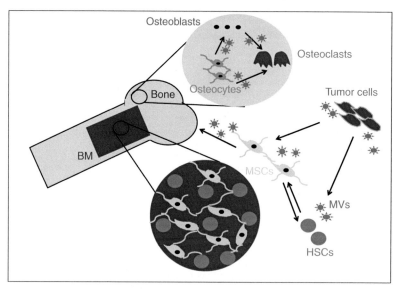

Figure 10.3 *Illustration Showing the Role of MVs in Bone Microenvironment.* Different cell types in the marrow (MSCs and HSCs) as well as bone tissue (osteoblasts, osteocytes, osteoclasts) secrete MVs and bring about physiological responses in other cell types. Direction of arrow indicates the direction of transfer of MVs from parent to the target cell.

the surface of these MVs. Thus, MVs represent an important means of cell–cell communication between different cell types in the bone tissue (Figure 10.3). Outside the bone tissue, MVs secreted from other cells also influence many processes in the bone marrow microenvironment. Murine BM-derived HSCs when cultured in the presence of MVs derived from murine embryonic stem (ES) cells showed improved survival and expansion coupled with upregulation of various early pluripotent genes such as octamer-binding transcription factor 4, Rex-1, and Nanog and many of the HSC markers such as Scleroderma, Homeobox protein B4, and GATA-binding protein 2. These MVs were also highly enriched in Wnt-3 protein as well as mRNA for many pluripotent transcription factors that are later on translated into their respective proteins in host cells [43]. Similarly, platelet-derived MVs are implicated in improved homing of HSCs in their microenvironment. It is seen that HSCs that interacted with platelet-derived particles display increased adherence to endothelial cells and immobilized SDF-1 as compared to normal cells, thus providing a significant potential in the optimization of stem cell transplantations [44].

4 EXTRACELLULAR VESICLES AND CANCER METASTASIS TO BONE

Bone metastasis is commonly seen in prostate cancer patients, and tumor cells represent another cell type that is reported to secrete a large number of EVs (Figure 10.3). These tumor cell-derived MVs are also known to affect a number of processes in bone physiology and are critical for bone disease progression. Exosomes derived from hormone refractory prostate cancer cells enhance the differentiation of osteoblasts via the delivery of erythroblastosis virus E26 oncogene homolog 1 (Ets1). Ets1 acts as a transcriptional factor for differentiation of osteoblasts facilitating bone metastasis during prostate cancer [45]. Similarly, MVs derived from chronic lymphocytic leukemia (CLL) stimulate BMSCs by activating a number of signaling cascades. Also, these stromal cells are shown to secrete more VEGF as compared to their normal counterparts, which may play some role in facilitating survival of a tumor microenvironment for CLL (Ghosh et al., 2010). The exact molecular mechanism by which these MVs activate the stromal cells is, however, unclear. A number of studies are now focused toward identifying the oncogenic factors present in these MVs that activate the downstream metastatic pathway. Identification of these molecules will significantly improve the potential of these MVs to be target different tumors. For example, MVs derived from K562 leukemia cells accumulate breakpoint cluster region-Abelson leukemia gene human homolog 1 (BCR-ABL1) mRNA. These BCR-ABL1-positive MVs cause transformation of hematopoietic transplants by inducing genomic instability via a number of mechanisms [46]. Similarly, exosomes derived from melanoma cells are known to permanently direct and mobilize the bone marrow cells by horizontal transfer of Met oncoprotein thereby promoting metastatic progression [47].

Once bone marrow metastasis occurs, there is another feedback loop between tumor cells in the bone and other tissues, which further results in tumor progression. One example is multiple myeloma where the exosomes released from BMSCs support tumor development [48]. These exosomes derived from multiple myeloma BMSCs contain lower amounts of tumor suppressor miR-15a and higher levels of oncogenic proteins, cytokines, and adhesion molecules thereby promoting multiple myeloma growth [49]. Similarly, exosomes derived from BMSCs from myelodysplastic patients were seen to upregulate 21 out of 384 miRNAs, amongst which were miR-10a and miR-132 that are involved in apoptosis. Thus, these differential expressions of miRNAs incorporate into HSCs and thus affect downstream processes leading to myelodysplasticity [40].

5 CLINICAL AND THERAPEUTIC APPLICATIONS OF BONE-DERIVED EXTRACELLULAR VESICLES

Identification of mRNAs and miRNAs from MVs derived from specific cell types often serve as biomarkers to detect various diseases or pathological conditions [50]. For instance, hematopoietic stem (precursor) cells secrete MVs in the marrow. These MVs containing signature mRNAs and miRNAs move from the marrow into peripheral blood, thus making it possible to assess the marrow condition post-transplantation by simply analyzing these mRNAs and miRNAs from peripheral blood instead of having to perform frequent bone marrow aspirations [51]. Apart from biomarkers, the identification of specific peptide cargo within MVs also allows for earlier detection and subsequent treatment for primary bone disorders, such as Ewing's sarcoma/Friend leukemia integration 1 transcription factor mRNA found in MVs of Ewing's sarcoma [52].

miRNAs also provide a potential target for malignancies, most notably cancer since its etiology is rooted in abnormal cell signaling. Valencia et al. [53] observed that miR-192 was underexpressed in MVs derived from their mouse model of bone metastasis. When miR-192 expression was upregulated in this model, bone metastasis was reduced by inhibiting tumor angiogenesis. In order to verify the role of MVs in delivering miR-192, they injected the mouse model with MVs enriched with miR-192 and observed the same antiangiogenic, tumor inhibiting effects. This finding could have important clinical implications since a variety of cancers frequently metastasize to bone [53]. In another oncological application, Bruno et al. observed that when MVs derived from hBM-MSCs were injected into mice expressing cell lines of Kaposi's sarcoma, Skov-3 ovarian tumors, and HepG2 hepatoma, tumor growth was inhibited. While *in vitro* studies determined these MVs were associated with an activation of negative cell cycle regulators, their exact mechanism *in vivo* is unknown [54].

Bone marrow transplants are commonly used to treat primary hematologic neoplasms. During the procedure, granulocyte colony stimulating factor (G-CSF) stimulates HSC mobilization from the bone marrow to the peripheral circulation so that the cells can be isolated and transplanted. Unfortunately, mobilization of HSCs is not an efficient process because HSCs make up a small portion of the mobilized cells [55]. However, Salvucci et al. have demonstrated that miR-126, which is concentrated in MVs of HSCs and stromal cells of bone marrow upon G-CSF administration, plays a pivotal role in this process by antagonizing the effects of vascular cell adhesion molecule 1 (VCAM1). VCAM1 binds and retains HSCs in the bone marrow environment. These findings are further supported by the fact that miR-126 knockout mouse models have little response to G-CSF and less mobilization [56]. Therefore, future therapy utilizing the efficiency of miR-126 could make bone marrow transplants more cost effective and less time intensive.

Oncology is not the only field that can benefit from MV-based therapy. For example, Bian et al. [57] demonstrated that when hBMSCs are injected into ischemic mouse myocardium, they have the capability of stimulating angiogenesis under hypoxic conditions via their release of cell mediators contained within MVs. These mediators not only stimulate angiogenesis but also reduce the myocardial infarct size and improve cardiac function when compared to controls. While many potential mediators of angiogenesis have been proposed, the exact cellular mediators contained within these MVs has not been determined, which is crucial for developing potential therapies [57]. In contrast, the role of MVs derived from hBMSCs in aiding the kidneys recover from acute kidney injury (AKI) is better understood [58]. For example, Bruno et al. compared the effects of injecting an acutely injured mouse kidney with hBMSCs versus only injecting MVs from the same hBMSCs. The MVs alone were found to be just as effective as hBMSCs in aiding recovery and limiting tubular damage. Interestingly, when RNase was added to the MV preparation, the recovery capacity was lost, suggesting a key mRNA mediator is present in the MVs [59].

6 SUMMARY AND CONCLUSIONS

The bone marrow microenvironment contains a variety of cell types that interact in complex ways to mediate tissue regeneration, bone metabolism, and disease progression. Communication among these different cells is key to the regulation of bone formation, bone resorption, and cell homing and engraftment. Exosomes and microvesicles, together referred to as EVs, represent a new and quite poorly understood mechanism by which cells communicate and interact. These EVs shuttle microRNAs, mRNAs, and proteins among different cell types, and can in this way impact the epigenetic reprogramming of target cells. In bone these EVs are likely to play a role in the differentiation of BMSCs and HSCs, maintenance of the stem cell niche in bone, and metastasis of cancer to bone.

Current efforts are exploring the possibilities and methods of how EVs can be used to deliver genetic and pharmacologic therapy to specific target tissues. EV carriers are advantageous because they have fewer nontarget tissue effects and evade immunological degradation better than current liposomal carriers, which are artificially derived EVs [60]. EVs have the potential to be used for gene therapy [61] as well as vaccines [62]. Furthermore, therapeutic activity of these EVs has been demonstrated in animal models of AKI and myocardial ischemia/reperfusion injury [54,63]. However, to be able to use them clinically for clinical applications, it is extremely important to be able to identify and purify the factors delivered by these EVs to their target cells. Currently, a number of biochemical purification techniques are used to purify these EVs and further identify the cargoes that are delivered by these vesicles. For instance, differential centrifugation followed by protein purification methods, quantitative mass spectrometry, immunological-based methods, flow cytometry, and bioinformatics are being employed to first isolate these membrane-bound vesicles and further to correctly identify and characterize the protein, RNA, and miRNA contents of these EVs depending on the parent cell type [64,65]. It is likely that further investigation into the basic biology of EVs in bone will lead to novel clinical applications, possibly including new targets to prevent cancer metastasis and progression, as well as biomarkers that may be of predictive value in bone-related disorders.

ABBREVIATIONS

AKI	Acute kidney injury
BM	Bone marrow
BMSC	Bone marrow stromal (mesenchymal) stem cell
BCR-ABL1	Breakpoint cluster region-Abelson leukemia gene human homolog 1
CLL	Chronic lymphocytic leukemia
CXCR4	C-X-C chemokine receptor type 4
ES	Embryonic stem cell
Ets1	Erythroblastosis virus E26 oncogene homolog 1
EVs	Extracellular vesicles
G-CSF	Granulocyte colony stimulating factor
hBM-MSCs	Human bone marrow mesenchymal stem cells
HSC	Hematopoietic stem cell
MV	Microvesicle
NF-κB	Nuclear factor kappa-light-chain-enhancer of activated B cells
PTH	Parathyroid hormone
PTHrP	Parathyroid hormone-related protein
RANK	Receptor activator of nuclear factor kappa-B
RANK-L	Receptor activator of nuclear factor kappa-B ligand
SDF-1	Stromal-derived factor 1
VCAM1	Vascular cell adhesion molecule 1
VEGF	Vascular endothelial growth factor

REFERENCES

[1] Gawande A. The way we age now. The New Yorker 2007;4:30.

[2] Frost HM. Bone's mechanostat: a 2003 update. Anat Rec A Discov Mol Cell Evol Biol 2003;275(2):1081–101.

[3] Karsenty G. Convergence between bone and energy homeostases: leptin regulation of bone mass. Cell Metab 2006;4:341–8.

[4] Baldock P, Lin S, Zhang L, Karl T, Shi Y, Driessler F, et al. Attenuates stress-induced bone loss through suppression of noradrenaline circuits. J Bone Miner Res 2014;29:2238–49.

[5] Iwaniec UT, Boghossian S, Lapke PD, Turner RT, Kalra SP. Central leptin gene therapy corrects skeletal abnormalities in leptin-deficient ob/ob mice. Peptides 2007;28: 1012–19.

[6] Bartell SM, Rayalam S, Ambati S, Gaddam DR, Hartzell DL, Hamrick M, et al. Central (ICV) leptin injection increases bone formation, bone mineral density, muscle mass, serum IGF-1, and the expression of osteogenic genes in leptin-deficient ob/ob mice. J Bone Miner Res 2011;26(8):1710–20.

[7] Vignaux G, Besnard S, Ndong J, Philoxène B, Denise P, Elefteriou F. Bone remodeling is regulated by inner ear vestibular signals. J Bone Miner Res 2013;28:2136–44.

[8] Pedersen BK. Muscles and their myokines. J Exp Biol 2011;214(Pt 2):337–46.

[9] Hamrick MW. A role for myokines in muscle–bone interactions. Exerc Sport Sci Rev 2011;39(1):43–7.

[10] Lee NK, Sowa H, Hinoi E, Ferron M, Ahn JD, Confavreux C, et al. Endocrine regulation of energy metabolism by the skeleton. Cell 2007;130(3):456–69.

[11] Brennan-Speranza TC, Henneicke H, Gasparini SJ, Blankenstein KI, Heinevetter U, Cogger VC, et al. Osteoblasts mediate the adverse effects of glucocorticoids on fuel metabolism. J Clin Invest 2012;122(11):4172–89.

[12] Mödder UI, Roforth MM, Nicks KM, Peterson JM, McCready LK, Monroe DG, et al. Characterization of mesenchymal progenitor cells isolated from human bone marrow by negative selection. Bone 2012;50(3):804–10.

[13] Mendez-Ferrer S, Michurina TV, Ferraro F, Mazloom AR, Macarthur BD, Lira SA, et al. Mesenchymal and haematopoietic stem cells form a unique bone marrow niche. Nature 2010;466:829–34.

[14] Nuttall M, Gimble J. Is there a therapeutic opportunity to either prevent or treat osteopenic disorders by inhibiting marrow adipogenesis? Bone 2000;27:177–84.

[15] Minaire P, Meunier P, Edouard C, Berbard J, Courpron J, Bourret J. Quantitative histological data on disuse osteoporosis. Calcif Tissue Res 1974;13:371–82.

[16] Ahdjoudj S, Lasmoles F, Holy X, Zerath E, Marie P. Transforming growth factor beta-2 inhibits adipocyte differentiation induced by skeletal unloading in rat bone marrow stroma. J Bone Miner Res 2002;17:668–77.

[17] Jung Y, Song J, Shiozawa Y, Wang J, Wang Z, et al. Hematopoietic stem cells regulate mesenchymal stromal cell induction into osteoblasts thereby participating in the formation of the stem cell niche. Stem Cells 2008;26:2042–51.

[18] Jackson K, Majka S, Wulf GG, Goodell MA. Stem cells: a minireview. J Cell Biochem Suppl 2002;38:1–6.

[19] Korbling M, Estrov Z. Adult stem cells for tissue repair – a new therapeutic concept? N Engl J Med 2003;349:570–82.

[20] Lin F, Cordes K, Li L, Hood L, Couser WG, Shankland SJ, et al. Hematopoietic stem cells contribute to the regeneration of renal tubules after renal ischemia-reperfusion injury in mice. J Am Soc Nephrol 2003;14:1188–99.

[21] Taichman R, Emerson S. Human osteoblasts support hematopoiesis through the population of granulocyte colony-stimulating factor. J Exp Med 1994;179:1677–82.

[22] Taichman RS, Reilly MJ, Emerson SG. Human osteoblasts support human hematopoietic progenitor cells in vitro bone marrow cultures. Blood 1996;87:518–24.

[23] Calvi L, Adams GB, Weibrecht KW, Weber JM, Olson DP, Knight MC, et al. Osteoblastic cells regulate the haematopoietic stem cell niche. Nature 2003;425:841–6.

[24] Zhang J, Niu C, Ye L, Huang H, He X, Tong WG, et al. Identification of the haematopoietic stem cell niche and control of the niche size. Nature 2003;425:836–41.

[25] Taichman RS. Blood and bone: two tissues whose fates are intertwined to create the hematopoietic stem-cell niche. Blood 2005;105:2631–9.

[26] Yin T, Li L. The stem cell niches in bone. J Clin Invest 2006;116:1195–201.

[27] Wilson A, Trumpp A. Bone-marrow haematopoietic-stem-cell niches. Nat Rev Immunol 2006;6:93–106.

[28] Fried S, Bunkin D, Greenberg A. Omental and subcutaneous adipose tissues of obese subjects release interleukin-6: depot difference and regulation by glucocorticoid. J Clin Endocrinol Metab 1998;83:847–50.

[29] Maurin A, Chavassieux P, Frappart L, Delmas P, Serre C, Meunier P. Influence of mature adipocytes on osteoblast proliferation in human primary cocultures. Bone 2000;26:485–9.

[30] Dar A, Kollet O, Lapidot T. Mutual, reciprocal SDF-1/CXCR4 interactions between hematopoietic and bone marrow stromal cells regulate human stem cell migration and development in NOD/SCID chimeric mice. Exp Hematol 2006;34(8):967–75.

[31] Valadi H, Ekström K, Bossios A, Sjöstrand M, Lee JJ, Lötvall JO. Exosome-mediated transfer of mRNAs and microRNAs is a novel mechanism of genetic exchange between cells. Nat Cell Biol 2007;9:654–9.

[32] Camussi G, Deregibus MC, Bruno S, Grange C, Fonsato V, Tetta C. Exosome/microvesicle-mediated epigenetic reprogramming of cells. Am J Cancer Res 2011;1:98–110.

[33] Mathivanan S, Ji H, Simpson R. Exosomes: extracellular organelles important in intercellular communication. J Proteomics 2010;73:1907–20.

[34] Vickers KC, Remaley AT. Lipid-based carriers of microRNAs and intercellular communication. Curr Opin Lipidol 2012;23:91–7.

[35] Collino F, Deregibus MC, Bruno S, Sterpone L, Aghemo G, Viltono L, et al. Microvesicles derived from adult human bone marrow and tissue specific mesenchymal stem cells shuttle selected pattern of miRNAs. PLoS One 2010;5:e11803.

[36] Xu L, Yang BF, Ai J. MicroRNA transport: a new way in cell communication. J Cell Physiol 2013;228:1713–19.

[37] Lee Y, El Andaloussi S, Wood MJ. Exosomes and microvesicles: extracellular vesicles for genetic information transfer and gene therapy. Hum Mol Genet 2012;21(R1): R125–34.

[38] Raposo G, Stoorvogel W. Extracellular vesicles: exosomes, microvesicles, and friends. J Cell Biol 2013;200(4):373–83.

[39] Villarroya-Beltri C, Gutierrez-Vazquez C, Sanchez-Cabo F, Perez-Hernandez D, Vazquez J, Martin-Cofreces N, et al. Sumoylated hnRNPA2B1 controls the sorting of miRNAs into exosomes through binding to specific motifs. Nat Commun 2013;4:2980.

[40] Muntión S, Ramos T, Paiva B, Roson B, Sarasquete ME, Diez-Campelo M, et al. Bone marrow mesenchymal stem cell (BM-MSC) release microvesicles/exosomes that incorporate into hematopoietic cells from MDS patients and may modify their behaviour. Blood 2013;21:863.

[41] Kordelas L, Rebmann V, Ludwig AK, Radtke S, Ruesing J, Doeppner TR, et al. MSC-derived exosomes: a novel tool to treat therapy-refractory graft-versus-host disease. Leukemia 2014;28:970–3.

[42] Fu Q, Deng L, Peng Y, Wu Y, Ding Y, Yang M, et al. Microvesicles released from stromal/osteoblast facilitate osteoclast formation via RANK/RANKL OPG pathway. Am Soc Bone Miner Res 2013;SA0204.

[43] Ratajczak J, Miekus K, Kucia M, Zhang J, Reca R, Dvorak P, et al. Embryonic stem cell-derived microvesicles reprogram hematopoietic progenitors: evidence for horizontal transfer of mRNA and protein delivery. Leukemia 2006;20(5):847–56.

[44] Janowska-Wieczorek A, Majka M, Ratajczak J, Ratajczak MZ. Autocrine/paracrine mechanisms in human hematopoiesis. Stem Cells 2001;19(2):99–107.

[45] Itoh T, Ito Y, Ohtsuki Y, Ando M, Tsukamasa Y, Yamada N, et al. Microvesicles released from hormone-refractory prostate cancer cells facilitate mouse pre-osteoblast differentiation. J Mol Histol 2012;43(5):509–15.

[46] Zhu X, You Y, Li Q, Zeng C, Fu F, Guo A, et al. BCR-ABL1-positive microvesicles transform normal hematopoietic transplants through genomic instability: implications for donor cell leukemia. Leukemia 2014;28:1666–75.

[47] Peinado H, Aleckovic M, Lavotshkin S, Matei I, Costa-Silva B, Moreno-Bueno G, et al. Melanoma exosomes educate bone marrow progenitor cells toward a pro-metastatic phenotype through MET. Nat Med 2012;18(6):883–91.

[48] Zhu W, Huang L, Li Y, Zhang X, Gu J, Yan Y, et al. Exosomes derived from human bone marrow mesenchymal stem cells promote tumor growth *in vivo*. Cancer Lett 2012;315(1):28–37.

[49] Roccaro AM, Sacco A, Maiso P, Azab AK, Tai YT, Reagan M, et al. BM mesenchymal stromal cell-derived exosomes facilitate multiple myeloma progression. J Clin Invest 2013;123(4):1542–55.

[50] D'Souza-Schorey C, Clancy JW. Tumor-derived microvesicles: shedding light on novel microenvironment modulators and prospective cancer biomarkers. Genes Dev 2012;26(12):1287–99.

[51] Aoki J, Ohashi K, Mitsuhashi M, Murakami T, Oakes M, Kobayashi T, et al. Posttransplantation bone marrow assessment by quantifying hematopoietic cell-derived mRNAs in plasma exosomes/microvesicles. Clin Chem 2014;60:675–82.

[52] Tsugita M, Yamada N, Noguchi S, Yamada K, Moritake H, Shimizu K, et al. Ewing sarcoma cells secrete EWS/Fli-1 fusion mRNA via microvesicles. PLoS One 2013;8 (10):4.

[53] Valencia K, Luis-Ravelo D, Bovy N, Antón I, Martínez-Canarias S, Zandueta C, et al. miRNA cargo within exosome-like vesicle transfer influences metastatic bone colonization. Mol Oncol 2014;8:689–703.

[54] Bruno S, Collino F, Deregibus MC, Grange C, Tetta C, Camussi G. Microvesicles derived from human bone marrow mesenchymal stem cells inhibit tumor growth. Stem Cells Dev 2013;22(5):758–67.

[55] Ryan MA, Nattamai KJ, Xing E, Schleimer D, Daria D, Sengupta A, et al. Pharmacological inhibition of EGFR signaling enhances G-CSF-induced hematopoietic stem cell mobilization. Nat Med 2010;16(10):1141–6.

[56] Salvucci O, Jiang K, Gasperini P, Maric D, Zhu J, Sakakibara S, et al. MicroRNA126 contributes to granulocyte colony-stimulating factor-induced hematopoietic progenitor cell mobilization by reducing the expression of vascular cell adhesion molecule 1. Haematologica 2012;97(6):818–26.

[57] Bian S, Zhang L, Duan L, Wang X, Min Y, Yu H. Extracellular vesicles derived from human bone marrow mesenchymal stem cells promote angiogenesis in a rat myocardial infarction model. J Mol Med 2014;92:387–97.

[58] Biancone L, Bruno S, Deregibus MC, Tetta C, Camussi G. Therapeutic potential of mesenchymal stem cell-derived microvesicles. Nephrol Dial Transplant 2012;(8):3037–42.

[59] Bruno S, Grange C, Deregibus MC, Calogero RA, Saviozzi S, Collino F, et al. Mesenchymal stem cell-derived microvesicles protect against acute tubular injury. J Am Soc Nephrol 2009;20:1053–67.

[60] Sun D, Zhuang X, Xiang X, Liu Y, Zhang S, Liu C, et al. A novel nanoparticle drug delivery system: the anti-inflammatory activity of curcumin is enhanced when encapsulated in exosomes. Mol Ther 2009;18:1606–14.

[61] Lai CP, Breakefield XO. Role of exosomes/microvesicles in the nervous system and use in emerging therapies. Front Physiol 2012;3:228.

[62] Thery C, Duban L, Segura E, Veron P, Lantz O, Amigorena S. Indirect activation of naive CD4+ T cells by dendritic cell-derived exosomes. Nat Immunol 2002;3(12):1156–62.

[63] Lai RC, Arslan F, Lee MM, Sze NS, Choo A, Chen TS, et al. Exosome secreted by MSC reduces myocardial ischemia/reperfusion injury. Stem Cell Res 2010;4:214–22.

[64] Simpson RJ, Jensen SS, Lim JW. Proteomic profiling of exosomes: current perspectives. Proteomics 2008;8(19):4083–99.

[65] Simpson RJ, Lim JW, Moritz RL, Mathivanan S. Exosomes: proteomic insights and diagnostic potential. Expert Rev Proteomics 2009;6(3):267–83.

[66] Ghosh AK, Secreto CR, Knox TR, Ding W, Mukhopadhyay D, Kay NE. Circulating microvesicles in B-cell chronic lymphocytic leukemia can stimulate marrow stromal cells: implications for disease progression. Blood 2010;115(9):1755–64.

[67] Oury F, Ferron M, Huizhen W, Confavreux C, Xu L, Lacombe J, et al. Osteocalcin regulates murine and human fertility through a pancreas-bone-testis axis. J Clin Invest 2013;123(6):2421–33.

[68] Reis LA, Borges FT, Simões MJ, Borges AA, Sinigaglia-Coimbra R, Schor N. Bone marrow-derived mesenchymal stem cells repaired but did not prevent gentamicin-induced acute kidney injury through paracrine effects in rat. PLoS One 2012;7(9):e44092.

[69] Xie Y, Yin T, Wiegraebe W, et al. Detection of functional haematopoietic stem cell niche using real-time imaging. Nature 2009;457:97–101.

CHAPTER 11

Diagnostic and Prognostic Applications of MicroRNA-Abundant Circulating Exosomes

Baron Arnone*, Xiaoqi Zhao**, Zhipeng Zou†, Gangjian Qin*, Min Cheng**

*Department of Medicine – Cardiology, Feinberg Cardiovascular Research Institute, Northwestern University Feinberg School of Medicine, Chicago, IL, USA
**Department of Cardiology, Union Hospital of Huazhong University of Science and Technology, Tongji Medical College, Wuhan, Hubei, China
†Department of Neurosurgery, Wuhan Iron & Steel Corporation General Hospital, Wuhan, Hubei, China

Contents

1 INTRODUCTION

Exosomes are a collective of bioactive lipid bilayer-based microvesicles secreted by a variety of cell types and have been demonstrated to play important roles as an alternative mechanism of biomolecular transport between distant cells [1]. This transport is characterized by an intracellular packaging process whereby various proteins and other gene products such as micro (mi)RNAs are loaded into exosomal vesicles and subsequently released into the extracellular environment [2]. The lipid bilayer of exosomes ensures the stability of their contents by protecting them from degradation by circulating proteases and nucleases.

miRNAs have recently been identified as novel biomarkers for various diseases and numerous studies have described the important predictive association between aberrant expression profiles of specific miRNAs and the disease status of various tissues or cells in which they are expressed [3]. These studies demonstrate the promising future value of using miRNA – particularly exosomal miRNA – as a powerful screening tool within the clinical setting. In this chapter, we summarize the latest research progress regarding the diagnostic and prognostic potential of exosomes and exosomal miRNA. Furthermore, we discuss the future direction for research in this area, especially as it relates to the development of prognostic strategies in the field of stem cell therapy.

2 EXOSOMES

2.1 Molecular and Biophysical Characteristics of Exosomes

A widely investigated cellular phenomenon is the vesicular packaging of proteins and/or other gene products (e.g., DNA, mRNAs, and miRNAs) into "bags" (microvesicles) whereby, under certain conditions, the contents of these microvesicles are released into the extracellular environment for the purposes of intercellular transport and signaling [1]. At present, the canonical view recognizes three main types of microvesicles: apoptotic bodies, microparticles, and exosomes bodies which, in turn, are primarily subclassified according to their size in diameter [4,5]. Apoptotic bodies (1–4 mm) are small, membranous particles released via blebbing from the surface of the cell membrane during the process of apoptosis [6]. Microparticles, ranging in size from 100 nm to 1 mm, occur as simple membrane fragments produced upon cellular activation [4]. Exosomes are small (30–100 nm), lipid, bilayer-enclosed particles of heterogeneous composition [1].

A diversity of cell types including B cells, T cells, mast cells, dendritic cells, platelets, neurons, and epithelial cells have demonstrated the capacity to produce and secrete exosomes into their extracellular environment. Additionally, free-floating exosomes have also been shown to be present in blood and other bodily fluids including urine, saliva, sputum, and cerebrospinal fluid (CSF) [7,8]. The contents of exosomes are composed of a wide variety of biomolecules including proteins, small signaling peptides, DNA fragments, mRNAs, miRNAs, and lipids. Their localization is characteristically diffuse in that their cellular targets can be either proximal or distal with respect to their cellular origin [9].

Characterization studies have shown that all exosomes possess some redundant features such as structure, size, and a panel of commonly shared proteins. The latter include cytoplasmic proteins (such as tubulin, actin, and Rab proteins), signaling proteins (such as protein kinases and G-proteins), major histocompatibility complex (MHC) molecules, and heat shock proteins [3]. Additionally, the tetraspanin transmembrane superfamily proteins are uniquely associated with exosomes and are typically employed as an exosome marker [10]. Exosomes consist of a bilayer lipid membrane and lack cellular organelles such as mitochondria, lysosomes, endoplasmic reticulum, or Golgi apparatus.

Interestingly, comparative genetic and proteomic analyses of the exosomal contents from various cell types, including both normal and diseased, have identified distinctive composition features that can be exploited as a means for discriminating between differing cell types including their disease status [3]. The powerful specificity of these discriminatory features, coupled with their relative availability and ease of collection, makes exosomes a highly valuable biomarker for the diagnosis and prognosis of various diseases [1,8].

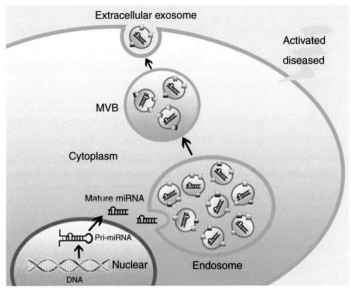

Figure 11.1 *Specific Activation or Cytopathogenic Stimulation of Cells Induces the Formation and Secretion of Exosomes.* The sequence of exosomal formation is initiated by a ceramide-triggered inward budding of the plasma membrane into early endosomes. Next, cytoplasmic molecules (e.g., miRNAs) are packaged by endosomal sorting complexes on the cytoplasmic surface of the early endosome, whereupon they are internalized via another inward budding step within the early endosome to form exosomes. The progressive accumulation of exosomes within the early endosome leads to the formation of an MVB. Homeostatic maintenance of MVB levels within the cell is dynamically regulated by directed fusion with lysosomal complexes leading to degradation. Both normal and pathological sources of external stimuli alter this equilibrium by inhibiting degradation and promoting the fusion of MVBs with the cell membrane, which leads to the budding and extracellular release of both MVBs and the individual exosomes.

2.2 Formation of Exosomes

In contrast with microparticles and apoptotic bodies, which are directly shed from the cell membrane by blebbing, exosomes are formed through an intracellular pathway involving budding of the cellular membrane after the aggregate fusion of endosome-derived multivesicular bodies (MVBs) within the plasma membrane (Figure 11.1) [1]. Specific activation or cytopathogenic stimulation of cells induces the formation of exosomes. The sequence of exosomal formation is initiated by a ceramide-triggered process characterized by an inward budding of the plasma membrane into early endosomes [11]. Next, cytoplasmic DNA fragments, mRNAs, miRNAs, and functional proteins are packaged by endosomal sorting complexes on the cytoplasmic surface of the early endosome, whereupon they are internalized via another inward budding step within the early endosome to form exosomes. The sorting and packaging of the exosomal contents at this stage is nonrandom and cell specific. For example, more than 1300 mRNAs and 100 miRNAs have been identified that are uniquely specific to mast cell exosomes [3].

The progressive accumulation of exosomes within the early endosome leads to the formation of an MVB [11]. Homeostatic maintenance of MVB levels within the cell is dynamically regulated by directed fusion with lysosomal complexes leading to degradation. Both normal and pathological sources of external stimuli alter this equilibrium by inhibiting degradation and promoting the fusion of MVBs with the cell membrane, which ultimately leads to the budding and extracellular release of both MVBs and the individual exosomes contained within [2]. RNA species that are transported between cells via exosomes are collectively referred to as "exosomal shuttle RNAs" [12]. Cellular activation status in response to either normal or pathogenic stimuli can significantly influence not only the production and secretion of exosomes but also their specific compositional profiles [1]. Exosomal composition is largely determined during the initial early endosome formation and packaging stages. Accordingly, research has focused primarily on these stages in an effort to better understand the molecular mechanisms that regulate and determine exosomal composition within a variety of cellular and pathophysiological contexts.

2.3 Models of Exosomal Delivery

It has been well established that exosomes can act as a vector for the transfer of mRNA, miRNA, and/or effector proteins to recipient cells [1]. Data from both *in vitro* and *in vivo* experiments have resulted in the postulation of at least three models to explain the mechanism(s) by which exosomes undergo fusion with their recipient cells [13]. The first model hypothesizes that exosomes directly fuse with the cellular membranes of target cells, thereby releasing their contents into the cytoplasm in a relatively passive manner. The second model stipulates that exosomes are absorbed into the target cell by a process of active endocytosis, while the third proposes that exosomal binding and uptake is mediated via specific chemotactic receptor proteins present on the surface of target cells [14,15]. It is reasonable to assume the existence of each of these models and that one may play a more dominant role depending the stimulatory context, cellular source, or target.

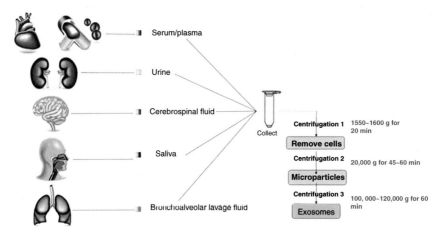

Figure 11.2 *Exosomes can be Isolated From Many Types of Bodily Fluids.* The most commonly used procedure is serial high-speed centrifugations: the first centrifugation removes cells, platelets, and cell debris; the second centrifugation prepares microparticles, from which the third centrifugation obtains exosomes.

2.4 Exosome Isolation and Detection Strategies

Various techniques have been developed to collect exosomes from bodily fluids such as serum or cell culture supernatants (Figure 11.2) [16]. The most common procedure utilizes high-speed centrifugation or magnetic activated cell sorting (MACS) using antibodies directed toward specific exosomal surface antigens [17]. High-speed centrifugation strategies take advantage of the relatively low density of exosomes compared to other lipid-based cellular structures. In this method, a progressive series of centrifugations (ranging from 10,000g to 100,000g) are applied to a sample, whereby cellular debris and other high-density vesicles are removed via pelleting. Exosomes remain within the supernatant fraction and are subsequently pelleted via ultracentrifugation at approximately 120,000g [18].

Recent improvements in isolation strategies with regard to both recovery time and yield have focused on taking advantage of other unique properties of exosomes such as size (nanomembrane-based filtration), surface antigens (MACS), and affinity characteristics (microfluidic immunoaffinity). Continued improvement of these strategies will rely upon a more comprehensive understanding and characterization of surface markers unique to exosomes and their cellular origins. Detailed morphological characterization of various exosome populations typically involves the use of electron microscopy while analysis of the internal molecular constituents relies upon the utilization of immunoelectron microscopy as well as other fluorescence-based techniques [17].

3 REGULATION OF GENE EXPRESSION BY MicroRNAs

3.1 MicroRNAs

MicroRNAs (miRNA) are derived from an evolutionarily conserved gene family whose mature gene products consist of small (18–25 nucleotide), noncoding RNAs. MiRNAs play a critical role in regulating gene expression throughout a wide range of normal biological and pathophysiological processes by binding to specific mRNA transcripts and thereby targeting them for degradation [19].The miRBase database (http://www.mirbase.org) has summarized information on a total of more than 30,000 mature miRNA products from 206 species, of which more than 2,000 miRNAs are present in *Homo sapiens* [20].

The maturation process for miRNAs consists of several posttranscriptional regulatory steps [21]. Primary miRNA transcripts (pri-miRNA) are first generated from DNA mainly by RNA polymerase II, resulting in the adoption of a long hairpin structure and, similar to protein-coding mRNA, is characterized by the addition of a 59 cap and a 39 poly (A) tail. In the second step, these pri-miRNAs are cleaved and spliced into precursor miRNA (pre-miRNA) in 70–110 nt dsRNA fragments by the RNase III enzyme complex Drosha-DGCR8 (Pasha). These pre-miRNAs contain a conserved 39 2–4-nucleotide overhang motif that enables interaction with exportin 5 and facilitates translocation of the pre-miRNA to the cytoplasm in a guanosine-59-triphosphate-dependent manner. Upon release from the export complex in the cytoplasm, the pre-miRNA is further processed by another RNase III enzyme "Dicer," resulting in cleavage to a mature 22mer single-stranded miRNA [22]. The mature miRNA is then loaded into the RNA-induced silencing complex (RISC)/Argonaut complex. This complex is then able to bind to and cleave mRNA containing a 7 nt sequence in its 39 untranslated region that is complementary to the miRNA, leading to the reduction in the stability of the mRNA and/or inhibition of protein translation and resulting in decreased expression of the target protein [23].

The discovery of miRNA has contributed enormously to researchers' understanding of the molecular mechanisms that regulate gene expression and protein stability [24]. Likewise, it has also become well established that dysregulation of this critical biological process (e.g., mutations affecting the gene sequence of miRNA genes or any of the processing enzymes that contribute to miRNA maturation) is known to be associated with a variety of disease pathologies including the development of cancer [25].

3.2 Circulating miRNAs

In 1931, Javillier and Fabrykant first reported the discovery of circulating nucleic acids. Since then, many researches have focused on the development of methods and applications for the use of circulating nucleic acids as biomarkers of disease. However, the production and regulation of circulating nucleic acids remains poorly understood [26]. Since the initial discovery of large amounts of circulating ribonucleases in the blood by Reddi and Holland in 1976, not until recently has substantial evidence accumulated that also demonstrates the stable existence of circulating miRNAs – not only in the blood but also in other bodily fluids, such as serum, plasma, ascites, saliva, breast milk, urine, and CSF [27]. Considering the crucial regulatory role of miRNAs in mediating biological processes, it is not surprising that physiological and pathological changes can induce alterations in the circulating miRNA expression profile [24]. miRNAs exhibit high sequence conservation across species, display consistent expression specificity to tissues or biological stages and provide easy detection by quantitative PCR through high-precision signal amplification. As such, these features establish miRNAs as ideal biomarkers for a variety of applications [19].

For example, researchers have reported a panel of placenta-specific miRNAs found within the serum of pregnant individuals that eventually disappear after childbirth, suggesting that circulating miRNAs could be an indicator of physiological status [28]. Likewise, expression levels of cancer-associated miRNAs (almost all derived from malignant cells) are prominently higher in serum from cancer patients than in healthy donors, indicating that circulating miRNAs can be used as biomarkers for the existence of cancer in patients [29]. Up to now, the aberrant expression of circulating miRNAs has been observed in many kinds of diseases, such as immune, cardiovascular, respiratory, digestive, urinary, and neurological diseases as well as a variety of cancers.

3.3 Exosomal miRNAs

miRNAs can remain stable in the extracellular environment by associating with either protein transporter complexes or microvesicles such as exosomes [30]. Interestingly, although a substantial portion of circulating miRNAs are present within exosomes, a recent study has shown that they are predominantly associated with the RISC complex protein Argonaute (Ago), suggesting that such miRNAs are deliberately secreted as primed complexes capable of directly binding to mature mRNAs [31].

With regards to stability, when stored at 4°C for up to 96 h, or −70°C for up to 28 days, microarray analysis confirmed that exosomal miRNAs remained quite stable when subjected to storage in either condition [32]. In addition, researchers have recently identified the presence of miRNAs in breast milk with particularly high levels of immune-associated miRNAs, indicating that miRNAs are very stable in acidic conditions [33].

As mentioned earlier, researchers have collectively identified up to 1300 mRNAs and 120 miRNAs in exosomes derived from the mast cell line MC/9 [34]. Interestingly, many of these RNAs were detectable only in exosomes and nowhere else within the various cellular fractions. The theoretically estimated nucleic acid carrying capacity for a single exosome has been calculated to be approximately 10,000 nucleotides − equivalent to about 500 miRNAs per exosome [3]. The website http://exocarta.org/ is a manually curated database providing informational resources concerning proteins, RNAs, and lipids that have been discovered in exosomes from a variety of organisms. Version 3.1 of ExoCarta includes information or over 10,000 proteins, 2,000 mRNAs, and 700 miRNAs obtained from over 100 exosomal studies [35]. In addition to their diagnostic and prognostic potential, exosomal miRNAs may also show promise as therapeutic targets [36].

3.4 Regulation of miRNA Expression and Exosomal Packaging

The multistep maturation of miRNA is orchestrated by a variety of regulatory factors, the imbalance and/or dysregulation of which can lead to a breakdown in miRNA expression and result in the progression of various diseases. Genomic mutations and polymorphisms of miRNA genes can cause sequence variation and structural rearrangements in miRNA transcripts that can negatively impact their expression and function [37]. Additionally, epigenetic modifications proximal to miRNA genomic loci, such as hyper- or hypomethylation of their promoter regions also plays a crucial role in regulating expression of miRNA [25]. Certain miRNAs located within the intronic regions of protein-coding genes can both influence and be influenced by the expression of such genes [38]. Moreover, inhibiting the translation of enzymes that are involved in epigenetic regulatory pathways can significantly influence the production of miRNA. Interestingly, research suggests that miRNAs themselves may also participate in recruiting and directing epigenetic factors or modifications, respectively, at various promoter regions.

The regulation of exosomal-packaged miRNA is best illustrated by studies involving the inhibition of microvesicle formation. Because the initial formation of exosomes is regulated by a ceramide-dependent mechanism, disruption of sphingomyelinase 2 activity, an enzyme responsible for ceramide production, can inhibit the cleavage of sphingomyelin into ceramide and lead to attenuation of packaging of exosomal miRNAs [39]. Likewise, suppression of sphingomyelinase 2 expression by small interfering (si)RNA leads to a reduction in exosomal miRNA levels, while overexpression produces an increase in exosomal miRNA levels [26]. Although these studies have significantly contributed to what is known about the mechanisms that regulate packaging of exosomal miRNA, little is known about these mechanisms with respect to pathophysiological effects such as infectious diseases and cancer.

4 DIAGNOSTIC AND PROGNOSTIC APPLICATIONS OF MicroRNAs IN EXOSOMES

Under the guidance of the National Institute for Health (NIH), the Biomarkers and Surrogate Endpoint Working Group has formulated crucial definitions for biomarkers of diseases as follows: (i) a characteristic that is objectively measured and evaluated as an indicator of normal biological processes, pathogenic processes or pharmacological responses to a therapeutic intervention; (ii) a characteristic or variable that reflects how a patient feels, functions, or survives; and (iii) a biomarker intended to substitute for a clinical endpoint with the potential for predicting the clinical benefit or harm (or lack of benefit or lack of harm) on the basis of epidemiological, therapeutic, pathophysiological, or other scientific evidence [40].

Exosomal miRNAs have received much attention regarding their diagnostic and prognostic potential. According to the above guidelines, exosomal miRNAs are ideal disease biomarkers for the following reasons: (i) Stability and tissue specificity: miRNA is arguably the more reliable nucleic acid biomarker due to its greater stability compared to other circulating nucleic acids such as mRNA and DNA, providing greater consistency and reproducibility for any clinical assays that utilize miRNAs as biomarkers. Exosomal miRNA also displays a high degree of tissue specificity, a feature that also comports well with NIH criteria. (ii) Diagnostic indication: genetic abnormalities, epigenetic alterations, as well as dysfunctional miRNA biogenesis can result in abnormal exosomal miRNA expression profiles, which, in turn, can provide meaningful correlative information about disease status and/or clinical endpoints. (iii) Easily obtained: techniques for rapid and easy collection of exosomes are well established for a variety of human bodily fluids such as plasma, ascites, saliva, breast milk, urine, and CSF [41]. Additionally, tissue-specific detection, separation, and analysis of exosomes is readily available via standard methods including flow cytometry and reverse transcription polymerase chain reaction (RT-PCR) [42].

Recent strides in various molecular biology-related technologies such as improved genomic and transcriptomic profiling have also increased the diagnostic and prognostic power of miRNAs. As such, these newly refined molecular approaches have altered fundamental diagnostic paradigms whereby disease pathogenesis is no longer viewed as originating from a single insult or event but rather the result of multiple physiologic alternations. In this respect, exosomal miRNA has been developed as a particularly insightful and useful biomarker for a variety of diseases. In the proceeding sections we provide a detailed summary of the current literature as it relates the development of such biomarkers for both tumor and nontumor diseases.

4.1 Tumor Diseases

A growing compendium of evidence increasingly supports the usefulness of miRNAs as novel disease biomarkers, particularly with regard to tissue specificity. This notion was first observed in 2002 when researchers demonstrated downregulation of miRNA – specifically, miR-15 and miR-16 – in chronic lymphocytic leukemia. This observation and other subsequent analyses led to the striking revelation that the locations of many miRNA genes were uniquely positioned at chromosomal regions associated with genomic instability and/or cancer [43]. Later, in 2005, another research group reported that miRNA expression profiles were associated with staging, diagnosis, prognosis, progression, and response to treatment of tumor diseases, and were more precise than mRNA expression profiles for the purposes of prediction and classification of malignant tumors by developmental lineage and differentiation state [44]. These results initiated an immediate increase in interest in miRNA within the field of oncology.

The discovery and characterization of tumor-derived vesicles (including exosomes and microvesicles) in a variety of malignancies such as breast, lung, ovarian, prostate, colorectal, and gastric cancers has led to a commensurate increasing interest in their value as diagnostic and prognostic biomarkers [29]. In 1979, tumor-derived exosomes were first discovered in the peripheral circulation of patients with cancer. A strong correlation was observed between increased exosomal content within the blood or other bodily fluids and developmental progression in cancer stage [45]. This phenomenon was the first indication of the potential of circulating tumor-derived exosomes as biomarkers for cancer, as evidenced by the variety of malignant cells from which they could be isolated and their amenableness to profiling analysis by virtue of their abundantly unique genetic contents such as mRNA and miRNA [46]. The fact that the miRNA expression profiles of these circulating tumor derived-exosomes were similar to the expression profiles of their parent tumor tissues indicated that circulating exosomal miRNA profiles could be utilized as diagnostic criteria for cancers in the absence of tissue selection [47]. This was later confirmed in studies where miRNAs known to be associated with specific types of cancer (such as let-7i, miR-16, miR-21, and miR-214) were found to be present in circulating exosomes and positively predicted both prognosis and clinical outcomes such as resistance to therapy [48].

Some data have led researchers to hypothesize that the presence and secretion of exosomal miRNAs by tumor cells points to the strong possibility that these miRNAs serve a pathological function to potentiate tumorigenicity such as by promoting invasiveness or inhibiting the antitumor effects of immune cells [49]. Thus, exploring the biological effects and mechanisms of tumor-derived exosomes and the miRNA contained within them should lead to a better understanding of their pathological role in cancer and increase their diagnostic and prognostic value as biomarkers for such diseases [45].

4.1.1 Ovarian Cancer

Ovarian cancer is the third-most occurring gynecologic malignancy, with an estimated 225,000 new diagnoses each year. Notably, it is the eighth-highest cause of death for women, due largely to its aggressive late-stage pathophysiology and detection difficulty resulting in late diagnosis that limit the beneficial effects of therapeutic intervention [50]. Four miRNAs – miR-141, miR-200a, miR-200b, and miR-200c – have been identified as being aberrantly overexpressed in ovarian cancer specimens compared with normal ovary tissue. They all belong to the same miR-200 subfamily and share similar genomic localization with miR-200a and miR-200b on chromosome 1p36.33 and miR-141 and miR-200c on chromosome 12p13.31 [51]. Being members of the miR-200 subfamily, they are highly related in sequence, especially in the mRNA recognition motifs of the guide strand, indicating that they likely share similar mRNA targets and exert similar biological effects. Recently, genomic profiling assays performed on circulating and tumor-derived exosomes collected from ovarian cancer patients identified four additional miRNAs (miR-21, miR-203, miR-205, and miR-214) whose expression profiles were found to be similar for both sample types [52]. Interestingly, exosomal miRNA profiles for cells derived from malignant and benign ovarian tumors exhibited elevated expression relative to one another, supporting earlier observations that exosomal miRNA expression levels are correlated with severity of cancer stage progression [53].

4.1.2 Nonsmall-Cell Lung Cancer

Lung cancer is the leading global cause of cancer-related deaths among both men and women [54]. Despite advances in surgical procedures and chemoradiotherapy, 5-year survival rates for lung cancer patients average just 15% [55]. Increasing importance has been placed on the need to develop earlier detection and diagnostic strategies for lung cancer in order to increase patient survival by maximizing therapeutic intervention. To this end, the development of exosomal miRNA as a biomarker for early-stage lung cancer has received significant attention. Transcriptional profiles from normal and diseased lung tissue at various stages of cancer reveal miRNA expression profiles that are markedly different from one another. One study identified a total of 12 miRNAs (miR-17-3p, miR-21, miR-106a, miR-146, miR-155, miR-191, miR-192, miR-203, miR-205, miR-210, miR-212, and miR-214) associated with nonsmall-cell lung cancer (NSCLC) compared with normal lung tissue [56]. Another study specifically identified overexpression of miR-155 and low expression of let-7a-2 with a poor prognosis for NSCLC patients [43].

Additionally, miRNA expression profiles of both circulating and noncirculating (i.e., tumor-derived) exosomes collected from lung cancer patients also display significant differences compared to normal controls [54].

4.1.3 Urologic Malignancies

Tumors arising from bladder, kidney, prostate, and testes are the most common forms of urological cancer in western countries [57]. Among them, prostate cancer is the most frequently diagnosed cancer within the male-exclusive cancer cohort. Accumulating evidence suggests that aberrant miRNA expression profiles in urological malignancies are strong predictors for disease progression. For example, one study identified high plasma levels of miR-141 as a diagnostic biomarker for prostate cancer with a reported

60% sensitivity and 100% specificity [58]. Similarly, researchers have reported miR-141 and, additionally, miR-375 as the most common urologic tumor-related circulating miRNAs associated with poor prognosis of prostate cancers [59]. Upregulated expression of both miR-141 and miR-375 has been proposed as a biomarker for metastatic prostate cancer and is correlated with a low level of relapse-free survival [57]. Overexpression of four exosomal miRNAs (miR-107, miR-141, miR-375, and miR-574-3p) has been observed as a distinguishing feature in serum of metastatic prostate cancer patients compared to nonmetastatic controls [60].

In addition to serum and plasma, urine is an ideal bodily fluid for sample collection because it is noninvasively derived and comes directly from potential tumor sources within the urological tract. As such, expression profiles of miRNAs strongly predictive of urological cancers have been found to be present in both tumors and their circulating exosomal counterparts [61].

Gene expression analysis of urine specimens collected from patients with urothelial cell carcinoma – the most common form of bladder cancer – revealed a panel of miRNAs including miR-24 that were specifically downregulated when compared to normal controls. In addition, differential expression of urine-derived miR-1224-3p is reported to possess diagnostic potential in bladder cancer, with a specificity of 83% and a sensitivity of 77% [62]. Furthermore, miR-15b, miR-135b, and miR-1224-3p provide a combinatorial predictive power with a sensitivity of 94.1% and specificity of 51%, as does the ratio of miR-126/miR-152 with a specificity of 82% and a sensitivity of 72% [57]. To date, exosomal miRNAs have not been investigated regarding their potential as predictive biomarkers for bladder cancer.

Urinary miR-15a is significantly upregulated in renal cell carcinoma (RCC) patients compared to controls, indicating urinary miR-15a as a biomarker for diagnosis of RCC. Moreover, combinatorial upregulation of serous miR-378 and miR-451 is associated with RCC, with a diagnostic sensitivity of 81% and a specificity of 83% [63]. Other potential biomarkers for RCC include miR-1233 and miR-210, though none so far have been investigated within the context of exosomes.

4.1.4 Melanoma

Melanoma is the most common form of invasive skin cancer with a characteristically high mortality rate for dermatological diseases [64]. Recent mRNA- and miRNA-based microarray profiling studies have identified thousands of exosomal mRNAs and miRNAs associated with the progressive and metastatic potential of melanoma. For example, differential expression of several of these miRNAs, including miR-31, miR-185, and miR-34b, have also been shown to be strongly correlated with the invasiveness of melanoma [65]. Other candidate exosomal miRNAs that show promise as diagnostic and prognostic biomarkers for melanoma include let-7a, miR-182, miR-221, miR-222, miR-31, miR-19b-2, miR-20b, miR-92a-2, miR-21, miR-15b, miR-210, miR-30b, miR-30d, and miR-532-5p, as determined by expression analysis performed in the A375 melanoma cell line compared to normal melanocytes [66].

4.1.5 Breast Cancer

Breast cancer is the most common cancer diagnosed in women and the leading cause of cancer-related deaths in women around the world [67]. It exerts a vastly disparate impact on women, with only 1% of all diagnosed cases occurring in men. Like other highly invasive and potentially lethal cancers, developing more sensitive, noninvasive biomarkers for breast cancer is important for achieving earlier detection and better prognoses [68]. To date, a total of three circulating miRNAs – miR-141, let-7a, and miR-195 – have been identified and found to be specifically upregulated ($P < 0.001$) in serum from recently diagnosed breast cancer patients, compared with age- and sex-matched healthy volunteers [69,70]. These studies did not elucidate the specific source of the miRNAs and so it remains to be determined whether they exist in the circulation as complexes within proteins or are predominantly transported via microvesicles such as exosomes – or both [67].

4.1.6 Glioblastoma Multiforme (GBM)

GBM is the most common primary brain cancer in adults. Despite recent advances in surgical procedures and combined chemoradiotherapy, the prognosis of GBM remains very poor, with a median survival time of less than 1 year [71]. Noninvasive as well as highly sensitive and specific biomarkers for early diagnosis of GBM have been widely explored [72]. A large-scale, genome-wide study recently identified 55 upregulated and 29 downregulated miRNAs in malignant gliomas [73]. Further studies confirmed that a combinatorial panel of 23 miRNAs is able to distinguish GBM from anaplastic astrocytoma with 95% sensitivity. A combination of specific mRNA and miRNA expression signatures has been established to designate five GBM subclasses. Compared with other upregulated miRNAs, miR-196a and miR-196b – both members of the miR-196 miRNA subfamily – exhibit particularly high expression levels in GBM patients, which, in turn, are strongly associated with shorter overall survival times [74]. Both miR-196a and miR-196b are known to be involved in the malignant transformation of gliomas [73].

Exosomal miRNA expression profiles have also been investigated for their diagnostic potential in GBM. A combined expression signature of exosomal miR-320, miR-574-3p, and a small nuclear RNA (RNU6-1) has been proposed as a serum biomarker for GBM patients [75]. In this study, results demonstrated that either RNU6-1 alone or the combination of miR-320/miR-574-3p/RNU6-1 (with different respective predictive indices) were able to distinguish GBM patients from healthy controls. Recent studies have also revealed the existence of exosomes and exosomal miRNA in CSF), thus raising the possibility for developing CSF-derived/specific exosomes and/or exosomal miRNAs as diagnostic and prognostic biomarkers for intracranial tumors such as GBM [76].

4.1.7 Head and Neck Cancer (HNC)

HNCs include malignant tumors of the nasal cavity, paranasal sinuses, oral cavity, nasopharynx, oropharynx, hypopharynx, and larynx. HNC is the fifth-most common cancer worldwide, with 650,000 new cases per year. After initial diagnosis more than 50% of HNC patients present with metastatic disease, associated with a relatively poor 5-year overall survival rate [77].

Oropharyngeal carcinoma (OPC) and oral cavity squamous cell carcinoma are two of the most prevalent forms of HNC [78]. Both of these diseases are highly associated with lifestyle risk factors such as smoking and alcohol intake [79]. However, infection from viruses such as human papilloma virus and Epstein–Barr virus (EBV) is also an important risk factor for OPC as well as another HNC, nasopharyngeal carcinoma (NPC) [80]. It is well known that exosomes serve as a primary transmission vector for virally infected cells to regulate their microenvironment and infect distant cells. Exosomes isolated from saliva, serum, and plasma in EBV-infected NPC patients have been shown to contain signal transduction-associated ligands such as LMP1 as well as EBV-related miRNAs, suggesting that the content analysis of circulating exosomes can serve as an etiological and diagnostic biomarker for HNC diseases [81].

4.1.8 Pancreatic Cancer

Pancreatic adenocarcinoma (PaCa) is the fourth-leading cause of cancer-related deaths around the world and is regarded as the deadliest cancer with an overall 5-year survival rate less than 1% and a mean survival time of 4–6 months. Significant progress has been made in the development of more effective surgical procedures and chemoradiotherapy for patients with early-stage PaCa. However, poor diagnostic options and early metastatic spread continue to serve as significant barriers to promoting better prognosis and survival outcomes [82]. Thus, continuing efforts addressing PaCa outcomes have focused on developing either noninvasive or minimally invasive biomarkers for early diagnosis and treatment. In a recent study, a combination of four serous miRNAs (miR-21, miR-210, miR-155, and miR-196a) were identified and proposed as biomarkers for PaCa, where miR-155 was particularly associated with early-stage PaCa and miR-196a with more progressive stages [83]. Another study reported elevated levels of miR-20a, miR-21, miR-24, miR-25, miR-99a, miR-185, and miR-191 in the serum of PaCa patients compared with healthy controls [84].

Exosomes and exosomal miRNA profiles have not been extensively investigated in PaCa patients but future studies that address these features are likely to improve the specificity and effectiveness of miRNA-based biomarkers for early diagnosis of PaCa [85].

4.2 Nontumor Diseases

Unlike cancer, which tends to originate from specific tissues or cells and whose causation and progression is typically rooted in a relatively limited number of molecular factors, nontumor diseases are both determined and driven by a large host of pathological variables (e.g., infection, inflammation, apoptosis, necrosis, proliferation, and senescence), which, in turn, can affect a multitude of physiological systems such as cardiovascular, respiratory, digestive, urological, and nervous systems. Significant progress has been made in the development of technologies aimed at identifying and targeting such diseases. Research focused on uncovering and characterizing exosomal miRNA expression profiles and exosomal proteomics will serve as an important contributor to further improving this technology by enhancing the ability to provide tissue-specific diagnoses, facilitate more accurate prognoses and maximize the predictive power of targeted therapeutic responses.

4.2.1 Cardiovascular Disease

Cardiovascular disease (CVD) is the leading cause of global mortality and one of the major contributors to morbidity around the world [86]. The clinical endpoint of CVD is heart failure (HF), typically prompted by myocardial infarction (MI), and is caused by irreversible loss of cardiac tissue as a result of either acute or chronic ischemia, of which coronary artery disease is the primary contributor. A rapid and accurate diagnosis of CVD sequelae is critical for successful therapeutic interventions in these patients. Current diagnostic strategies for CVD include electrocardiogram as well as a serum panel for the MI biomarkers creatinine phosphokinase-muscle band, troponin-T, and troponin I.

Recent evidence from several animal model studies shows that cardiac-specific miRNAs are released into the circulation upon coronary artery ligation-induced MI, indicating a potential for the development of miRNA-based biomarkers for CVD detection [87]. The cardiac-specific miRNAs identified were miR-1, miR-133, miR-208, and miR-499. Other MI-related miRNAs such as miR-1 and miR-208 have also been reported to be present in urine. In the latter experiments, researchers confirmed an exosomal origin of miR-208.

Advances in primary percutaneous coronary intervention strategies and newly-developed antithrombotic drugs have resulted in a significant reduction in mortality rates due to acute MI but these strategies have done little to mitigate the long-term mortality rates in these patients due to the effects of progressive ischemic HF in these patients [88]. Thus, it is necessary to establish diagnostic biomarkers indicative of HF development after MI in order to optimize management and treatment strategies. Although cardiac-specific miRNAs such as miR-1, miR-133a, miR-208a, and miR-499 have been shown to be upregulated in the circulation of MI patients, these miRNAs display relatively short half-lives in the blood and are unsuitable for use as biomarkers for HF development in post-MI patients. Recently, researchers identified several circulating p53-responsive miRNAs (miR-192, miR-194, and miR-34a) that were upregulated in patients who survived MI but experienced development of HF within 1 year [89]. Furthermore, miR-194 and miR-34a were shown to be predominantly localized to circulating exosomes and their levels specifically correlated with left ventricular diastolic function in the recuperative stage of MI.

Hypertension is one of the primary risk factors for CVD. Blood pressure salt sensitivity (SS), resulting in elevated blood pressure upon high dietary salt intake, is associated with an increased risk for mortality in CVD patients with hypertension. Conversely, inverse salt sensitivity (ISS) resulting in a decline in blood pressure after high salt intake is also associated with increased mortality of CVD patients [90]. Unfortunately, the accuracy and reproducibility of currently available diagnostic tests for both SS and ISS are not ideal. Researchers recently identified 45 exosomal miRNAs (out of a total panel of 194) that were associated with either SS or ISS [90]. These miRNAs have been implicated as effectors in known regulatory pathways involved in hypertension, such as PPARγ, EGFR, TGFβ-1, and PTEN/PI3K signaling. Further confirmational studies may lead to the successful development of any number of these miRNAs as biomarkers for individuals with SS or ISS who may be at risk for CVD.

4.2.2 Respiratory Diseases

Asthma is a chronic inflammatory respiratory disease characterized by reversible airway narrowing in response to various inhaled environmental stimuli such as allergens, microorganisms, and air pollutants. The etiological factors associated with the risk for developing asthma are determined by both genetic and environmental considerations. As such, though the clinical presentation of asthma can be fairly consistent between individuals, because of the disease's etiological complexity it continues to prove difficult to accurately determine causation and provide optimally effective treatment regimens for individual patients [91].

Exosomes derived from both normal and inflammatory cells within the lung have been shown to play an important role in the sensitization to allergens and the development of asthma. In one study, researchers observed significant phenotypic and functional differences between exosomes derived from bronchoalveolar lavage fluid (BALF) of normal and asthmatic patients [92]. Additionally, aberrant expression profiles were also observed for miRNAs isolated from these exosomes. Specifically, a total of 24 miRNAs were found to be significantly associated with changes in forced expiratory volume. Among these, members of the let-7 and miRNA-200 families were found to display the most significant differential variation and all 24 miRNAs were exclusively downregulated relative to healthy controls. The miR-200 family is known to play a role in epithelial-to-mesenchymal transitions for pulmonary epithelial cells and has also been implicated in the regulation of airway remodeling processes in asthma.

In addition to BALF, exhaled breath condensate (EBC) is another respiratory tract-specific bodily fluid known to contain miRNA. RT-PCR analysis of EBC recently identified at least three miRNAs (miR-574-5p, miR-516a-5p, and miR-421) that were differentially expressed in asthma patients relative to healthy controls and that have also been previously reported to play crucial roles in inflammation-related pathways [93].

4.2.3 Urologic Diseases

Chronic kidney disease (CKD) is another global leading cause of death whose incidence and prevalence continues to rise. The primary, long-term pathophysiological outcome of CKD is tubulointerstitial fibrosis, which promotes progressive loss of kidney function and eventually leads to renal failure [94]. Worthy of note is autosomal-dominant polycystic kidney disease (ADPKD) Types 1 and 2, which are the most common heritable kidney genetic diseases and typically progress to CKD. Current standard diagnostic and prognostic strategies for CKD involve the assessment of tubulointerstitial fibrosis using ultrasound-guided renal biopsy. In addition to its particularly invasive nature, considerations such as lack of accessibility and high cost necessitate the development of alternative procedures that mitigate these issues while still providing the same reliability.

A variety of epithelial cell lineages including podocytes, renal tubule cells, and transitional epithelial cells line the urinary tract and are able to release exosomes into the urine for collection [95]. Exosomes from these cells contain a broad range of tissue- and cell-specific proteins, mRNAs, and miRNAs [96] and collection and analysis of these exosomes may provide advantages that lead to the establishment of biomarkers critical for the diagnosis and treatment of CKD. For example, protein levels of polycystin-1 and -2 (whose mutant forms are principally responsible for the development of ADPKD) are typically undetectable in kidney tissue homogenate but are easily detected in urinary exosomes [97]. Mutations in the thiazide-sensitive Na–Cl co-transporter are responsible for the development of Gitelman's syndrome, another form of CKD. Mutant forms of this co-transporter can be detected in urinary exosomes released from the distal convoluted tubule via immunoblot assays. Similarly, absence of the sodium–potassium–chloride co-transporter 2 normally present in urinary exosomes can be used as a biomarker to identify patients with Bartter syndrome type I [98]. Other studies have shown that upregulated levels of activating transcription factor 3 (ATF3) and Wilms' tumor 1 (WT-1) are associated with acute kidney injury (AKI) and podocyte injury, respectively. Levels of urinary exosomal ATF3 and WT-1 also show increases in a small number of patients with AKI and focal segmental glomerulosclerosis [99]. Thus, establishing quantitative benchmarks for elevated levels of exosomal ATF3 may serve as a novel diagnostic biomarker for AKI, and exosomal WT-1 for injury. Moreover, the presence of urinary exosomal fetuin-A has been demonstrated as a prognostic biomarker for AKI, as well as the absence of urinary exosomal aquaporin-1 as a novel biomarker for ischemia reperfusion injury [99].

miRNAs have also been shown to be indispensable regulators in the development and homeostasis of the renal system. Aberrant miRNA expression has been observed in kidney fibrosis animal models. Dicer is a key enzyme associated with the maturation of miRNAs and the absence of Dicer in podocytes has been shown to promote proteinuria and glomerulosclerosis. Recent studies suggest that urinary exosomal miRNAs can be easily detected in various kidney diseases such as diabetic nephropathy, focal segmental glomerulosclerosis, membranous nephropathy, and IgA nephropathy [96]. Specifically, researchers observed that levels of miR-29 and miR-200 were downregulated in urinary exosomes collected from CKD patients, and these miRNAs have been shown to have a protective effect in renal disease [100]. Based on the examples listed above it is clear that both the protein and miRNA constituents of urinary exosomes show significant promise as biomarker targets for the diagnosis and prognosis of urological diseases.

4.2.4 Nervous System Diseases

Central nervous system (CNS) diseases can arise from a complex variety of ischemic, hemorrhagic, neurodegenerative, inflammatory, and developmental disorders [101]. The multitude of physiological and etiological variables that contribute to such disorders as well as the sensitive nature of the CNS itself necessitate diagnostic techniques that provide high specificity and are noninvasive. Therefore, the development of highly specific surrogate biomarkers for CNS diseases will be important in establishing diagnostic and prognostic criteria such as patient subclassification, reliable therapeutic response assessments as well as contributing to a better understanding of CNS pathological features.

Growing evidence demonstrates the important role miRNA plays in the regulation of both CNS physiological homeostasis and the pathogenesis of CNS diseases, indicating also that miRNAs are ideal for use as CNS-specific biomarker targets [101]. Sources for miRNA in the CNS include blood mononulear cells, serum, and CSF. For example, specific CSF-derived miRNAs have been used as biomarkers for Alzheimer's disease and multiple sclerosis [102]. In addition, a single study established a panel of CSF-derived miRNAs as a biomarker for diagnosing brain injury in patients with neurotrauma.

4.2.5 Digestive System Diseases

Drug induced liver injury (DILI) is the most common drug-related disease whose mortality and morbidity rates have been rising in recent years [103].

Diagnostic and prognostic methods remain insufficient, leading a call for the development of more specific and effective biomarkers for the disease [103]. RNA-based profiling analyses have demonstrated the existence of several liver-specific miRNAs. Recently, miR-122 and miR-155 have received considerable attention as potential biomarkers for liver injury and it has been suggested that they may even be more effective than the traditional diagnostic marker for DILI, alanine transaminase [104]. In alcohol-induced and lipopolysaccharide-induced murine liver injury models, the plasma levels of miR-122 and miR-155 were significantly upregulated and almost exclusively localized to circulating exosomes [104]. However, in the acetaminophen-induced liver injury model, though their plasma levels were elevated, an inverse distribution was observed in which these miRNAs were localized primarily to nonexosomal fractions. The latter study highlights the possibility that the specific distribution patterns of circulating miRNAs may reflect the differing pathological aspects of various DILI-related diseases.

4.2.6 Immunological Disorders

Immunological effector cells can be categorized as belonging to either the "innate" or "adaptive" immune compartments. The innate immune compartment is constituted by effector cells (and their molecular products) that lack antigen specificity (such as neutrophils, monocytes, macrophages, complement, and acute phase proteins) and generally provide protection against exposure to acute pathogenic factors. Conversely, cellular effectors of the adaptive immune compartment (such as T and B lymphocytes) demonstrate high antigen specificity and promote the establishment of immunological memory to various pathogens by coordinating the actions of both the adaptive and the innate compartments [105].

Numerous studies have demonstrated the crucial role that exosomes play in regulating the function of and/or coordination between the innate and adaptive immune compartments. It was first reported that the membrane protein profiles of exosomes originating from and secreted by B lymphocytes exhibited were wholly distinct from their counterpart parent membranes. Interestingly, it was soon after discovered that these exosomes contained MHC class II molecules and were able to induce antigen-specific T cell responses. Exosomes derived from T lymphocytes after T-cell receptor triggering have also been reported.

Exosomally derived miRNAs are widely recognized for their important roles in determining both the extent of activation and specificity of immune-mediated responses via control of gene expression programs of immune effector cells. For example, one study demonstrated that exosomes serve as transmission vectors for tRNA and miRNA between antigen presenting cells and T cells to regulate gene expression in recipient cells [105]. In this study, 20 intracellular miRNAs were identified in several distinct subpopulations of CD4$^+$ T cells [106]. It was subsequently determined that two of these miRNAs were exclusively present in CD4$^+$ T cell-derived exosomes compared to B cells.

Of these, miR-150 has recently been shown to play an important role in the development of the adaptive lymphoid and innate myeloid cell lineage of the immune system. In particular, miR-150 is almost undetectable in B- and T-progenitor cells but highly upregulated in mature lymphocytes. Interestingly, when mature T cells differentiate into activated effector T cells, the level of intracellular miR-150 is downregulated again. This post-activation reduction in intracellular levels of miR-150 is partly achieved by release of miR-150-containing exosomes into the circulation, thus providing a novel and unique method for determining immune activation status. For example, a recent study observed that decreasing serum levels of miR-150 was associated with a poor outcome in critically ill patients with sepsis.

A primary feature of the adaptive immune system is the ability to recognize and distinguish between self and non-self antigens. Reactivity to self-antigens can result in the development of autoimmune disease. Various autoimmune diseases display a heterogeneity of clinical presentations, which increases the difficulty in obtaining accurate diagnoses and/or prognoses for such diseases [107]. Recent studies in patients with autoimmune diseases have shown promising potential for the use of circulating exosomal miRNAs as sensitive and effective biomarkers for determining disease status, such as miR107 for lupus erythematosus [108].

4.2.7 Metabolic Diseases

Development of metabolic disorders is determined by two fundamental physiological contributors – excess energy intake and insufficient energy expenditure – leading to clinical outcomes characterized by dysfunctional sugar, lipid, or protein metabolism [109]. Chronic inflammation and low-level oxidative stress are both regarded as stimuli significantly associated with the origination and progression of metabolic diseases [40].

A specific subset of circulating exosomal miRNAs has been found to be upregulated in certain patients suffering from metabolic diseases. For example, miR-133a, an important regulator of the NFATc4 protein associated with cardiac hypertrophy shows elevated levels in plasma-derived exosomes of patients with cardiometabolic diseases. In addition, differential expression of five other exosomal miRNAs – miR-15a, miR-29b, miR-126, miR-223, and miR-28-3p – has been detected in the plasma of type II diabetes patients compared to healthy age- and sex-matched controls. [109]

4.2.8 Aging

Increasing age is an important risk factor in the development of a variety of diseases and disorders such as CVD, metabolic diseases, nervous disorders, and cancer and results in the breakdown of a wide range of cellular, molecular, and physiological functions. Insights into the specific mechanisms that regulate the aging process and account for individual differences within the population reveal that cellular senescence is a major determinant of aging and age-related diseases. A recent study demonstrated that specific miRNA known to be upregulated in elderly patients (including the miR-17-92 cluster, let-7, and miR-34a) were also capable of inducing cellular senescence. Overexpression or downregulation of these miRNAs likely promoted senescence by acting on genes of the p53–p21 and p16–pRB pathways.

4.2.9 Obstetrics and Gynecology

Maintenance of placental homeostasis is crucial for proper fetomaternal nutrition exchange and disruption of this homeostasis during pregnancy leads to significantly adverse clinical outcomes such as impaired fetal growth and diminished maternal health. Such considerations underline the importance of developing noninvasive and accurate biomarkers for monitoring fetomaternal health during pregnancy [110]. Several miRNAs have been identified as potential biomarkers for placental homeostasis. For example, trophoblast-related, exosome-derived, and protein-bound miRNAs are constantly released into the extracellular environment (collected from plasma or amniotic fluid), and can be steadily detected in the maternal circulation during gestation. Specifically, analysis of miRNA isolated from maternal plasma indicates that elevated levels of miR-127, miR-134, and C19MC are associated with pregnancy. C19MC miRNAs are the largest cluster of miRNAs in the human genome and are predominantly expressed in the placenta. Interestingly, several studies have shown the progressive upregulation of C19MC miRNAs in placental trophoblasts from the first to the third trimester during pregnancy, indicating that C19MC levels could be used as biomarker for both pregnancy status and pregnancy-related diseases [110].

4.2.10 Infectious Diseases

Prion diseases including Creutzfeldt–Jakob disease, fatal familial insomnia, Gerstmann–Sträussler–Scheinker syndrome and Kuru are transmissible, fatal neurodegenerative disorders. These diseases originate from a molecular defect in protein folding characterized by conformational transformation of the host-encoded cellular prion protein (PrP^C) into the pathogenic isoform (PrP^{Sc}) [111]. It has been demonstrated that release of both PrP^C and PrP^{Sc} by prion-containing cells is facilitated via exosomal packaging and secretion, thus enabling a highly infectious mechanism of transmission for prion proteins in promoting neurodegenerative disease. Exosomes derived from prion-infected neuronal cells show an enrichment for several specific miRNAs whose expression levels are significantly upregulated including let-7b, let-7i, miR-128a, miR-21, miR-222, miR-29b, miR-342-3p, and miR-424 as well as miR-146a, which is downregulated. Importantly, the expression variation pattern for these exosomal miRNAs was also confirmed in terminally-infected mouse and primate models of prion disease. Results indicated that let-7b, miR-128a, and miR-342-3p were significantly upregulated in prion-infected mouse brains, while miR-128a and miR-342-3p were significantly upregulated in brain tissue from primate models infected with BSE prion strains.

As noted earlier, virally infected cells release exosomal miRNAs that can be targeted to and internalized by peripheral cells to actively repress viral target genes. This suggests that exosomal miRNAs derived from prion-infected cells may be involved in the regulation of prion protein propagation in and dissemination from these cells and contribute to the spread of infection [111]. Similarly, these exosomal miRNAs could be regarded as novel biomarkers for diagnosis and prognosis of prion infection. In addition to prion-based infectious diseases, exosomes have also been developed as biomarkers for parasitic infections such as *Leishmania, Cryptococcus*, and *Trypanosoma* [112].

5 CONCLUSIONS

Exosomes are actively secreted into the extracellular environment by a variety of cell types in a wide range of activation states. Current and past research has led to the development of methods to isolate exosomes from various kinds of bodily fluids such as plasma, serum, saliva, CSF, and urine. Exosomes contain an abundance of proteins and nucleic acids, particularly miRNAs. Due to their tissue specificity and strong association with pathological processes miRNAs have come to be regarded as important early- and late-stage biomarkers for many kinds of diseases. In this chapter, we summarized the biological characteristics of exosomes and exosomal miRNAs, with an emphasis on current research and future directions into the development of exosomal miRNAs as diagnostic and prognostic bio-markers for both cancerous and noncancerous diseases.

Exosomal miRNAs are ideal biomarkers for diagnosis and prognosis of diseases due to their stability in circulation, tissue specificity, active se-cretion, direct or indirect association with disease status, and their relative ease of procurement. A list of many currently identified exosomal miRNA biomarkers for both cancerous and noncancerous diseases is provided (Table 11.1). In order to optimize the diagnostic and prognostic power of

Table 11.1 Exosomal miRNA as biomarkers for disease

Diseases	Exosomal miRNA biomarker candidate(s)	References
Ovarian cancer	Expression levels of 8 microRNAs (miR-21, miR-141, miR-200a, miR-200c, miR-200b, miR-203, miR-205, and miR-214) were similar between tumor cells and circulating exosomes	[52]
NSCLC	Strong correlation in expression levels for 12 specific miRNAs (miR-17-3p, miR-21, miR-106a, miR-146, miR-155, miR-191, miR-192, miR-203, miR-205, miR-210, miR-212, and miR-214) between NSCLC tissue and periph-eral blood-derived exosomal miRNAs	[56]
Prostate cancer	Overexpression of four exosomal miRNAs (miR-107, miR-141, miR-375, and miR-574-3p) in serum of metastatic prostate cancer patients	[60]
Melanoma	Upregulated miRNAs in melanoma patients (let-7a, miR-182, miR-221, miR-222, miR-31, miR-19b-2, miR-20b, miR-92a-2, miR-21, miR-15b, miR-210, miR-30b, miR-30d, and miR-532-5p) as determined by exosomal miRNA profiling	[66]
GBM	Two exosomal miRNAs (miR-320 and miR-574-3p) combined with one small nuclear RNA (RNU6-1)	[75]
HNC	Latent EBV-infected nasopharyngeal carcinoma cells release exosomes containing LMP1 and EBV-related miRNAs; exosomal BART-miR-NAs exist in plasma of nasopharyngeal carci-noma patients	[81,113]

Table 11.1 Exosomal miRNA as biomarkers for disease *(cont.)*

Diseases	Exosomal miRNA biomarker candidate(s)	References
CVD	Exosomal miR-1 is upregulated in plasma of AMI patients and also detected in urine; circulating miR-194 and miR-34a exist particularly in exosomal fraction and were correlated with heart failure in the recuperative stage of AMI	[89,114]
Asthma	Variation of exosomal let-7 family (a–e) and the miRNA-200 family (miR-200b and miR-141) derived from bronchoalveolar lavage fluid most pronounced between asthma patients and healthy controls	[92]
CKD	Levels of miR-29 and miR-200 downregulated in urinary exosomes released from CKD patients	[100]
DILI	Two liver-abundant miRNAs (miR-122 and miR-155) mainly present in circulating exosomes have been regarded as more effective biomarkers than alanine transaminase alone in diagnosis of DILI	[104]
Metabolic disease	Aberrant expression profiles of exosomal miRNAs (including downregulated miR-15a, miR-29b, miR-126, miR-223, and upregulated miR-28-3p) detected in type II diabetes patients	[109]
Prion diseases	Exosomes derived from prion-infected neuronal cells contain miRNAs significantly upregulated (let-7b, let-7i, miR-128a, miR-21, miR-222, miR-29b, miR-342-3p, and miR-424) and downregulated (miR-146a)	[111]

exosomal miRNAs as biomarkers, future research should focus on the specific mechanisms by which they exert their biological effects and attempt to address how these effects contribute to the pathogenesis of disease.

ABBREVIATIONS

ADPKD	Autosomal-dominant polycystic kidney disease
AKI	Acute kidney injury
ATF3	Activating transcription factor 3
BALF	Bronchoalveolar lavage fluid
CKD	Chronic kidney disease
CNS	Central nervous system
CVD	Cardiovascular disease
CSF	Cerebrospinal fluid
DILI	Drug-induced liver injury

EBC	Exhaled breath condensate
EBV	Epstein–Barr virus
EGFR	Epidermal growth factor receptor
GBM	Glioblastoma multiforme
HF	Heart failure
HNC	Head and neck cancer
HPV	Human papilloma virus
MI	Myocardial infarction
MVB	Multivesicular body
MHC	Major histocompatibility complex
MACS	Magnetic activated cell sorting
NIH	National Institute for Health
NPC	Nasopharyngeal carcinoma
NSCLC	Nonsmall-cell lung cancer
PaCa	Pancreatic adenocarcinoma
PI3K	Phosphatidylinositol-4,5-bisphosphate 3-kinase
PPARγ	Peroxisome proliferator-activated receptor gamma
pre-miRNA	Precursor miRNA
pri-miRNA	Primary miRNA transcripts
PrP^C	Prion protein
PrP^{Sc}	Pathogenic isoform of PrP^C
PTEN	Phosphatase and tensin homolog
OPC	Oropharyngeal carcinoma
RCC	Renal cell carcinoma
RISC	RNA-induced silencing complex
RT-PCR	Reverse transcription polymerase chain reaction
SS	Salt sensitivity
TGFβ-1	Transforming growth factor beta 1
WT-1	Wilms' tumor 1

ACKNOWLEDGMENTS

This work was supported by the National Science Foundation of China (Young Investigator Grant# 81100084 to M.C.); and the National Institute of Health (R01 Grants HL093439 and HL113541 to G.Q.); and the American Diabetes Association (Grant# 1-15-BS-148 to G.Q.).

REFERENCES

[1] Vlassov AV, Magdaleno S, Setterquist R, Conrad R. Exosomes: current knowledge of their composition, biological functions, and diagnostic and therapeutic potentials. Biochim Biophys Acta 2012;1820(7):940–8.

[2] Keller S, Sanderson MP, Stoeck A, Altevogt P. Exosomes: from biogenesis and secretion to biological function. Immunol Lett 2006;107(2):102–8.

[3] Valadi H, Ekstrom K, Bossios A, Sjostrand M, Lee JJ, Lotvall JO. Exosome-mediated transfer of mRNAs and microRNAs is a novel mechanism of genetic exchange between cells. Nat Cell Biol 2007;9(6):654–9.

[4] Fang DY, King HW, Li JY, Gleadle JM. Exosomes and the kidney: blaming the messenger. Nephrol (Carlton) 2013;18(1):1–10.

[5] Raposo G, Stoorvogel W. Extracellular vesicles: exosomes, microvesicles, and friends. J Cell Biol 2013;200(4):373–83.

[6] Proudfoot D, Skepper JN, Hegyi L, Bennett MR, Shanahan CM, Weissberg PL. Apoptosis regulates human vascular calcification *in vitro*: evidence for initiation of vascular calcification by apoptotic bodies. Circ Res 2000;87(11):1055–62.

[7] Michael A, Bajracharya SD, Yuen PS, Zhou H, Star RA, Illei GG, Alevizos I. Exosomes from human saliva as a source of microRNA biomarkers. Oral Dis 2010;16(1):34–8.

[8] Keller S, Ridinger J, Rupp AK, Janssen JW, Altevogt P. Body fluid derived exosomes as a novel template for clinical diagnostics. J Transl Med 2011;9:86.

[9] Smalheiser NR. Exosomal transfer of proteins and RNAs at synapses in the nervous system. Biol Direct 2007;2:35.

[10] Nazarenko I, Rana S, Baumann A, McAlear J, Hellwig A, Trendelenburg M, Lochnit G, Preissner KT, Zoller M. Cell surface tetraspanin Tspan8 contributes to molecular pathways of exosome-induced endothelial cell activation. Cancer Res 2010;70(4):1668–78.

[11] Fevrier B, Raposo G. Exosomes: endosomal-derived vesicles shipping extracellular messages. Curr Opin Cell Biol 2004;16(4):415–21.

[12] Wahlgren J, De LKT, Brisslert M, Vaziri Sani F, Telemo E, Sunnerhagen P, Valadi H. Plasma exosomes can deliver exogenous short interfering RNA to monocytes and lymphocytes. Nucleic Acids Res 2012;40(17):e130.

[13] Akao Y, Iio A, Itoh T, Noguchi S, Itoh Y, Ohtsuki Y, Naoe T. Microvesicle-mediated RNA molecule delivery system using monocytes/macrophages. Mol Ther 2011;19(2):395–9.

[14] Hagiwara K, Ochiya T, Kosaka N. A paradigm shift for extracellular vesicles as small RNA carriers: from cellular waste elimination to therapeutic applications. Drug Deliv Transl Res 2014;4:31–7.

[15] Lee C, Mitsialis SA, Aslam M, Vitali SH, Vergadi E, Konstantinou G, Sdrimas K, Fernandez-Gonzalez A, Kourembanas S. Exosomes mediate the cytoprotective action of mesenchymal stromal cells on hypoxia-induced pulmonary hypertension. Circulation 2012;126(22):2601–11.

[16] Rani S, O'Brien K, Kelleher FC, Corcoran C, Germano S, Radomski MW, Crown J, O'Driscoll L. Isolation of exosomes for subsequent mRNA, microRNA, and protein profiling. Methods Mol Biol 2011;784:181–95.

[17] Taylor DD, Zacharias W, Gercel-Taylor C. Exosome isolation for proteomic analyses and RNA profiling. Methods Mol Biol 2011;728:235–46.

[18] van der Pol E, Hoekstra AG, Sturk A, Otto C, van Leeuwen TG, Nieuwland R. Optical and non-optical methods for detection and characterization of microparticles and exosomes. J Thromb Haemost 2010;8(12):2596–607.

[19] Etheridge A, Lee I, Hood L, Galas D, Wang K. Extracellular microRNA: a new source of biomarkers. Mutat Res Fund Mol Mech Mut 2011;717(1–2):85–90.

[20] Griffiths-Jones S, Saini HK, van Dongen S, Enright AJ. miRBase: tools for microRNA genomics. Nucleic Acids Res 2008;36(Database issue):D154–8.

[21] Zeng Y. Principles of micro-RNA production and maturation. Oncogene 2006;25(46):6156–62.

[22] Yeom KH, Lee Y, Han J, Suh MR, Kim VN. Characterization of DGCR8/Pasha, the essential cofactor for Drosha in primary miRNA processing. Nucleic Acids Res 2006;34(16):4622–9.

[23] Cai Y, Yu X, Hu S, Yu J. A brief review on the mechanisms of miRNA regulation. Genom Proteom Bioinform 2009;7(4):147–54.

[24] Ajit SK. Circulating microRNAs as biomarkers, therapeutic targets, and signaling molecules. Sensors (Basel) 2012;12(3):3359–69.

[25] Eulalio A, Huntzinger E, Nishihara T, Rehwinkel J, Fauser M, Izaurralde E. Deadenylation is a widespread effect of miRNA regulation. RNA 2009;15(1):21–32.

[26] Kosaka N, Yoshioka Y, Hagiwara K, Tominaga N, Katsuda T, Ochiya T. Trash or treasure: extracellular microRNAs and cell-to-cell communication. Front Genet 2013;4:173.

[27] Lasser C. Identification and analysis of circulating exosomal microRNA in human body fluids. Methods Mol Biol 2013;1024:109–28.

[28] Moldovan L, Batte K, Wang Y, Wisler J, Piper M. Analyzing the circulating microRNAs in exosomes/extracellular vesicles from serum or plasma by qRT-PCR. Methods Mol Biol 2013;1024:129–45.

[29] Kosaka N, Iguchi H, Ochiya T. Circulating microRNA in body fluid: a new potential biomarker for cancer diagnosis and prognosis. Cancer Sci 2010;101(10):2087–92.

[30] Hu G, Drescher KM, Chen XM. Exosomal miRNAs: biological properties and therapeutic potential. Front Genet 2012;3:56.

[31] Peters L, Meister G. Argonaute proteins: mediators of RNA silencing. Mol Cell 2007;26(5):611–23.

[32] Lasser C, Eldh M, Lotvall J. Isolation and characterization of RNA-containing exosomes. J Vis Exp 2012;(59):e3037.

[33] Lasser C, Alikhani VS, Ekstrom K, Eldh M, Paredes PT, Bossios A, Sjostrand M, Gabrielsson S, Lotvall J, Valadi H. Human saliva, plasma and breast milk exosomes contain RNA: uptake by macrophages. J Transl Med 2011;9:9.

[34] Record M, Subra C, Silvente-Poirot S, Poirot M. Exosomes as intercellular signalosomes and pharmacological effectors. Biochem Pharmacol 2011;81(10):1171–82.

[35] Mathivanan S, Fahner CJ, Reid GE, Simpson RJ. ExoCarta 2012: database of exosomal proteins, RNA and lipids. Nucleic Acids Res 2012;40(Database issue):D1241–4.

[36] Kosaka N, Yoshioka Y, Hagiwara K, Tominaga N, Ochiya T. Functional analysis of exosomal microRNA in cell-cell communication research. Methods Mol Biol 2013;1024:1–10.

[37] Boon RA, Vickers KC. Intercellular transport of microRNAs. Arterioscler Thromb Vasc Biol 2013;33(2):186–92.

[38] Guil S, Esteller M. DNA methylomes, histone codes and miRNAs: tying it all together. Int J Biochem Cell Biol 2009;41(1):87–95.

[39] Graham TR, Kozlov MM. Interplay of proteins and lipids in generating membrane curvature. Curr Opin Cell Biol 2010;22(4):430–6.

[40] Muller G. Microvesicles/exosomes as potential novel biomarkers of metabolic diseases. Diabetes Metab Syndr Obes 2012;5:247–82.

[41] Taylor DD, Gercel-Taylor C. The origin, function, and diagnostic potential of RNA within extracellular vesicles present in human biological fluids. Front Genet 2013;4:142.

[42] Gallo A, Alevizos I. Isolation of circulating microRNA in saliva. Methods Mol Biol 2013;1024:183–90.

[43] Calin GA, Croce CM. MicroRNA–cancer connection: the beginning of a new tale. Cancer Res 2006;66(15):7390–4.

[44] Schwarzenbach H, Hoon DS, Pantel K. Cell-free nucleic acids as biomarkers in cancer patients. Nat Rev Cancer 2011;11(6):426–37.

[45] Kahlert C, Kalluri R. Exosomes in tumor microenvironment influence cancer progression and metastasis. J Mol Med 2013;91(4):431–7.

[46] Kogure T, Patel T. Isolation of extracellular nanovesicle microRNA from liver cancer cells in culture. Methods Mol Biol 2013;1024:11–18.

[47] Wendler F, Bota-Rabassedas N, Franch-Marro X. Cancer becomes wasteful: emerging roles of exosomes in cell-fate determination. J Extracell Vesicles 2013; 2:10.3402/jev.v2i0.22390.

[48] Hummel R, Hussey DJ, Haier J. MicroRNAs: predictors and modifiers of chemo- and radiotherapy in different tumour types. Eur J Cancer 2010;46(2):298–311.

[49] Zhang HG, Grizzle WE. Exosomes and cancer: a newly described pathway of immune suppression. Clin Cancer Res 2011;17(5):959–64.

[50] Davidson BA, Secord AA. Profile of pazopanib and its potential in the treatment of epithelial ovarian cancer. Int J Womens Health 2014;6:289–300.

[51] Korpal M, Lee ES, Hu G, Kang Y. The miR-200 family inhibits epithelial-mesenchymal transition and cancer cell migration by direct targeting of E-cadherin transcriptional repressors ZEB1 and ZEB2. J Biol Chem 2008;283(22):14910–4.

[52] Taylor DD, Gercel-Taylor C. MicroRNA signatures of tumor-derived exosomes as diagnostic biomarkers of ovarian cancer. Gynecol Oncol 2008;110(1):13–21.

[53] Li J, Sherman-Baust CA, Tsai-Turton M, Bristow RE, Roden RB, Morin PJ. Claudin-containing exosomes in the peripheral circulation of women with ovarian cancer. BMC Cancer 2009;9:244.

[54] Rabinowits G, Gercel-Taylor C, Day JM, Taylor DD, Kloecker GH. Exosomal microRNA: a diagnostic marker for lung cancer. Clin Lung Cancer 2009;10(1):42–6.

[55] Lwin Z, Riess JW, Gandara D. The continuing role of chemotherapy for advanced non-small cell lung cancer in the targeted therapy era. J Thorac Dis 2013;5(Suppl. 5):S556–64.

[56] Molina-Pinelo S, Suarez R, Pastor MD, Nogal A, Marquez-Martin E, Martin-Juan J, Carnero A, Paz-Ares L. Association between the miRNA signatures in plasma and bronchoalveolar fluid in respiratory pathologies. Dis Markers 2012;32(4):221–30.

[57] Huang X, Liang M, Dittmar R, Wang L. Extracellular microRNAs in urologic malignancies: chances and challenges. Int J Mol Sci 2013;14(7):14785–99.

[58] DeVere White RW, Vinall RL, Tepper CG, Shi XB. MicroRNAs and their potential for translation in prostate cancer. Urol Oncol 2009;27(3):307–11.

[59] Selth LA, Townley S, Gillis JL, Ochnik AM, Murti K, Macfarlane RJ, Chi KN, Marshall VR, Tilley WD, Butler LM. Discovery of circulating microRNAs associated with human prostate cancer using a mouse model of disease. Int J Cancer 2012;131(3):652–61.

[60] Hessvik FNP, Sandvig K, Llorente A. Exosomal miRNAs as biomarkers for prostate cancer. Front Genet 2013;4:36.

[61] Wang G, Szeto CC. Methods of microRNA quantification in urinary sediment. Methods Mol Biol 2013;1024:211–20.

[62] Miah S, Dudziec E, Drayton RM, Zlotta AR, Morgan SL, Rosario DJ, Hamdy FC, Catto JWF. An evaluation of urinary microRNA reveals a high sensitivity for bladder cancer. Br J Cancer 2012;107(1):123–8.

[63] Redova M, Poprach A, Nekvindova J, Iliev R, Radova L, Lakomy R, Svoboda M, Vyzula R, Slaby O. Circulating miR-378 and miR-451 in serum are potential biomarkers for renal cell carcinoma. J Transl Med 2012;10:55.

[64] Voskoboynik M, Arkenau HT. Combination therapies for the treatment of advanced Melanoma: a review of current evidence. Biochem Res Int 2014;2014:307059.

[65] Xiao D, Ohlendorf J, Chen Y, Taylor DD, Rai SN, Waigel S, Zacharias W, Hao H, McMasters KM. Identifying mRNA, microRNA and protein profiles of melanoma exosomes. PLoS One 2012;7(10):e46874.

[66] Gajos-Michniewicz A, Duechler M, Czyz M. MiRNA in melanoma-derived exosomes. Cancer Lett 2014;347(1):29–37.

[67] Friel AM, Corcoran C, Crown J, O'Driscoll L. Relevance of circulating tumor cells, extracellular nucleic acids, and exosomes in breast cancer. Breast Cancer Res Treat 2010;123(3):613–25.

[68] Shupe MP, Graham LJ, Schneble EJ, Flynt FL, Clemenshaw MN, Kirkpatrick AD, Stojadinovic A, Peoples GE, Shumway NM. Future directions for monitoring treatment responses in breast cancer. J Cancer 2014;5(1):69–78.

[69] Corcoran C, Friel AM, Duffy MJ, Crown J, O'Driscoll L. Intracellular and extracellular microRNAs in breast cancer. Clin Chem 2011;57(1):18–32.

[70] Wittmann J, Jack HM. Serum microRNAs as powerful cancer biomarkers. Biochim Biophys Acta 2010;1806(2):200–7.

[71] Redzic JS, Ung TH, Graner MW. Glioblastoma extracellular vesicles: reservoirs of potential biomarkers. Pharmgenom Pers Med 2014;7:65–77.

[72] De Smaele E, Ferretti E, Gulino A. MicroRNAs as biomarkers for CNS cancer and other disorders. Brain Res 2010;1338:100–11.

[73] Mizoguchi M, Guan Y, Yoshimoto K, Hata N, Amano T, Nakamizo A, Sasaki T. Clinical implications of microRNAs in human glioblastoma. Front Oncol 2013;3:19.

[74] Karsy M, Arslan E, Moy F. Current progress on understanding microRNAs in Glioblastoma Multiforme. Genes Cancer 2012;3(1):3–15.

[75] Manterola L, Guruceaga E, Gallego Perez-Larraya J, Gonzalez-Huarriz M, Jauregui P, Tejada S, Diez-Valle R, Segura V, Sampron N, Barrena C, Ruiz I, Agirre A, Ayuso A, Rodriguez J, Gonzalez A, Xipell E, Matheu A, Lopez de Munain A, Tunon T, Zazpe I, Garcia-Foncillas J, Paris S, Delattre JY, Alonso MM. A small noncoding RNA signature found in exosomes of GBM patient serum as a diagnostic tool. Neuro Oncol 2014;16(4):520–7.

[76] Akers JC, Ramakrishnan V, Kim R, Skog J, Nakano I, Pingle S, Kalinina J, Hua W, Kesari S, Mao Y, Breakefield XO, Hochberg FH, Van Meir EG, Carter BS, Chen CC. MiR-21 in the extracellular vesicles (EVs) of cerebrospinal fluid (CSF): a platform for glioblastoma biomarker development. PLoS One 2013;8(10):e78115.

[77] Zheng M, Li L, Tang Y, Liang XH. How to improve the survival rate of implants after radiotherapy for head and neck cancer? J Periodontal Implant Sci 2014;44(1):2–7.

[78] Iriti M, Varoni EM. Chemopreventive potential of flavonoids in oral squamous cell carcinoma in human studies. Nutrients 2013;5(7):2564–76.

[79] Majchrzak E, Szybiak B, Wegner A, Pienkowski P, Pazdrowski J, Luczewski L, Sowka M, Golusinski P, Malicki J, Golusinski W. Oral cavity and oropharyngeal squamous cell carcinoma in young adults: a review of the literature. Radiol Oncol 2014;48(1):1–10.

[80] Xu T, Tang J, Gu M, Liu L, Wei W, Yang H. Recurrent nasopharyngeal carcinoma: a clinical dilemma and challenge. Curr Oncol 2013;20(5):e406–19.

[81] Keryer-Bibens C, Pioche-Durieu C, Villemant C, Souquere S, Nishi N, Hirashima M, Middeldorp J, Busson P. Exosomes released by EBV-infected nasopharyngeal carcinoma cells convey the viral latent membrane protein 1 and the immunomodulatory protein galectin 9. BMC Cancer 2006;6:283.

[82] Reznik R, Hendifar AE, Tuli R. Genetic determinants and potential therapeutic targets for pancreatic adenocarcinoma. Front Physiol 2014;5:87.

[83] Greither T, Grochola LF, Udelnow A, Lautenschlager C, Wurl P, Taubert H. Elevated expression of microRNAs 155, 203, 210 and 222 in pancreatic tumors is associated with poorer survival. Int J Cancer 2010;126(1):73–80.

[84] Liu R, Chen X, Du Y, Yao W, Shen L, Wang C, Hu Z, Zhuang R, Ning G, Zhang C, Yuan Y, Li Z, Zen K, Ba Y, Zhang CY. Serum microRNA expression profile as a biomarker in the diagnosis and prognosis of pancreatic cancer. Clin Chem 2012;58(3):610–18.

[85] Zoller M. Pancreatic cancer diagnosis by free and exosomal miRNA. World J Gastrointest Pathophysiol 2013;4(4):74–90.

[86] Kuwabara Y, Ono K, Horie T, Nishi H, Nagao K, Kinoshita M, Watanabe S, Baba O, Kojima Y, Shizuta S, Imai M, Tamura T, Kita T, Kimura T. Increased microRNA-1 and microRNA-133a levels in serum of patients with cardiovascular disease indicate myocardial damage. Circ Cardiovasc Genet 2011;4(4):446–54.

[87] Corsten MF, Dennert R, Jochems S, Kuznetsova T, Devaux Y, Hofstra L, Wagner DR, Staessen JA, Heymans S, Schroen B. Circulating MicroRNA-208b and MicroRNA-499 reflect myocardial damage in cardiovascular disease. Circ Cardiovasc Genet 2010;3(6):499–506.

[88] Taylor DA, Zenovich AG. Cardiovascular cell therapy and endogenous repair. Diabetes Obes Metab 2008;10(Suppl. 4):5–15.

[89] Matsumoto S, Sakata Y, Suna S, Nakatani D, Usami M, Hara M, Kitamura T, Hamasaki T, Nanto S, Kawahara Y, Komuro I. Circulating p53-responsive microRNAs are predictive indicators of heart failure after acute myocardial infarction. Circ Res 2013;113(3):322–6.

[90] Gildea JJ, Carlson JM, Schoeffel CD, Carey RM, Felder RA. Urinary exosome miRNome analysis and its applications to salt sensitivity of blood pressure. Clin Biochem 2013;46(12):1131–4.

[91] Papaiwannou A, Zarogoulidis P, Porpodis K, Spyratos D, Kioumis I, Pitsiou G, Pataka A, Tsakiridis K, Arikas S, Mpakas A, Tsiouda T, Katsikogiannis N, Kougioumtzi I, Machairiotis N, Siminelakis S, Kolettas A, Kessis G, Beleveslis T, Zarogoulidis K. Asthma-chronic obstructive pulmonary disease overlap syndrome (ACOS): current literature review. J Thorac Dis 2014;6(Suppl. 1):S146–51.

[92] Levanen B, Bhakta NR, Torregrosa Paredes P, Barbeau R, Hiltbrunner S, Pollack JL, Skold CM, Svartengren M, Grunewald J, Gabrielsson S, Eklund A, Larsson BM, Woodruff PG, Erle DJ, Wheelock AM. Altered microRNA profiles in bronchoalveolar lavage fluid exosomes in asthmatic patients. J Allergy Clin Immunol 2013;131(3):894–903.

[93] Bajaj P. Exhaled breath condensates as a source for biomarkers for characterization of inflammatory lung diseases. J Analyt Sci Methods Instrumen 2013;3(1):17–29.

[94] Afsar B, Turkmen K, Covic A, Kanbay M. An update on coronary artery disease and chronic kidney disease. Int J Nephrol 2014;2014:767424.

[95] Hoorn EJ, Pisitkun T, Zietse R, Gross P, Frokiaer J, Wang NS, Gonzales PA, Star RA, Knepper MA. Prospects for urinary proteomics: exosomes as a source of urinary biomarkers. Nephrol (Carlton) 2005;10(3):283–90.

[96] Alvarez ML, Khosroheidari M, Kanchi Ravi R, DiStefano JK. Comparison of protein, microRNA, and mRNA yields using different methods of urinary exosome isolation for the discovery of kidney disease biomarkers. Kidney Int 2012;82(9):1024–32.

[97] Hogan MC, Manganelli L, Woollard JR, Masyuk AI, Masyuk TV, Tammachote R, Huang BQ, Leontovich AA, Beito TG, Madden BJ, Charlesworth MC, Torres VE, LaRusso NF, Harris PC, Ward CJ. Characterization of PKD protein-positive exosome-like vesicles. J Am Soc Nephrol 2009;20(2):278–88.

[98] van Balkom BW, Pisitkun T, Verhaar MC, Knepper MA. Exosomes and the kidney: prospects for diagnosis and therapy of renal diseases. Kidney Int 2011;80(11):1138–45.

[99] Zhou H, Cheruvanky A, Hu X, Matsumoto T, Hiramatsu N, Cho ME, Berger A, Leelahavanichkul A, Doi K, Chawla LS, Illei GG, Kopp JB, Balow JE, Austin III HA, Yuen PS, Star RA. Urinary exosomal transcription factors, a new class of biomarkers for renal disease. Kidney Int 2008;74(5):613–21.

[100] Lv LL, Cao YH, Ni HF, Xu M, Liu D, Liu H, Chen PS, Liu BC. MicroRNA-29c in urinary exosome/microvesicle as a biomarker of renal fibrosis. Am J Physiol Renal Physiol 2013;305(8):F1220–7.

[101] Rao P, Benito E, Fischer A. MicroRNAs as biomarkers for CNS disease. Front Mol Neurosci 2013;6:39.

[102] Cheng L, Quek CY, Sun X, Bellingham SA, Hill AF. The detection of microRNA associated with Alzheimer's disease in biological fluids using next-generation sequencing technologies. Front Genet 2013;4:150.

[103] Han D, Dara L, Win S, Than TA, Yuan L, Abbasi SQ, Liu ZX, Kaplowitz N. Regulation of drug-induced liver injury by signal transduction pathways: critical role of mitochondria. Trends Pharmacol Sci 2013;34(4):243–53.

[104] Yang X, Weng Z, Mendrick DL, Shi Q. Circulating extracellular vesicles as a potential source of new biomarkers of drug-induced liver injury. Toxicol Lett 2014;225(3):401–6.

[105] Villarroya-Beltri C, Gutierrez-Vazquez C, Sanchez-Madrid F, Mittelbrunn M. Analysis of microRNA and protein transfer by exosomes during an immune synapse. Methods Mol Biol 2013;1024:41–51.

[106] de Candia P, Torri A, Pagani M, Abrignani S. Serum microRNAs as biomarkers of human lymphocyte activation in health and disease. Front Immunol 2014;5:43.

[107] Alevizos I, Illei GG. MicroRNAs as biomarkers in rheumatic diseases. Nat Rev Rheumatol 2010;6(7):391–8.

[108] Alevizos I, Illei GG. MicroRNAs in Sjogren's syndrome as a prototypic autoimmune disease. Autoimmun Rev 2010;9(9):618–21.

[109] Zampetaki A, Kiechl S, Drozdov I, Willeit P, Mayr U, Prokopi M, Mayr A, Weger S, Oberhollenzer F, Bonora E, Shah A, Willeit J, Mayr M. Plasma microRNA profiling reveals loss of endothelial miR-126 and other microRNAs in type 2 diabetes. Circ Res 2010;107(6):810–7.

[110] Ouyang Y, Mouillet JF, Coyne CB, Sadovsky Y. Review: placenta-specific microRNAs in exosomes – good things come in nano-packages. Placenta 2014;35(Suppl.):S69–73.

[111] Bellingham SA, Coleman BM, Hill AF. Small RNA deep sequencing reveals a distinct miRNA signature released in exosomes from prion-infected neuronal cells. Nucleic Acids Res 2012;40(21):10937–49.

[112] Barteneva NS, Maltsev N, Vorobjev IA. Microvesicles and intercellular communication in the context of parasitism. Front Cell Infect Microbiol 2013;3:49.

[113] Principe S, Hui AB, Bruce J, Sinha A, Liu FF, Kislinger T. Tumor-derived exosomes and microvesicles in head and neck cancer: implications for tumor biology and biomarker discovery. Proteomics 2013;13(10–11):1608–23.

[114] Cheng Y, Wang X, Yang J, Duan X, Yao Y, Shi X, Chen Z, Fan Z, Liu X, Qin S, Tang X, Zhang C. A translational study of urine miRNAs in acute myocardial infarction. J Mol Cell Cardiol 2012;53(5):668–76.

SUBJECT INDEX

A

Printed in the United States
By Bookmasters